THEORY

OF

DIFFERENTIAL EQUATIONS.

THEORY

OF

DIFFERENTIAL EQUATIONS.

PART II.

ORDINARY EQUATIONS, NOT LINEAR.

BY

ANDREW RUSSELL FORSYTH,

Sc.D. (CAMB.), HON. Sc.D. (DUBL.), F.R.S.,

SADLERIAN PROFESSOR OF PURE MATHEMATICS,
FELLOW OF TRINITY COLLEGE, CAMBRIDGE.

VOL. III.

CAMBRIDGE:

AT THE UNIVERSITY PRESS.

1900

CAMBRIDGE UNIVERSITY PRESS
Cambridge, New York, Melbourne, Madrid, Cape Town,
Singapore, São Paulo, Delhi, Tokyo, Mexico City

Cambridge University Press
The Edinburgh Building, Cambridge CB2 8RU, UK

Published in the United States of America by Cambridge University Press, New York

www.cambridge.org
Information on this title: www.cambridge.org/9781107630123

First published 1890
First paperback edition 2011

A catalogue record for this publication is available from the British Library

ISBN 978-1-107-63012-3 Paperback

CONTENTS.

CHAPTER XI.

REDUCED FORMS OF SYSTEMS OF EQUATIONS OF THE FIRST ORDER IN THE VICINITY OF SINGULARITIES OF THE DERIVATIVES.

CHAPTER XII.

THE INTEGRALS OF THE REDUCED FORMS OF A SYSTEM OF EQUATIONS, CHIEFLY OF TWO DEPENDENT VARIABLES.

CHAPTER XIII.

SYSTEMS OF EQUATIONS WITH MULTIFORM VALUES OF THE DERIVATIVES ; SINGULAR SOLUTIONS.

CHAPTER XIV.

EQUATIONS OF THE SECOND ORDER AND THE FIRST DEGREE.

CHAPTER XV.

EQUATIONS OF THE SECOND ORDER AND ANY DEGREE.

CHAPTER XVI.

EQUATIONS OF THE SECOND ORDER WITH SUB-UNIFORM INTEGRALS : WITH SOME GENERAL CONSIDERATIONS.

CHAPTER XI.

SYSTEMS OF EQUATIONS OF THE FIRST ORDER AND THEIR REDUCED FORMS IN THE VICINITY OF SINGULARITIES OF THE DERIVATIVES.

145. IN most of the preceding investigations, the discussion has been restricted to a single equation of the first order. It is clear that many of the processes, with appropriate modifications and obvious extensions, can be applied to a system of equations of the first order: and accordingly it is possible, in making such applications, to deal more briefly both with the explanations and the details of the processes.

When the number of dependent variables is two, an appropriate geometrical illustration is provided by the skew curves in ordinary space which satisfy two differential equations of the first order. The full development of this illustration requires the restriction that the variables, dependent and independent, shall have real values. As this restriction prevents the full consideration of the functional relation between the variables, the geometrical mode of regarding the variables will be adopted chiefly for purposes of illustration: largely for the same reason that, in the corresponding problems involving only one dependent variable, the geometrical association of plane curves with the equations was adopted for the subsidiary purpose of illustration.

The general theorem in Chap. II establishes the existence of the integrals of a system of equations of the first order, determined by the condition that the integrals shall assume assigned values when the independent variable acquires its initial value. It effectively gives an expression for the integral, in the form of power-series valid over a finite domain: there being a precedent hypothesis that the aggregate of values, assigned as initial values to all the variables, constitutes an ordinary combination for the functions, which are the values of the respective derivatives as given by the differential equations. But if the combination of initial values should not have the character of an ordinary combination for the functions in question, then further investigation is required in order to afford knowledge of the integrals for values of the variables in the vicinity of such a combination.

146. Consider, in the first instance, a set of equations involving $n-1$ dependent variables and one independent variable, say $x_1, x_2, \ldots, x_n$ in all; and let them be

$$\frac{dx_1}{X_1} = \frac{dx_2}{X_2} = \ldots = \frac{dx_n}{X_n},$$

where, for the present purpose, each of the functions $X_1, X_2, \ldots, X_n$ vanishes when $x_1 = 0$, $x_2 = 0$, $\ldots$, $x_n = 0$. Let t denote a new parametric variable, such that the common value of the n fractions is

$$= \frac{dt}{t}.$$

Taking only the simplest case, so that in each of the functions X the terms of lowest dimensions which occur, when the functions are expanded in power-series, are of the first order, let

$$X_s = \sum_{i=1}^{n} a_{si} x_i + \text{terms of higher orders.}$$

Change the variables from x to y by means of the homogeneous linear substitution

$$y_r = \sum_{j=1}^{n} c_{rj} x_j, \qquad (r = 1, 2, \ldots, n),$$

where the n^2 coefficients c are disposable constants. Thus

$$t\frac{dy_r}{dt} = \sum_{j=1}^{n} c_{rj} X_j$$

$$= \sum_{j=1}^{n}\sum_{i=1}^{n} c_{rj} (a_{ji} x_i + \text{terms of higher order}),$$

for all values of r. If possible, let a set of constants c be chosen so that

$$\sum_{j=1}^{n}\sum_{i=1}^{n} c_{rj} a_{ji} x_i = \lambda y_r$$

$$= \sum_{j=1}^{n} \lambda c_{rj} x_j;$$

this choice can be made if the constants c satisfy the equations

$$\sum_{j=1}^{n} c_{rj} a_{js} = \lambda c_{rs},$$

for $s = 1, 2, \ldots, n$. These n equations are linear and homogeneous in the n constants $c_{r1}, c_{r2}, \ldots, c_{rn}$: in order that they may be consistent, we must have

$$A = \begin{vmatrix} a_{11} - \lambda, & a_{21}, & a_{31}, & \ldots \\ a_{12}, & a_{22} - \lambda, & a_{32}, & \ldots \\ a_{13}, & a_{23}, & a_{33} - \lambda, & \ldots \\ \hdotsfor{4} \end{vmatrix} = 0,$$

an equation in λ of degree n.

The importance of this equation does not lie mainly in the comparative simplicity of the forms of differential equations which are suggested as possible equivalents of the original system, but is rather constituted by an invariantive property which it possesses, viz. whatever be the intermediate linear transformations effected upon the variables x, the roots λ of the determinantal equation are entirely independent of those intermediate transformations. The formal proof of this property is practically identical with the formal proof of the invariantive character of the fundamental equation appertaining to a singularity of an ordinary linear equation of order n; it is as follows [*].

Let a new set of variables be given by the relations

$$z_m = \sum_{s=1}^{n} g_{ms} x_s,$$

where the determinant

$$\Gamma = \begin{vmatrix} g_{11}, & g_{12}, & \ldots \\ g_{21}, & g_{22}, & \ldots \\ \hdotsfor{3} \end{vmatrix}$$

is different from zero; and let the new equations be

$$\frac{dz_1}{Z_1} = \frac{dz_2}{Z_2} = \ldots = \frac{dz_n}{Z_n} = \frac{dt}{t},$$

where

$$Z_s = \sum_{i=1}^{n} b_{si} z_i + \text{terms of higher orders.}$$

---

[*] This is similar to the proof given by Hamburger, *Crelle*, t. LXXVI (1873), p. 115, for the corresponding result connected with linear equations.

The corresponding equation, for the reduction to the simpler form with a multiplier λ, is

$$B = \begin{vmatrix} b_{11} - \lambda, & b_{21}, & b_{31}, & \ldots \\ b_{12}, & b_{22} - \lambda, & b_{32}, & \ldots \\ b_{13}, & b_{23}, & b_{33} - \lambda, & \ldots \\ \cdots\cdots\cdots\cdots\cdots\cdots\cdots \end{vmatrix} = 0.$$

It is clear that the highest term in powers of λ, viz. $(-\lambda)^n$, is the same in A as in B.

We have

$$\sum_{p=1}^{n} g_{mp} X_p = t \sum_{p=1}^{n} g_{mp} \frac{dx_p}{dt}$$

$$= t \frac{dz_m}{dt}$$

$$= Z_m$$

$$= \sum_{i=1}^{n} b_{mi} z_i + \text{terms of higher orders}$$

$$= \sum_{i=1}^{n} b_{mi} \sum_{s=1}^{n} g_{is} x_s + \text{terms of higher orders};$$

whence, on equating to one another the coefficients of the first powers of the variables x on the two sides of the equation thus obtained, we find

$$\sum_{p=1}^{n} g_{mp} a_{pt} = \sum_{p=1}^{n} b_{mp} g_{pt} = f_{mt},$$

say, for all values of m and t. Now

$$\Gamma A = \begin{vmatrix} g_{11}, & g_{12}, & g_{13}, & \ldots \\ g_{21}, & g_{22}, & g_{23}, & \ldots \\ g_{31}, & g_{32}, & g_{33}, & \ldots \\ \cdots\cdots\cdots\cdots\cdots\cdots \end{vmatrix} \begin{vmatrix} a_{11} - \lambda, & a_{21}, & a_{31}, & \ldots \\ a_{12}, & a_{22} - \lambda, & a_{32}, & \ldots \\ a_{13}, & a_{23}, & a_{33} - \lambda, & \ldots \\ \cdots\cdots\cdots\cdots\cdots\cdots\cdots\cdots \end{vmatrix}$$

$$= \begin{vmatrix} f_{11} - g_{11}\lambda, & f_{12} - g_{12}\lambda, & f_{13} - g_{13}\lambda, & \ldots \\ f_{21} - g_{21}\lambda, & f_{22} - g_{22}\lambda, & f_{23} - g_{23}\lambda, & \ldots \\ f_{31} - g_{31}\lambda, & f_{32} - g_{32}\lambda, & f_{33} - g_{33}\lambda, & \ldots \\ \cdots\cdots\cdots\cdots\cdots\cdots\cdots\cdots \end{vmatrix};$$

and

$$B\Gamma = \begin{vmatrix} b_{11}-\lambda, & b_{12}, & b_{13}, & \cdots \\ b_{21}, & b_{22}-\lambda, & b_{23}, & \cdots \\ b_{31}, & b_{32}, & b_{33}-\lambda, & \cdots \\ \cdots\cdots\cdots\cdots & \cdots\cdots\cdots\cdots \end{vmatrix} \begin{vmatrix} g_{11}, & g_{21}, & g_{31}, & \cdots \\ g_{12}, & g_{22}, & g_{32}, & \cdots \\ g_{13}, & g_{23}, & g_{33}, & \cdots \\ \cdots\cdots\cdots\cdots \end{vmatrix}$$

$$= \begin{vmatrix} f_{11}-g_{11}\lambda, & f_{12}-g_{12}\lambda, & f_{13}-g_{13}\lambda, & \cdots \\ f_{21}-g_{21}\lambda, & f_{22}-g_{22}\lambda, & f_{23}-g_{23}\lambda, & \cdots \\ f_{31}-g_{31}\lambda, & f_{32}-g_{32}\lambda, & f_{33}-g_{33}\lambda, & \cdots \\ \cdots\cdots\cdots\cdots\cdots\cdots\cdots\cdots \end{vmatrix} = \Gamma A,$$

so that $B = A$. The equation $A = 0$ is therefore absolutely invariantive for all intermediate linear transformations of the variables.

The invariance of this fundamental equation makes it possible to construct a canonical equivalent of the original system of equations.

147. If ω be a simple root of this equation, some (or it may be any) $n-1$ of the preceding n homogeneous linear equations determine the ratios of the n coefficients $c_{r1}, ..., c_{rn}$; so that, taking one of the coefficients arbitrarily, say c_{rr}, the others are known, and the consequent value of y is obtained.

First, suppose that the roots of the determinantal equation of degree n are distinct from one another; denote them by $\lambda_1, \lambda_2, ...,$ λ_n, so that each of them is simple. After the above explanation, each root determines a substitution (c_{rj}) and a consequent new variable y; let y_r be the variable determined by λ_r, for $r = 1, ..., n$. Conversely, the variables $x_1, ..., x_n$ are linearly expressible in terms of the variables $y_1, ..., y_n$; so that an aggregate of terms of any order in x becomes, after substitution, an aggregate of terms of the same order in y. Moreover, the expression for $t\dfrac{dy_r}{dt}$ is

$$\frac{dy_r}{dt} = \lambda_r y_r + \text{terms of order higher than the first in } x;$$

hence after substitution, the system of differential equations is

$$\frac{dy_1}{Y_1} = \frac{dy_2}{Y_2} = ... = \frac{dy_n}{Y_n} = \frac{dt}{t},$$

where $$Y_r = \lambda_r y_r + \eta_r,$$

and η_r is an aggregate of terms of order higher than the first in the variables $y_1, \ldots, y_n$.

This is the canonical form when the n roots of the determinantal equation are distinct.

Next, suppose that a root of the determinantal equation is repeated, say $\lambda = \omega$, and that its multiplicity is κ. Then the equations

$$\sum_{j=1}^{n} c_{rj} a_{js} = \omega c_{rs}$$

are consistent with one another, and they provide a set of co-efficients c, though the set may not be uniquely determined. Taking them in some definite form, let them be associated with a variable y_1; and transform the variables from $x_1, x_2, \ldots, x_n$ to $y_1, x_2, \ldots, x_n$. The differential equations then take the form

$$t\frac{dy_1}{dt} = \omega y_1 + \eta_1{}',$$

$$t\frac{dx_s}{dt} = \beta_s y_1 + \sum_{j=2}^{n} a_{sj}{}' x_j + \eta_s{}', \qquad (s = 2, \ldots, n),$$

where $\eta_1{}', \eta_2{}', \ldots, \eta_n{}'$ are the aggregates of terms of order higher than the first on the respective sides. The determinantal equation now is

$$\begin{vmatrix} \omega - \lambda, & 0, & 0, & \ldots \\ \beta_2, & a_{22}{}' - \lambda, & a_{32}{}', & \ldots \\ \beta_3, & a_{23}{}', & a_{33}{}' - \lambda, & \ldots \\ \cdots\cdots\cdots\cdots\cdots\cdots \end{vmatrix} = 0;$$

as the equation is invariantive, the root ω is of multiplicity κ, and therefore the equation

$$\begin{vmatrix} a_{22}{}' - \lambda, & a_{32}{}', & \ldots \\ a_{23}{}', & a_{33}{}' - \lambda, & \ldots \\ \cdots\cdots\cdots\cdots\cdots \end{vmatrix} = 0$$

has the root ω of multiplicity $\kappa - 1$. Hence equations

$$\sum_{j=2}^{n} c_{rj}{}' a_{js}{}' = \omega c_{rs}{}' \qquad \begin{matrix} (s = 2, \ldots, n), \\ (r = 2, \ldots, n), \end{matrix}$$

are consistent with one another; and they provide a set of coefficients c', though the set may not be uniquely determined. Taking them in some definite form, let them be associated with a variable y_2; and transform the variables from x_2, x_3, $\ldots$, x_n to y_2, x_3, $\ldots$, x_n. The differential equations then take the form

$$t\frac{dy_2}{dt} = \beta_2 y_1 + \omega y_2 + \eta_2'',$$

$$t\frac{dx_s}{dt} = \beta_s y_1 + \gamma_s y_2 + \sum_{j=3}^{n} a_{sj}'' x_j + \eta_s'', \qquad (s = 3, \ldots, n),$$

where η_2'', η_3'', $\ldots$, η_n'' are the aggregates of terms of order higher than the first on the respective sides.

Proceeding in this way, we see that κ variables y_1, $\ldots$, y_κ can be associated with a root $\lambda = \omega$, of multiplicity κ in the determinantal equation, in such a way as to replace κ variables x_1, $\ldots$, x_κ, and to transform the original system of differential equations to the form

$$t\frac{dy_1}{dt} = \omega y_1 + \eta_1,$$

$$t\frac{dy_2}{dt} = a_{21} y_1 + \omega y_2 + \eta_2,$$

$$t\frac{dy_3}{dt} = a_{31} y_1 + a_{32} y_2 + \omega y_3 + \eta_3,$$

$$\ldots\ldots\ldots\ldots\ldots\ldots\ldots\ldots\ldots\ldots\ldots\ldots\ldots\ldots\ldots\ldots$$

$$t\frac{dy_\kappa}{dt} = a_{\kappa 1} y_1 + a_{\kappa 2} y_2 + \ldots + a_{\kappa,\,\kappa-1} y_{\kappa-1} + \omega y_\kappa + \eta_\kappa,$$

so far as concerns the derivatives of the κ variables indicated.

Taking the simple roots in turn, the transformation is effected for a variable associated with each of them; and the expression for the corresponding derivative is monomial in terms of the first order on the right-hand side. Taking the multiple roots of the determinantal equation in turn, with each of them is associated a group of variables in number equal to the multiplicity of the root; when the transformation is effected, the expressions for the derivatives of the corresponding variables are of the above form.

148. The reduction indicated is effective if, in the expressions for the regular functions X_1, $\ldots$, X_n, which are valid in the vicinity of $0, 0, \ldots, 0$, a common zero of all of them, the terms of the first order occur in quite general form. We shall not discuss the case in which the terms of the first order occur only in specialised forms: nor shall we discuss, for an unrestricted number of variables, the case in which the terms of lowest order that occur

are of dimensions higher than the first. For this investigation, it will be sufficient to refer to Königsberger's treatise *.

The simplest instance of the preceding class of equations is furnished by a system of n equations in n dependent variables, which occur only linearly and homogeneously in the equations. When the system is of the form

$$\frac{dy_{n-1}}{dx} = \sum_{r=0}^{n-1} P_{n-r} y_r,$$

$$\frac{dy_r}{dx} = y_{r+1}, \qquad (r = 0, 1, \ldots, n-2),$$

it can immediately be replaced by a single linear equation of order n. A complete discussion of such an equation and of the properties of its integrals is reserved for a separate section.

The next most important instance is that in which the system has the form

$$\frac{dy_r}{dx} = \sum_{i=1}^{n} A_{ir} y_i,$$

where the coefficients A_{ij} are functions of x alone. An ample discussion of the properties and characteristics of the integrals of such a system, leading to results associated with the ordinary linear equation of order n to which reference has just been made, will be found in various papers by Sauvage; a general exposition of the results is given by him in a separately published volume†. Among these, the most important are:—

    (i)   The establishment of the existence of integrals of the equation in the vicinity of an ordinary point of the coefficients A;

---

* *Lehrbuch der Theorie der Differentialgleichungen mit einer unabhängigen Variabeln*, (Leipzig, Teubner, 1889); see, in particular, Chapter v.

    Upon this subject, reference may also be made to the following authorities:

        Picard, *Cours d'Analyse*, t. iii, chap. i.

        Horn, *Crelle*, t. cxvi (1896), pp. 265—306, *ib.* t. cxvii (1897), pp. 104—128, 254—266.

        Poincaré, *Inaugural Dissertation*, (1879).

        Bendixson, *Stockh. Öfv.*, t. li (1894), pp. 141—151.

    Further references will be found in the works quoted.

    † *Théorie générale des systèmes d'équations différentielles linéaires et homogènes*, (Paris, Gauthier-Villars, 1895).

    A memoir by Grünfeld, *Wiener Akad. Denkschr.* t. liv, ii Abth., (1888), pp. 93—104, in which he follows the earlier work of Sauvage, dispensing with the use of

(ii) A proof that equations, which have **regular*** integrals in the vicinity of $x = 0$, are of the form

$$x\frac{dy_r}{dx} = \sum_{j=1}^{n} a_{ri}y_i, \qquad (r = 1, \ldots, n),$$

where a_{ri} is regular in the vicinity of the point $x = 0$. This is the canonical form for such equations in the vicinity of the point $x = 0$; and it is the form which must be possessed by the equations in the vicinity of every non-regular point of the coefficients A;

(iii) The construction of the solution when the tests for regularity are satisfied;

(iv) A generalisation, to the class of equations considered, of the main properties established by Fuchs and others for the ordinary linear differential equation of the nth order.

The investigations will not be repeated here, as they follow somewhat closely the development of the theory of ordinary linear equations, which the present writer hopes to discuss in detail in another volume.

FORM OF EQUATIONS IN THE VICINITY OF NON-ORDINARY COMBINATIONS OF VALUES FOR THE DERIVATIVES.

149. In order to give some indication of the mode of obtaining integrals of simultaneous equations, involving more than one dependent variable, it is desirable to shew how the processes, applied in preceding chapters to the discussion of the integral of a single equation, can be extended so as to be effective for such a purpose. Accordingly, we shall discuss in some detail, though without attempting the same completeness as in the case of a single equation, the case when the system involves one independent variable z and two dependent variables u and v, and when it

---

Weierstrass's theory of bilinear forms, and also a paper by the same author, Schlömilch's *Zeitschrift*, t. xxxvi (1891), pp. 21—33, may be consulted.

The form of the equations which have "regular" integrals, and the characteristics of the regular integrals of such equations, are discussed by Königsberger in his treatise (quoted p. 8, note) pp. 441—469: also in two memoirs by Horn, *Math. Annalen*, t. xxxix (1891), pp. 391—408, *ib.* t. xl (1892), pp. 527—550.

Reference may also be made to Picard, *Cours d'Analyse*, t. iii, ch. i, ii; and to Jordan, *Cours d'Analyse*, t. iii, ch. ii, Section 1.

* In the same sense as in § 30.

therefore contains two independent equations. The general discussion of Chapter I shews that, when the equations are given in
the form

$$F\left(\frac{du}{dz}, \frac{dv}{dz}, u, v, z\right) = 0$$
$$G\left(\frac{du}{dz}, \frac{dv}{dz}, u, v, z\right) = 0$$

then, except in the vicinity of what may be called branch-values
for $\dfrac{du}{dz}$ and $\dfrac{dv}{dz}$ given by these two algebraical equations, we may
take

$$\frac{du}{dz} = f(u, v, z), \qquad \frac{dv}{dz} = g(u, v, z),$$

as representing the equations. The nature of the integrals in
the vicinity of any assigned values depends upon the character
of the functions f and g in such a vicinity.

Cauchy's existence-theorem gives the character of the integrals
of the system

$$\frac{du}{dz} = f(u, v, z), \qquad \frac{dv}{dz} = g(u, v, z),$$

which are regular functions of $z - c$ in the vicinity of c, and acquire
values α and β respectively when $z = c$: provided $u = \alpha$, $v = \beta$, $z = c$
are an ordinary combination of values for the functions f and g.
It therefore is necessary to discuss the character of the integrals
in the vicinity of values, which constitute a non-ordinary combination of those functions. For this purpose, we take $u - \alpha$,
$v - \beta$, $z - c$ as new variables and, assuming the corresponding
transformations effected, we can consider $0, 0$ as initial values
assigned to the dependent variables, when the independent variable
vanishes.

Now $u = 0$, $v = 0$, $z = 0$, when not an ordinary combination of
values for both f and g, can give rise to several alternatives.
Denoting

(i)   an ordinary combination for one function by O:

(ii)   an accidental singularity of the first kind for one function
       by A_1:

(iii) an accidental singularity of the second kind for one
       function by A_2:

(iv)   an essential singularity by E:

then the possible alternatives are $0, A_1$; $0, A_2$; $0, E$; $\boldsymbol{A}_1, \boldsymbol{A}_1$; A_1, A_2; A_1, E; A_2, A_2; A_2, E; E, E; where, for instance, $0, \boldsymbol{A}_1$ signifies that the values $0, 0, 0$ constitute an ordinary combination for one of the functions and an accidental singularity of the first kind for the other.

Of these, we shall discuss only the combinations which are free from essential singularities. All that was said in Chapter VII, as to the exiguity of results that are obtainable concerning an equation the form of which gives rise to an essential singularity, applies with at least equal force to systems of equations. Accordingly, we shall discuss only the five (non-ordinary) combinations represented by $0, A_1$; $0, A_2$; A_1, A_1; A_1, A_2; A_2, A_2, taking them in this order.

Case I.

150. $0, A_1$. Let the combination be ordinary for f; so that f can be expanded as a regular function of u, v, z in the vicinity, the appropriate power-series being

$$f = f_0 + f_1 u + f_2 v + f_3 z + \dots.$$

Let the combination be an accidental singularity of the first kind for g, so that there exists some power-series $P_1(u, v, z)$, regular in the vicinity, such that

$$P_1(u, v, z)\, g(u, v, z) = \text{regular function not vanishing at } 0, 0, 0$$

$$= 1 + Q(u, v, z),$$

say, where Q is a regular function that does vanish at $0, 0, 0$. Then

$$\frac{dz}{dv} = \frac{1}{g(u, v, z)}$$

$$= \frac{P_1(u, v, z)}{1 + Q(u, v, z)}$$

$$= P(u, v, z),$$

where $P(u, v, z)$ is a regular function that vanishes at $0, 0, 0$. Now

$$\frac{du}{dz} = f(u, v, z),$$

and therefore

$$\frac{du}{dv} = f(u, v, z)\, P(u, v, z).$$

Thus the new form of the equations is such that the values of $\dfrac{dz}{dv}$ and $\dfrac{du}{dv}$ are regular functions in the vicinity of the origin: and therefore, by Cauchy's theorem, regular integrals exist, determined uniquely by the condition that the functions of v, which represent the values of z and u, vanish when $v = 0$.

The actual expansions depend upon the forms of the functions f and P.

Taking the simplest case, let

$$f = f_0 + f_1 u + f_2 v + f_3 z + f_4 v^2 + \ldots,$$

$$P = \alpha u + \beta v + \gamma z + \delta v^2 + \ldots;$$

then

$$\frac{dz}{dv} = P, \quad \frac{du}{dv} = fP.$$

It is easy to prove that

$$z = \tfrac{1}{2}\beta v^2 + \tfrac{1}{3}\left(\delta + \tfrac{1}{2}\gamma\beta + \tfrac{1}{2}f_0\alpha\beta\right)v^3 + \ldots,$$

$$u = \tfrac{1}{2}f_0\beta v^2 + \tfrac{1}{3}\left\{f_0\left(\delta + \tfrac{1}{2}\gamma\beta + \tfrac{1}{2}f_0\alpha\beta\right) + f_2\beta\right\}v^3 + \ldots.$$

From the former of these, we have

$$v = \left(\frac{2}{\beta}\right)^{\frac{1}{2}} z^{\frac{1}{2}} R\left(z^{\frac{1}{2}}\right),$$

where $R\left(z^{\frac{1}{2}}\right)$ is a regular function of $z^{\frac{1}{2}}$ such that $R(0) = 1$; and then

$$u = z f_0 + z^{\frac{3}{2}} S\left(z^{\frac{1}{2}}\right),$$

where $S\left(z^{\frac{1}{2}}\right)$ is a regular function of $z^{\frac{1}{2}}$ such that $S(0) = \tfrac{2}{3}\left(\dfrac{2}{\beta}\right)^{\frac{1}{2}} f_2$.

Clearly $z = 0$ is a branch-point of algebraical type, for each of the functions u and v, in this case.

For a less simple case, let

$$\frac{dz}{dv} = P_m + \ldots, \quad \frac{du}{dv} = f_n P_m + \ldots,$$

where P_m denotes the aggregate of terms in P of dimensions m in u, v, z, these being the terms of lowest dimension that occur; and f_n similarly denotes the aggregate of terms in f of dimensions n in u, v, z, these also being the terms of lowest dimension that occur;

and assume that the term v^m occurs in P_m: and that the term v^n occurs in f_n. Then

$$z = \lambda v^{m+1} + \ldots, \quad u = \mu v^{m+n+1} + \ldots,$$

so that

$$v = z^{\frac{1}{m+1}} R \left(z^{\frac{1}{m+1}} \right),$$

$$u = z^{1 + \frac{n}{m+1}} S \left(z^{\frac{1}{m+1}} \right),$$

where R and S denote regular functions of their arguments, these functions not vanishing with z. The point $z = 0$ clearly is an algebraical critical point for u and v.

If a term in v alone should not occur in P_m, or if a term in v alone should not occur in f_n, then the dimension of the lowest terms of z and u in powers of v is not obvious upon inspection: it may be obtainable by means of a space-diagram (§ 157).

Case II.

151. $0, A_2$.   Let the combination be ordinary for f, so that the function can be expanded in the form

$$f = f_0 + f_1 u + f_2 v + f_3 z + \ldots$$

in the vicinity of 0, 0, 0. Let the combination be an accidental singularity of the second kind for g, so that regular functions $P_1(u, v, z)$, $P_2(u, v, z)$ exist, such that

$$P_1(u, v, z) \, g(u, v, z) = P_2(u, v, z),$$

both the functions P_1, P_2 vanishing at 0, 0, 0.

In the simplest case of all, we have

$$P_2(u, v, z) = a'u + b'v + c'z + \ldots,$$

$$P_1(u, v, z) = au + bv + cz + \ldots;$$

and the initial conditions are that the integrals u and v are to vanish with z. The equations thus are

$$\frac{du}{dz} = f_0 + f_1 u + f_2 v + f_3 z + \ldots,$$

$$\frac{dv}{dz} = \frac{a'u + b'v + c'z + \ldots}{au + bv + cz + \ldots}.$$

Now as the integrals u and v are to vanish with z, the most important term in the power-series, which represent them in the immediate vicinity of the origin—supposing such power-series possible—must be some positive power of z. In the case when the coefficients are quite general, it is clear that, both for u and for v, this power is the first, so that we may write

$$u = \rho z + \dots, \quad v = \sigma z + \dots,$$

when the unexpressed terms have their modulus small compared with $|z|$ in the immediate vicinity of the origin. Substituting and taking account of terms of lowest order in

$$\frac{du}{dz} = f_0 + f_1 u + f_2 v + f_3 z + \dots,$$

$$(au + bv + cz + \dots)\frac{dv}{dz} = a'u + b'v + c'z + \dots,$$

we have

$$\rho = f_0,$$

$$(a\rho + b\sigma + c)\,\sigma = a'\rho + b'\sigma + c',$$

so that σ is a root of the equation

$$b\sigma^2 + \sigma\,(af_0 + c - b') - (a'f_0 + c') = 0.$$

Accordingly, new variables U and V are introduced, defined by the equations

$$u = z\,(\rho + U), \quad v = z\,(\sigma + V),$$

where U and V are to vanish with z: and then

$$z\,\frac{dU}{dz} = -U + \alpha z + \text{terms of second and higher orders in } U,\ V,\ z,$$

$$z\,\frac{dV}{dz} = (a' - \sigma a)\,U + (b' - a\rho - c)\,V + \beta z + \dots.$$

Taking a less simple case, let the terms of lowest order in the expression for f in the immediate vicinity of the combination $0, 0, 0$ be of dimension l, so that we can take

$$\frac{du}{dz} = (\ast \mathbb{Q} u,\, v,\, z)^l + \dots\ ;$$

and similarly suppose P_1 and P_2 to be such that we can take

$$\frac{dv}{dz} = \frac{(\ast \mathbb{Q} u,\, v,\, z)^m + \dots}{(\ast \mathbb{Q} u,\, v,\, z)^n + \dots},$$

both u and v vanishing with z: where $l,\ m,\ n \geqslant 1$.

If $m \geqslant n$, then in the expansion of u the most important term is a constant multiple of z^{l+1}; and in the expansion of v the most important term is a constant multiple of z^{m-n+1}. New dependent variables, defined by the relations

$$u = (\rho + U) z^{l+1},$$

$$v = (\sigma + V) z^{m-n+1},$$

are introduced, ρ and σ being determined so as to annihilate the lowest terms in the equation, and U and V being subject to the condition of vanishing with z; and the reduced form is

$$z \frac{dU}{dz} = -lU + kV + jz + \ldots,$$

$$z \frac{dV}{dz} = \alpha U + \beta V + \gamma z + \ldots,$$

where $\alpha = 0$ if $l > 1$, and $k = 0$ if $m > n$.

If $m < n$, the determination of the orders of the lowest terms can be effected by means of a space-diagram (§ 157).

Case III.

152. A_1, A_1. This case can be included, by transformation of the variables, in preceding cases. The form being

$$\frac{du}{dz} = f(u, v, z), \quad \frac{dv}{dz} = g(u, v, z),$$

where the functions f and g are such that their reciprocals are finite at 0, 0, 0, then

$$\frac{dz}{du} = \frac{1}{f(u, v, z)} = F(u, v, z),$$

$$\frac{dv}{du} = \frac{g(u, v, z)}{f(u, v, z)} = G(u, v, z).$$

For the function F, the combination 0, 0, 0 is ordinary: for the function G, it either is ordinary, or is accidental of the first kind, or is accidental of the second kind. The respective possibilities have been dealt with; and the inversion of the series in the respective cases leads to the corresponding results.

Case IV.

153. A_1, A_2. The form of the equations is

$$\frac{du}{dz} = f(u, v, z), \quad \frac{dv}{dz} = \frac{g_1(u, v, z)}{g_2(u, v, z)},$$

where both the functions g_1 and g_2 vanish at 0, 0, 0, and the reciprocal of f is finite there. Then

$$\frac{dz}{du} = \frac{1}{f(u, v, z)} = F(u, v, z),$$

$$\frac{dv}{du} = \frac{g_1(u, v, z)}{f(u, v, z)\, g_2(u, v, z)} = G(u, v, z).$$

The combination is ordinary for the function F; it is either ordinary, or accidental of the first kind, or accidental of the second kind, for the function G. Hence the equations now are of one of the types already discussed.

As an example, take the equations

$$\frac{du}{dz} = \frac{1 + \text{positive powers}}{au + bv + cz + \ldots},$$

$$\frac{dv}{dz} = \frac{a''u + b''v + c''z + \ldots}{au + bv + cz + \ldots},$$

so that

$$\frac{dz}{du} = au + bv + cz + hu^2 + kuv + luz + mu^3 + \ldots,$$

$$\frac{dv}{du} = a''u + b''v + c''z + h''u^2 + k''uv + l''uz + m''u^3 + \ldots,$$

the unexpressed terms being of an order that is easily seen from the analysis to be higher in powers of small quantities than those which are retained. As both the functions on the right-hand side have 0, 0, 0 for an ordinary combination, there exist integrals of the equations which are regular functions of u vanishing with z. Moreover, the slightest inspection shews that each of these integral functions, when expressed as a power-series in u, begins with a term in u^2; so that, if

$$v = A_1 u^2 + B_1 u^3 + C_1 u^4 + \ldots,$$

$$z = A_0 u^2 + B_0 u^3 + C_0 u^4 + \ldots,$$

then

$$A_0 = \tfrac{1}{2}a, \quad A_1 = \tfrac{1}{2}a'';$$

$$B_0 = \tfrac{1}{3}h + \tfrac{1}{6}(a''b + ac), \quad B_1 = \tfrac{1}{3}h'' + \tfrac{1}{6}(a''b'' + ac'');$$

$$C_1 = \tfrac{1}{4}\{m'' + b''B_1 + c''B_0 + k''A_1 + l''A_0\},$$

$$C_0 = \tfrac{1}{4}\{m + bB_1 + cB_0 + kA_1 + lA_0\};$$

and so on. Reversing the series between u and z, we have

$$u = \sqrt{\frac{2}{a}} z^{\frac{1}{2}} + z P(z^{\frac{1}{2}}),$$

where $P(z^{\frac{1}{2}})$ is a regular power-series of $z^{\frac{1}{2}}$, that does not vanish with z; and then, substituting this value in the series for v, we have

$$v = \frac{a''}{a} z + z^{\frac{3}{2}} Q(z^{\frac{1}{2}}),$$

where $Q(z^{\frac{1}{2}})$ is a regular power-series of $z^{\frac{1}{2}}$, that does not vanish with z. Thus $z = 0$ is a simple branch-point of the integral functions determined by the particular differential equations.

Case V.

154. A_2, A_2. Passing now to the case in which the combination $0, 0, 0$ is an accidental singularity of the second kind for each of the functions f and g, so that each of them is expressible in the form of quotients of regular power-series which vanish at $0, 0, 0$, we shall restrict ourselves to the case in which the denominator in the two quotients is the same*, so that we can take

$$f(u, v, z) = \frac{P(u, v, z)}{R(u, v, z)}, \quad g(u, v, z) = \frac{Q(u, v, z)}{R(u, v, z)},$$

where P, Q, R are regular functions of their arguments and vanish with u, v, z.

To obtain a more useful representation of these fractions, we use Weierstrass's theorem (Note, Chap. I) on the expression for a regular function of any number of variables in the immediate vicinity of a zero; we apply it to each of the functions P, Q, R in succession.

We begin with the function $R(u, v, z)$.

First, if $R(u, 0, 0)$ does not vanish for all values of u, let Cu^μ be the lowest power of u in its expression; then we know that $R(u, v, z)$ can be expressed in the form

$$(Cu^\mu + r_1 u^{\mu-1} + r_2 u^{\mu-2} + \ldots + r_\mu) e^{G(u, v, z)},$$

---

* The alternative possibility, viz. that each quotient has its own denominator without reference to any other, is discussed in Königsberger's treatise (cited in § 1), pp. 353 et seq. The hypothesis in the text is adopted, partly in connection with the reduction of a system of equations to a canonical form as in § 5, partly as exhibiting an adequate illustration of the method of dealing with other cases.

where G is a regular function vanishing with u, v, z; and where $r_1, r_2, \ldots, r_\mu$ are regular functions of v and z, which vanish with v and z.

Secondly, if $R(u, 0, 0)$ does vanish identically, but either $R(0, v, 0)$ or $R(0, 0, z)$ does not vanish identically, then there is a corresponding expression for $R(u, v, z)$ either in powers of v or in powers of z for the respective cases, the coefficients of powers of v or of powers of z being of the same character as $r_1, \ldots, r_\mu$ in the preceding case.

Thirdly, if $R(u, 0, 0)$, $R(0, v, 0)$, $R(0, 0, z)$ all vanish identically, (as would, for instance, be the case with a function equal to $uv + vz + zu$, or with any function every term in the regular expansion of which contains powers of not fewer than two of the variables), then new variables ξ, η, ζ are introduced, defined by the relations

$$(u, v, z) = \left(\begin{array}{ccc} a_1, & b_1, & c_1 \\ a_2, & b_2, & c_2 \\ a_3, & b_3, & c_3 \end{array} \right)(\xi, \eta, \zeta),$$

where the constants a, b, c are subject solely to restrictions of inequality

$$(a_1, \ b_2, \ c_3) \neq 0,$$

$$R(a_1, \ a_2, \ a_3) \neq 0.$$

If $R'(\xi, \eta, \zeta)$ be the transformed equivalent of $R(u, v, z)$, the properties of the transformation are such that $R'(\xi, 0, 0)$ is not identically zero. Let $C'\xi^\mu$ be the lowest power of ξ in $R'(\xi, 0, 0)$; then, as before, it is possible to obtain an expression for $R'(\xi, \eta, \zeta)$, that is, for $R(u, v, z)$, in the form

$$R(u, v, z) = (C'\xi^\mu + r_1'\xi^{\mu-1} + \ldots + r_\mu')e^{G(u, v, z)},$$

where $r_1', \ldots, r_\mu'$ are regular functions of η and ζ, which vanish when $\eta = 0$, $\zeta = 0$.

Similarly, the expressions for $P(u, v, z)$ and $Q(u, v, z)$ can be transformed. If $P(u, 0, 0)$ and $Q(u, 0, 0)$ do not vanish identically, then no transformation to the new variables ξ, η, ζ is necessary. If $P(u, 0, 0)$ vanish identically, the sole additional restriction of inequality is that

$$P(a_1, a_2, a_3) \neq 0;$$

and if $Q(u, 0, 0)$ vanish identically, the sole additional restriction of inequality is that

$$Q(a_1, a_2, a_3) \neq 0.$$

If then no one of the quantities $R(u, 0, 0)$, $P(u, 0, 0)$, $Q(u, 0, 0)$ should vanish, we have

$$f(u, v, z) = C_1 \frac{u^\lambda + p_1 u^{\lambda-1} + \ldots + p_\lambda}{u^\mu + r_1 u^{\mu-1} + \ldots + r_\mu} e^{H_1(u, v, z)},$$

$$g(u, v, z) = C_2 \frac{u^\nu + q_1 u^{\nu-1} + \ldots + q_\nu}{u^\mu + r_1 u^{\mu-1} + \ldots + r_\mu} e^{H_2(u, v, z)},$$

where the functions H_1 and H_2 are regular functions of u, v, z; and the coefficients p, q, r are regular functions of v and z, which vanish when $v = 0$, $z = 0$.

155. Now the objects of our discussion are the construction of the conditions of existence, and the determination of the forms, of integrals of the equations which vanish when $z = 0$: so that u and v vanish with z. If there are integrals which are regular functions of z or are quasi-regular functions, e.g. such as are regular functions of a fractional power of z, then some information as to the integrals will be furnished when the leading terms are known. These terms will be of the form

$$u = \alpha z^\rho, \quad v = \beta z^\sigma,$$

where ρ and σ (which will be found to be real quantities in the simpler cases) have their real parts positive, and α and β are non-vanishing constants. If such leading terms can once be obtained, then the equations will be transformed by the substitutions

$$u = (\alpha + U) z^\rho, \quad v = (\beta + V) z^\sigma,$$

where the new dependent variables U and V must vanish with z; and the solution of the equations will then depend upon the quantities U and V.

In order to obtain these leading terms, it is necessary to consider the terms in the numerator and the denominator of f and of g, which are of lowest order in the variables. In any coefficient, such as p_κ, these are obtainable by another application of Weierstrass's theorem. Thus p_κ is an analytical function of v

and z, which vanishes when $v = 0$ and $z = 0$; and therefore, as in § 35, it is of the form

$$v^{R_\kappa} z^{S_\kappa} q_\kappa (v, z) e^{G(v, z)},$$

where G is a regular function of v and z, the integers R_κ and S_κ are positive integers (including zero), and where

$$q_\kappa (v, z) = A_1 v^{m_\kappa} + Q_1 v^{m_\kappa - 1} + Q_2 v^{m_\kappa - 2} + \dots + Q_{m_\kappa},$$

the coefficients $Q_1, \dots, Q_{m_\kappa}$ being analytical functions of z which vanish with z. Let

$$Q_i = z^{n_i} Q_i' (z),$$

where $Q_i' (z)$ is a regular function of z such that $Q_i' (0)$ is not zero and n_i is an integer; then clearly the most important terms in p_κ for our purpose are

$$v^{R_\kappa} z^{S_\kappa} \sum_{i=0}^{m_\kappa} v^{m_\kappa - i} z^{n_i} Q_i' (0),$$

because all other neglected terms have their moduli small compared with the moduli of one or more of the retained terms. Similarly as regards the other coefficients p; and also as regards the coefficients r and q.

Consequently, we have expressions of the form

$$f(u, v, z) = \frac{\Sigma K_{mnp} u^m v^n z^p + \dots}{\Sigma K_{abc} \, u^a \, v^b \, z^c + \dots} e^{F(u, v, z)},$$

$$g(u, v, z) = \frac{\Sigma K_{\mu\nu\pi} \, u^\mu \, v^\nu \, z^\pi + \dots}{\Sigma K_{abc} \, u^a \, v^b \, z^c + \dots} e^{(u, v, z)},$$

where the denominator and each of the numerators possess terms in u alone independent of v and z, and may possess terms in each of the three variables alone: and the functions F and G are regular.

If, however, any one, or any two, or all three, of the quantities $R(u, 0, 0)$, $P(u, 0, 0)$, $Q(u, 0, 0)$ should vanish identically, then the variables ξ, η, ζ are introduced, subject to the restrictions indicated before: and we obtain expressions of the form

$$f(u, v, z) = \frac{\Sigma K'_{mnp} \xi^m \eta^n \zeta^p + \dots}{\Sigma K'_{abc} \, \xi^a \, \eta^b \, \zeta^c + \dots} e^{F(u, v, z)},$$

$$g(u, v, z) = \frac{\Sigma K'_{\mu\nu\pi} \xi^\mu \, \eta^\nu \, \zeta^\pi + \dots}{\Sigma K'_{abc} \, \xi^a \, \eta^b \, \zeta^c + \dots} e^{G(u, v, z)},$$

of the same general type as in the last case. When re-transformation to the variables u, v, z is effected upon these expressions as in § 37, the ultimate forms are

$$f(u, v, z) = u^M v^N z^P \frac{\Sigma K_{mnp} u^m v^n z^p + \ldots}{\Sigma K_{abc}\ u^a v^b z^c + \ldots}\ e^{F(u,v,z)},$$

$$g(u, v, z) = u^A v^B z^C \frac{\Sigma K_{\mu\nu\pi}\ u^\mu v^\nu z^\pi + \ldots}{\Sigma K_{abc}\ u^a v^b z^c + \ldots}\ e^{G(u,v,z)},$$

where M, N, P, A, B, C are integers, positive or negative or zero; and the other functions that occur have the same general significance as before.

The expressions for f and g, substituted in the differential equations, lead to forms of the latter that are convenient for our immediate purpose.

If however the re-transformation does not lead to forms of this type, as would be the case were the original numerator or denominator of such a form as $uv + vz + zu$ or $u^2v + v^2z + z^2u$ or the like, then we proceed as follows. We have

$$(\xi, \eta, \zeta) = \left(\begin{array}{ccc} A_1, & B_1, & C_1 \\ A_2, & B_2, & C_2 \\ A_3, & B_3, & C_3 \end{array}\right)(u, v, z),$$

where the matrix (A_1, B_2, C_3) is the inverse of (a_1, b_2, c_3); and then

$$\frac{d\xi}{d\zeta} = \frac{A_1 \dfrac{du}{dz} + B_1 \dfrac{dv}{dz} + C_1}{A_3 \dfrac{du}{dz} + B_3 \dfrac{dv}{dz} + C_3}$$

$$= \frac{A_1 P + B_1 Q + C_1 R}{A_3 P + B_3 Q + C_3 R};$$

and similarly

$$\frac{d\eta}{d\zeta} = \frac{A_2 P + B_2 Q + C_2 R}{A_3 P + B_3 Q + C_3 R}.$$

The functions P, Q, R, when expressed in terms of ξ, η, ζ, possess terms involving one variable only: thus the reduction to the preceding form is possible, the variables u, v, z being replaced by ξ, η, ζ.

156.   Consider, therefore, the equations in the form

$$(\Sigma K_{abc} u^a v^b z^c + \dots) \frac{du}{dz} = (\Sigma K_{mnp} u^m v^n z^p + \dots) e^F,$$

$$(\Sigma K_{abc} u^a v^b z^c + \dots) \frac{dv}{dz} = (\Sigma K_{\mu\nu\pi} u^\mu v^\nu z^\pi + \dots) e^G ;$$

so that M, N, P, A, B, C are all assumed to be zero. To obtain the leading terms in u and v, we substitute

$$u = \alpha z^\rho, \quad v = \beta z^\sigma,$$

where α, β, ρ, σ are to be determined; the condition, leading to the desired determination, is that the adopted substitutions make the lowest terms on both sides of each equation cancel one another. Owing to the fact that the coefficient of $\frac{du}{dz}$ in the first equation is the same as that of $\frac{dv}{dz}$ in the second, the lowest terms on the left-hand side in the first and the lowest terms on the left-hand side in the second arise from the same group of terms in that coefficient: the order in the first is $\rho - \sigma$ greater than the order in the second.

Suppose that the term

$$K_{abc} u^a v^b z^c$$

on the left-hand sides, and that the terms

$$K_{mnp} u^m v^n z^p, \quad K_{\mu\nu\pi} u^\mu v^\nu z^\pi$$

on the right-hand sides, are terms which, for the substitutions

$$u = \alpha z^\rho, \quad v = \beta z^\sigma,$$

give powers of z at least as low as all others, and that they lead to equal terms in lowest powers of z on the two sides. In order that the indices of these lowest terms may be the same on the two sides, we must have

$$\left. \begin{array}{l} \rho a + \sigma b + c + \rho - 1 = \rho m + \sigma n + p \\ \rho a + \sigma b + c + \sigma - 1 = \rho \mu + \sigma \nu + \pi \end{array} \right\} ,$$

and therefore

$$\rho (a + 1) + \sigma (b + 1) + c = \rho m + \sigma (n + 1) + p + 1$$
$$= \rho (\mu + 1) + \sigma \nu + \pi + 1$$
$$= N,$$

say. In order that the coefficients of the terms on the two sides may be the same, we must have

$$\left.\begin{array}{l} \alpha\rho \, \Sigma \, K_{abc}\alpha^a\sigma^b = F_0 \Sigma K_{mnp}\alpha^m\beta^n \\ \beta\sigma \, \Sigma \, K_{abc}\alpha^a\sigma^b = G_0 \Sigma K_{\mu\nu\pi} \; \alpha^\mu\beta^\nu \end{array}\right\},$$

where F_0, G_0 are the values of $e^{F(0,0,0)}$, $e^{G(0,0,0)}$ respectively: and in the various summations those coefficients are included, which belong to the terms that give rise to the lowest indices of z after the substitutions are made.

The quantity N, being $\rho + 1$ greater than the index of the lowest power in the first equation and $\sigma + 1$ greater than the index of the lowest power in the second equation, (which two excesses are independent of the indices of terms in the numerators and the denominator of f and g), is smaller for the substitution adopted than is the corresponding quantity N deduced from other terms in the equation.

SPACE-DIAGRAMS.

157. To determine the quantities ρ and σ, we construct an obvious generalisation* of the Puiseux diagram which was used in the corresponding determination when there is only a single dependent variable (§§ 39—42).

We take, in ordinary three-dimensional space, three perpendicular axes OU, OV, OZ. Referred to these axes, we mark all the points

$$a + 1, \quad b + 1, \quad c$$

associated with the common denominator of f and g; all the points

$$m, \quad n + 1, \quad p + 1$$

associated with the numerator of f, being the expression for $\dfrac{du}{dz}$; and all the points

$$\mu + 1, \quad \nu, \quad \pi + 1$$

---

* This is not adopted by Königsberger (*l.c.*, § 1) p. 354, in the case of n dependent variables: the corresponding diagram would then exist in a space of $n + 1$ dimensions. The first occasion on which use was made of the generalisation of the Puiseux diagram to space of three dimensions, corresponding to the case of two dependent variables, appears to be in Goursat's preliminary propositions in the discussion of the singular solutions of simultaneous equations, *Amer. Journ. Math.*, t. XI (1889), p. 341.

associated with the numerator of g, being the expression for $\dfrac{dv}{dz}$.

Through the origin draw a line in the octant of space where all these points are marked; let its polar coordinates be θ, its inclination to the positive direction of the axis OZ, and ϕ, its azimuth in the positive direction from the plane of UZ: and let

$$\rho = \tan \theta \cos \phi, \quad \sigma = \tan \theta \sin \phi,$$

so that ρ and σ are positive quantities. Then any plane perpendicular to this line is

$$\rho U + \sigma V + Z = \Theta,$$

the value of Θ depending upon the distance from the origin.

In connection with the substitution

$$u = \alpha z^\rho, \quad v = \beta z^\sigma,$$

take a plane, determined by ρ and σ as coordinates of direction and passing through the point $a+1,\ b+1,\ c$; its equation is

$$\rho U + \sigma V + Z = \rho(a+1) + \sigma(b+1) + c$$
$$= N.$$

It passes also through the point $m,\ n+1,\ p+1$; through the point $\mu+1,\ \nu,\ \pi+1$; and through all points associated with terms that, for the adopted substitution, give rise to the lowest powers of z in the differential equations. The perpendicular from the origin upon this plane is

$$N(1 + \rho^2 + \sigma^2)^{-\frac{1}{2}};$$

since the value of N is a minimum, this quantity is less than the perpendicular from the origin upon a parallel plane through any of the other points in the tableau. If a surface consisting of portions of planes can be constructed, so that it is always convex to the origin, and such that between the three coordinate planes and this surface no point of the triple system in the tableau is enclosed, then each of the various plane portions of the surface determines quantities ρ and σ that are positive. Each such portion furnishes the orders of the associable substitution.

The process of obtaining these planes is similar to that adopted for obtaining the broken line in the Puiseux diagram. When points occur lying in the three coordinate planes, some of them on the axes of reference, it is convenient to construct in each of the coordinate planes the broken line as for a Puiseux diagram in that plane. When there are no points in the tableau lying in the coordinate planes, it is convenient to select the point or points of lowest Z-ordinate: if there be several, through them draw a plane, which is parallel to the U-V plane, and in it construct the broken line as for a Puiseux diagram in that plane.

Through the initial point, if there be only one point of lowest Z-ordinate, take a plane parallel to the U-V plane: through each portion of the broken line in the plane of points of lowest Z-ordinates, if there be more than one point in this plane, take the plane parallel to the U-V plane. Round the point, or round each portion of the broken line, as the case may be, cause the plane to turn towards coincidence with the U-Z plane or the V-Z plane: and let it continue to turn, until it meets one point or more than one point in the space-tableau. In each such position, construct a broken line as for a Puiseux diagram in that position of the plane among the points which it contains: and round each portion of the broken line let the plane be turned in succession, until it meets other points in the space-tableau. Let the operation be continued, until the portions of broken line (if any) in the U-Z plane are made to lie in the moving plane in one or more than one of its positions: and so also with the portions of broken line (if any) in the V-Z plane.

It is clear that each portion of the broken surface thus constructed satisfies the conditions as to (i) the respective minimum values of N, (ii) the positiveness in sign of the quantities ρ and σ.

Further, it should be remarked that the description of the construction is applicable, not solely to the case immediately under consideration when M, N, P, A, B, C, all are zero, but also to other cases when several of them can differ from zero: and even in the cases when several of them are negative, it can be applied if, instead of beginning with the coordinate planes, we take planes parallel to them through the points of smallest ordinates. Also, it may happen that the broken line in one of the coordinate planes does not find itself subjected to pass through a point on an axis of coordinates in the plane: thus for the tableau of points $a+1,\ b+1,\ c$, all the points for which c is zero lie off the axis of U and the axis of V.

Now take a portion of a plane containing three points such as $a+1,\ b+1,\ c:\ m,\ n+1,\ p+1:\ \mu+1,\ \nu,\ \pi+1:$ its angular coordinates ρ and σ are given by the equations

$$\frac{\rho}{\begin{vmatrix} b+1, & c & , 1 \\ n+1, & p+1, & 1 \\ \nu & , \pi+1, & 1 \end{vmatrix}} = \frac{\sigma}{\begin{vmatrix} c & , a+1, & 1 \\ p+1, & m & , 1 \\ \pi+1, & \mu+1, & 1 \end{vmatrix}} = \frac{1}{\begin{vmatrix} a+1, & b+1, & 1 \\ m & , n+1, & 1 \\ \mu+1, & \nu & , 1 \end{vmatrix}}.$$

Each of the denominators is a determinant of integers and therefore each itself is an integer; let any factor common to all three denominator-values be removed, so that we have (say)

$$\rho = \frac{\theta}{\psi}, \quad \sigma = \frac{\phi}{\psi},$$

where $\theta,\ \phi,\ \psi$ are integers having no factor common to all three. Then we change the variables by the substitutions

$$z = t^{\psi}, \qquad u = t^{\theta} u_1 \qquad\qquad v = t^{\phi} v_1$$
$$= t^{\theta}(\alpha + U), \qquad = t^{\phi}(\beta + V),$$

where α and β are constants, and the quantities U and V are required to vanish with z and therefore with t.

REDUCED FORM OF THE EQUATION IN CASE V.

158. In order to obtain the explicit expression of the reduced typical forms, at least in the more important instances, we shall assume that the factors $e^{F},\ e^{G}$ are absorbed into the other parts of the expressions on the right-hand sides of the equations—an assumption that does not affect any of the results thus far obtained: consequently the equations may be taken in the form

$$\left.\begin{aligned} (\Sigma K_{abc} u^{a} v^{b} z^{c} + \ldots)\frac{du}{dz} &= \Sigma K_{mnp} u^{m} v^{n} z^{p} + \ldots \\[2mm] (\Sigma K_{abc} u^{a} v^{b} z^{c} + \ldots)\frac{dv}{dz} &= \Sigma K_{\mu\nu\pi} w^{u} v^{\nu} z^{\pi} + \ldots \end{aligned}\right\}.$$

When the preceding substitutions are effected, these equations become

$$\left(\theta u_1 + t\,\frac{du_1}{dt}\right) t^{\psi N - \phi - \psi}\left[\Sigma K_{abc} u_1{}^a v_1{}^b + t S_1\left(u_1,\,v_1,\,t\right)\right]$$
$$= \psi\left[\Sigma K_{mnp} u_1{}^m v_1{}^n + t S_2\left(u_1,\,v_1,\,t\right)\right] t^{\psi N - \phi - \psi},$$

$$\left(\phi v_1 + t\,\frac{dv_1}{dt}\right) t^{\psi N - \theta - \psi}\left[\Sigma K_{abc} u_1{}^a v_1{}^b + t S_1\left(u_1,\,v_1,\,t\right)\right]$$
$$= \psi\left[\Sigma K_{\mu\nu\pi} u_1{}^\mu v_1{}^\nu + t S_3\left(u_1,\,v_1,\,t\right)\right] t^{\psi N - \theta - \psi},$$

where S_1, S_2, S_3 are regular functions of u_1, v_1, t. On removing the common power of t from the respective sides of these equations, they become

$$\left(\theta u_1 + t\,\frac{du_1}{dt}\right)\{\Sigma K_{abc} u_1{}^a v_1{}^b + t S_1\left(u_1,\,v_1,\,t\right)\}$$
$$= \psi\{\Sigma K_{mnp} u_1{}^m v_1{}^n + t S_2\left(u_1,\,v_1,\,t\right)\},$$

$$\left(\phi v_1 + t\,\frac{dv_1}{dt}\right)\{\Sigma K_{abc} u_1{}^a v_1{}^b + t S_1\left(u_1,\,v_1,\,t\right)\}$$
$$= \psi\{\Sigma K_{\mu\nu\pi} u_1{}^\mu v_1{}^\nu + t S_3\left(u_1,\,v_1,\,t\right)\},$$

respectively.

The variables u_1 and v_1 are required to be distinct from zero when $t = 0$; their values are α and β respectively when $t = 0$. (If they were expressible as power-series in t, representing regular functions, the series would have α and β as the initial terms.) Hence, taking $t = 0$ in the above differential equations, α and β are determined by the two algebraical equations

$$\left.\begin{aligned}G\left(\alpha,\,\beta\right) &= \theta\alpha\Sigma K_{abc}\alpha^a\beta^b - \psi\Sigma K_{mnp}\alpha^m\beta^n = 0 \\ H\left(\alpha,\,\beta\right) &= \phi\beta\Sigma K_{abc}\alpha^a\beta^b - \psi\Sigma K_{\mu\nu\pi}\alpha^\mu\beta^\nu = 0\end{aligned}\right\};$$

simultaneous solutions of these equations—the solutions giving values different from zero—are suitable values of α and β.

Let α and β represent a simultaneous solution of these equations: the character of α and β, in regard to multiplicity, may be illustrated by reference to the theory of plane curves. The equations $G\left(\alpha,\,\beta\right) = 0$, $H\left(\alpha,\,\beta\right) = 0$ may be taken as representing two plane curves: α, β will then denote the coordinates of a point of intersection. If it be an ordinary point on each curve, then the order of the roots is simple when the curves have distinct tangents at the common point; it is duplex when the curves, having a common tangent at the point, have different curvatures: and so on. If the point be a multiple point on either curve or on both, the order of occurrence of the roots is determined (by the known results obtained for the intersections of curves) according to the character of the multiple points.

159. Taking as the first case that in which α, β is an ordinary point on each curve, let

$$u_1 = \alpha + U, \quad v_1 = \beta + V,$$

where U and V (if they exist) are functions of t that vanish with t; then the coefficient of $t\dfrac{du_1}{dt}$ is

$$\Sigma K_{abc}\alpha^a\beta^b + R(U, V, t),$$

where $R(U, V, t)$ is a regular function that vanishes with U, V, t. Let all the terms in the first equation, other than those which involve $t\dfrac{du_1}{dt}$, be transferred to the right-hand side; their aggregate is

$$-U\frac{\partial G}{\partial \alpha} - V\frac{\partial G}{\partial \beta} - \underset{M+N\geqslant 2}{\Sigma\Sigma} \frac{U^M V^N}{M!\,N!}\frac{\partial^{M+N}G}{\partial\alpha^M\partial\beta^N} - tS(U, V, t),$$

where S is a regular function of its arguments. Thus the equation for $t\dfrac{dU}{dt}$ is

$$t\frac{dU}{dt} = \frac{-U\dfrac{\partial G}{\partial \alpha} - V\dfrac{\partial G}{\partial \beta} - \ldots - tS(U, V, t)}{\Sigma K_{abc}\alpha^a\beta^b + R(U, V, t)};$$

and similarly the equation for $t\dfrac{dV}{dt}$ is

$$t\frac{dV}{dt} = \frac{-U\dfrac{\partial H}{\partial \alpha} - V\dfrac{\partial H}{\partial \beta} - \ldots - tT(U, V, t)}{\Sigma K_{abc}\alpha^a\beta^b + R(U, V, t)}.$$

Suppose that α and β are such as to make

$$\Sigma K_{abc}\alpha^a\beta^b \neq 0,$$

and let its value be Λ, which is a constant: then in the functions on the right-hand side, the denominator does not vanish when $U=0$, $V=0$, $t=0$; and consequently the functions can be expanded as regular power-series in U, V, t, so that the equations become

$$t \frac{dU}{dt} = \kappa_1 U + \lambda_1 V + i_1 t + \theta_1 (U, V, t) \Big|$$

$$t \frac{dV}{dt} = \kappa_2 U + \lambda_2 V + i_2 t + \theta_2 (U, V, t) \Big|,$$

where θ_1 and θ_2 are regular functions of U, V, t in the form of power-series, the lowest terms being of dimensions at least two, and the values of the coefficients κ and λ being

$$\Lambda \kappa_1 = -\frac{\partial G}{\partial \alpha} \Big| \qquad \Lambda \kappa_2 = -\frac{\partial H}{\partial \alpha} \Big|$$

$$\Lambda \lambda_1 = -\frac{\partial G}{\partial \beta} \Big| , \qquad \Lambda \lambda_2 = -\frac{\partial H}{\partial \beta} \Big| .$$

This form will be regarded as the typical reduced form: and in this form, the quantities κ_1 and λ_1 do not both vanish, and also the quantities κ_2 and λ_2 do not vanish. Manifestly the reduction is effective provided that Λ does not vanish.

If, however, while Λ does not vanish, κ_1 and λ_1 both vanish, or κ_2 and λ_2 both vanish, or both pairs vanish, then the equation for $t \frac{dU}{dt}$, or the equation for $t \frac{dV}{dt}$, or both equations, will have a different form.

Accordingly, as our second case, suppose that the lowest derivatives of G with regard to u_1 and v_1, which do not vanish for the values of α, β, are of order M_1 where $M_1 \geqslant 2$. The equation for $t \frac{dU}{dt}$ then is

$$t \frac{dU}{dt} = -\frac{1}{M_1!} \frac{1}{\Lambda} \left(U \frac{\partial}{\partial \alpha} + V \frac{\partial}{\partial \beta} \right)^{M_1} G + i_1 t + \Theta_1 (U, V, t),$$

where Θ_1 is a regular function of t vanishing with t, U, V, the lowest term in Θ_1 involving t alone being of index $\geqslant 2$, and the dimensions of the lowest terms in Θ_1 independent of t being of order $> M_1$.

As a third case, let the lowest derivatives of H with regard to u_1 and v_1, which do not vanish at α, β, be of order M_2 where $M_2 \geqslant 2$; the equation for $t \frac{dV}{dt}$ is

$$t \frac{dV}{dt} = -\frac{1}{M_2!} \frac{1}{\Lambda} \left(U \frac{\partial}{\partial \alpha} + V \frac{\partial}{\partial \beta} \right)^{M_2} H + i_2 t + \Theta_2 (U, V, t),$$

where Θ_2 is a regular function of t vanishing with t, U, V, the lowest term in Θ_2 involving t alone being of index $\geqslant 2$, and the dimensions of the lowest terms in Θ_2 independent of t being of order $> M_2$.

Similarly for a fourth case, viz. that in which the conditions for both the second and the third cases are satisfied, the corresponding change occurs in both equations. All the instances can be included in the form

$$
\left.
\begin{aligned}
t\frac{dU}{dt} &= -\frac{1}{M_1!}\frac{1}{\Lambda}\left(U\frac{\partial}{\partial\alpha} + V\frac{\partial}{\partial\beta}\right)^{M_1} G + i_1 t + \Theta_1(U,\,V,\,t) \\
t\frac{dV}{dt} &= -\frac{1}{M_2!}\frac{1}{\Lambda}\left(U\frac{\partial}{\partial\alpha} + V\frac{\partial}{\partial\beta}\right)^{M_2} H + i_2 t + \Theta_2(U,\,V,\,t)
\end{aligned}
\right\},
$$

where Θ_1 and Θ_2 are regular functions in which the lowest power of t has its index $\geqslant 2$, and the terms of lowest order in U and V independent of t are of dimensions $> M_1$, $> M_2$, respectively.

For the first case of all, $M_1 = 1$, $M_2 = 1$; for the second case, $M_1 \geqslant 2$, $M_2 = 1$; for the third case, $M_1 = 1$, $M_2 \geqslant 1$; for the fourth case, $M_1 \geqslant 2$, $M_2 \geqslant 2$.

160. If however Λ be zero, that is, if $u_1 = \alpha$, $v_1 = \beta$ make

$$\Sigma K_{abc} u_1{}^a v_1{}^b$$

vanish, then the preceding form does not arise as the resulting type. The coefficient of $t\dfrac{du_1}{dt}$ in the equations, before the substitutions $u_1 = \alpha + U$, $v_1 = \beta + V$ are made, is

$$\Sigma K_{abc} u_1{}^a v_1{}^b + tS_1(u_1,\,v_1,\,t).$$

Let

$$F(\alpha,\,\beta) = \Sigma K_{abc}\alpha^a\beta^b;$$

then, since $F = 0$, the value of the coefficient of $t\dfrac{du_1}{dt}$ after the transformations have been made is

$$\frac{1}{M_3!}\left(U\frac{\partial}{\partial\alpha} + V\frac{\partial}{\partial\beta}\right)^{M_3} F + i_3 t + \Theta_3(U,\,V,\,t),$$

where $M_3 \geqslant 1$, Θ_3 is a regular function vanishing with U, V, t in which the lowest power of t has its index $\geqslant 2$, and the terms of lowest order in U and V independent of t are of dimensions $> M_3$.

The equations thus become

$$t\frac{dU}{dt} = -\frac{\dfrac{1}{M_1!}\left(U\dfrac{\partial}{\partial\alpha}+V\dfrac{\partial}{\partial\beta}\right)^{M_1}G+i_1t+\Theta_1(U,V,t)}{\dfrac{1}{M_3!}\left(U\dfrac{\partial}{\partial\alpha}+V\dfrac{\partial}{\partial\beta}\right)^{M_3}F+i_3t+\Theta_3(U,V,t)}$$

$$t\frac{dV}{dt} = -\frac{\dfrac{1}{M_2!}\left(U\dfrac{\partial}{\partial\alpha}+V\dfrac{\partial}{\partial\beta}\right)^{M_2}H+i_2t+\Theta_2(U,V,t)}{\dfrac{1}{M_3!}\left(U\dfrac{\partial}{\partial\alpha}+V\dfrac{\partial}{\partial\beta}\right)^{M_3}F+i_3t+\Theta_3(U,V,t)}$$

The simplest case is manifestly that which is given by

$$M_3 = 1, \quad M_1 = 1, \quad M_2 = 1.$$

161. As already indicated, we are limiting our detailed investigations to the case when the number of dependent variables is two. The perfectly general character of the last result indicates what the corresponding form is when the number of dependent variables is n.

Let the equations be

$$(\Sigma K_{ab\ldots kl}u_1{}^a u_2{}^b \ldots u_n{}^k z^l + \ldots)\frac{du_r}{dz}$$

$$= \Sigma K_{m_r n_r \cdots q_r p}u_1{}^{m_r}u_2{}^{n_r}\ldots u_n{}^{q_r}z^p + \ldots,$$

for $r = 1, 2, \ldots, n$; determine quantities $\mu_1, \mu_2, \ldots, \mu_n$ by the process (either geometrically or analytically completed) used to determine ρ and σ in the preceding case, and suppose that they are expressed in the form

$$\mu_s = \frac{\theta_s}{\psi}, \qquad (s = 1, \ldots, n),$$

where the integers $\theta_1, \ldots, \theta_n, \psi$ have no factor common to all. Let

$$F_r(\alpha_1, \ldots, \alpha_n) = \psi\Sigma K_{m_r n_r \cdots q_r p}\alpha_1{}^{m_r}\alpha_2{}^{n_r}\ldots\alpha_n{}^{q_r}$$
$$- \theta_r\alpha_r\Sigma K_{ab\ldots kl}\alpha_1{}^a\alpha_2{}^b \ldots \alpha_n{}^k,$$

for $r = 1, \ldots, n$; and let

$$G = G(\alpha_1, \ldots, \alpha_n) = \Sigma K_{ab\ldots kl}\alpha_1{}^a\alpha_2{}^b \ldots \alpha_n{}^k.$$

Then let

$$z = t^\psi, \quad u_r = (\alpha_r + U_r)t^{\theta_r}, \qquad (r = 1, \ldots, n),$$

where the n quantities α are determined by the n equations

$$F_s(\alpha_1, \ldots, \alpha_n) = 0.$$

Then the equations for the quantities U are

$$t\frac{dU_r}{dt} = -\frac{\dfrac{1}{M_r!}\left\{\Sigma U_s\dfrac{\partial}{\partial\alpha_s}\right\}^{M_r} F_r + i_r t + \Theta_r(U_1, \ldots, U_n, t)}{\dfrac{1}{M!}\left\{\Sigma U_s\dfrac{\partial}{\partial\alpha_s}\right\}^{M} G + it + \Theta(U_1, \ldots, U_n, t)},$$

for $r = 1, \ldots, n$. In obtaining this result, the conditions are:—

(i) the derivatives of F_r with regard to $\alpha_1, \ldots, \alpha_n$ of lowest order which do not vanish are of order M_r, so that $M_r \geqslant 1$;

(ii) the derivatives of G with regard to $\alpha_1, \ldots, \alpha_n$ of lowest order which do not vanish are of order M. If $M = 0$, the fractions for $t\dfrac{dU_r}{dt}$ can be at once expressed as regular functions of $U_1, \ldots, U_n, t$; if $M > 0$, further transformations are necessary;

(iii) the functions Θ_r are regular; the lowest power of t alone occurring in them has its index > 1: and the lowest terms in the variables U independent of t are of dimensions $> M_r$. Likewise for the function Θ.

REMAINING FORMS.

162. Thus far we have discussed only the case in which the values of ρ and σ, being the orders of the leading terms of u and v in powers of z, arose from terms in the numerators of $\dfrac{du}{dz}$ and $\dfrac{dv}{dz}$ and from terms in their common denominator. It might, of course, happen for particular cases that values of ρ and σ were provided through the denominator alone; in which case the left-hand side of the preceding equations would be replaced by

$$t^{k+1}\frac{dV}{dt}, \quad t^{l+1}\frac{dV}{dt}, \qquad (k \geqslant 1, \ l \geqslant 1),$$

in the appropriate reduced forms. It might happen that values of ρ and σ were provided by the denominator and by only one of the numerators and not by the other, say by that of $\dfrac{du}{dz}$ but not by that of $\dfrac{dv}{dz}$; in which case the left-hand sides of the preceding equations would be replaced by

$$t\frac{dU}{dt}, \quad t^{l+1}\frac{dV}{dt}, \qquad\qquad (l \geqslant 1),$$

in the appropriate reduced forms.

Similarly for other possible combinations: we use the general method adopted for the preceding cases, in the succession which corresponds to that considered (Chapter V) for a single equation.

Again, the preceding reductions are effective for the forms indicated at the beginning of § 156; but these forms are only particular cases of the most general form given in § 155.

Without entering into the full discussion of all the alternatives connected with

$$\left.\begin{array}{l} \dfrac{du}{dz} = u^M v^N z^P \dfrac{\Sigma K_{mnp} u^m v^n z^p + \cdots}{\Sigma K_{abc} u^a v^b z^c + \cdots} e^{F\,(u,\,v,\,z)} \\[3mm] \dfrac{dv}{dz} = u^A v^B z^C \dfrac{\Sigma K_{\mu\nu\pi} u^\mu v^\nu z^\pi + \cdots}{\Sigma K_{abc} u^a v^b z^c + \cdots} e^{G\,(u,\,v,\,z)} \end{array}\right\},$$

that can arise according as the integers M, N, P, A, B, C are positive, zero, or negative in all the possible combinations, we shall merely indicate the initial stage. The detailed development bears the same broad relation to the development in the case when all these integers are zero as, in the case of a single equation, does the detailed development (§§ 50—58) bear to the development when the corresponding integers are zero (§§ 43—46).

If for sufficiently small values of $|z|$, the leading terms in u and v are

$$u = \alpha z^\rho + \ldots, \quad v = \beta z^\sigma + \ldots,$$

and if the lowest powers of z for these substitutions spring from the denominator and both the numerators, say from terms

$$K_{abc} u^a v^b z^c, \quad K_{mnp} u^m v^n z^p, \quad K_{\mu\nu\pi} u^\mu v^\nu z^\pi,$$

then, as before, we have

$$\rho - 1 + a\rho + b\sigma + c = M\rho + N\sigma + P + m\rho + n\sigma + p,$$

$$\sigma - 1 + a\rho + b\sigma + c = A\rho + B\sigma + C + \mu\rho + \nu\sigma + \pi,$$

that is,

$$(a+1)\rho+(b+1)\sigma+c=(m+M)\rho+(n+1+N)\sigma+p+1+P$$
$$=(\mu+1+A)\rho+(\nu+B)\sigma+\pi+1+C.$$

Taking the same three axes as before, and referring points to them as axes of reference, we mark all the points

$$a+1, \quad b+1, \quad c,$$

associated with the common denominator for the various values of a, b, c; all the points

$$m+M, \quad n+1+N, \quad p+1+P,$$

associated with the numerator of $\dfrac{du}{dz}$ for the various values of m, n, p; and all the points

$$\mu+1+A, \quad \nu+B, \quad \pi+1+C,$$

associated with the numerator of $\dfrac{dv}{dz}$ for the various values of μ, ν, π.

With all the points of this tableau, and by means of planes, we construct the same configuration as in the case when M, N, P, A, B, C all are zero. Then those plane portions of the broken surface thus constructed, which give positive values for their angular coordinates ρ and σ, correspond to appropriate values of leading terms in u and v; and they lead, by corresponding analysis, similar to that in the preceding instance, to the reduced typical forms.

As in the case, when M, N, P, A, B, C all are zero, it may be necessary to take account of instances in which terms of the lowest order for an appropriate substitution arise out of the numerators only, and not from the common denominator: similarly, also for the other possible combinations for the source of terms of the lowest orders.

Many of these forms are evidently only reduced in part to typical forms. The further reduction can be carried out by a repetition of the process adopted for the change from the original form to the form which has already been obtained: but as regards general investigations, the process merely leads to an accumulation of analytical forms of no special value for the present purpose. The actual details of such reductions in the case of a single equation, for sub-reduced forms, were exhibited for one case in Ex. 5, § 60:

the corresponding process can be carried out for the cases that have arisen in connection with a system of two equations.

Some particular examples of the general process will now be considered.

163. *Ex.* 1. Consider the equations

$$\left.\begin{array}{l}\dfrac{du}{dz} = \dfrac{a'u + b'v + c'z + \text{higher powers}}{au + bv + cz + \text{higher powers}} \\[3mm] \dfrac{dv}{dz} = \dfrac{a''u + b''v + c''z + \text{higher powers}}{au + bv + cz + \text{higher powers}}\end{array}\right\} .$$

The points in the tableau that need to be taken into account are the three points : $(2, 2, 1)$ from the common denominator, $(1, 2, 2)$ from the numerator of $\dfrac{du}{dz}$, and $(2, 1, 2)$ from the numerator of $\dfrac{dv}{dz}$; and no others. A plane through them is

$$U + V + Z = 5,$$

that is, we have $\rho = 1$, $\sigma = 1$; and therefore—no transformation of the dependent variable being necessary—the substitutions are

$$u = u_1 z, \quad v = v_1 z.$$

The equations for u_1 and v_1 become

$$\{c + au_1 + bv_1 + zP(u_1, v_1, z)\} z \frac{du_1}{dz} = \{c' + a'u_1 + b'v_1 + zQ(u_1, v_1, z)\}$$
$$- u_1 \{c + au_1 + bv_1 + zP(u_1, v_1, z)\},$$
$$\{c + au_1 + bv_1 + zP(u_1, v_1, z)\} z \frac{dv_1}{dz} = \{c'' + a''u_1 + b''v_1 + zR(u_1, v_1, z)\}$$
$$- v_1 \{c + au_1 + bv_1 + zP(u_1, v_1, z)\}.$$

Let a, β be the respective values of u_1, v_1, when $z = 0$; then

$$\frac{c' + a'a + b'\beta}{a} = \frac{c'' + a''a + b''\beta}{\beta} = c + aa + b\beta.$$

Denoting the common value of the three expressions by θ, we have

$$\begin{vmatrix} c - \theta, & a\ , & b \\ c'\ , & a' - \theta, & b' \\ c''\ , & a''\ , & b'' - \theta \end{vmatrix} = 0,$$

a cubic which in general has three roots distinct from one another. In connection with a root $\theta = \vartheta$, we have

$$\left.\begin{array}{l}(a' - \vartheta)\, a + b'\beta = -c' \\ a''a + (b'' - \vartheta)\,\beta = -c''\end{array}\right\},$$

which determine a and β uniquely unless

$$(a' - \vartheta)(b'' - \vartheta) - a''b' = 0 ;$$

with this exclusive condition, each root ϑ gives a determination of a and β. Hence, writing

$$u_1 = a + U, \quad v_1 = \beta + V,$$

for any such determination, we have

$$z\frac{dU}{dz} = \frac{(a'-aa-\vartheta)\,U + (b'-ba)\,V + \text{term in } z + zP_1\,(U,\,V,\,z)}{\vartheta + aU + bV + zP\,(u_1,\,v_1,\,z)}$$

$$= \frac{1}{\vartheta}\{(a'-aa-\vartheta)\,U + (b'-ba)\,V + \gamma z\} + zR_1\,(U,\,V,\,z)$$

$$z\frac{dV}{dz} = \frac{1}{\vartheta}\{(a''-a\beta)\,U + (b''-b\beta-\vartheta)\,V + \gamma'z\} + zR_2\,(U,\,V,\,z)$$

where R_1 and R_2 are regular functions of their arguments.

This reduction is satisfactory unless $\vartheta = 0$: and this can happen only when

$$\begin{vmatrix} c, & a, & b \\ c', & a', & b' \\ c'', & a'', & b'' \end{vmatrix} = 0.$$

If however this determinant does vanish, then the forms of the equations are

$$z\frac{dU}{dz} = \frac{a_1'U + b_1'V + c_1'z + \text{higher powers}}{a_1 U + b_1 V + c_1 z + \text{higher powers}},$$

$$z\frac{dV}{dz} = \frac{a_1''U + b_1''V + c_1''z + \text{higher powers}}{a_1 U + b_1 V + c_1 z + \text{higher powers}}.$$

Ex. 2.   Consider the equations

$$z\frac{du}{dz} = \frac{a'u + b'v + c'z + \ldots}{au + bv + cz + \ldots}$$

$$z\frac{dv}{dz} = \frac{a''u + b''v + c''z + \ldots}{au + bv + cz + \ldots},$$

which represent the simplest of the incompletely reduced forms in § 160, and also the special unreduced form in the last example when ϑ is zero.

Proceeding as by the general method, and in the first place assuming the possibility of terms of the lowest order arising from both the numerators and from the common denominator, we have $M = 0$, $N = 0$, $P = -1$, $A = 0$, $B = 0$, $C = -1$: the points to be taken account of in the tableau are $(2, 2, 1)$, $(1, 2, 1)$, $(2, 1, 1)$: so that the plane $\rho U + \sigma V + Z = N$ is $Z = 1$, that is, $\rho = 0$, $\sigma = 0$. This shews that no substitution of the required type exists in connection with the assumption that terms of the lowest order arise from the denominator and the numerators simultaneously : a result that is sufficiently obvious from independent inspection of the equation.   In fact, the equations can be satisfied (if at all) only by values of u and v, which make the terms of lowest order in the numerators vanish, without at the same time making the terms of that order vanish in the denominator.

Accordingly, we take two new variables U and V, vanishing with z, such that

$$a'u + b'v + c'z = Uz,$$

$$a''u + b''v + c''z = Vz,$$

and therefore

$$au + bv + cz = \frac{z}{a'b'' - a''b'}\begin{vmatrix} a', & b', & c'-U \\ a'', & b'', & c''-V \\ a, & b, & c \end{vmatrix},$$

a result which satisfies the first condition if the determinant $(a'b''c)$ does not vanish : this will be assumed. Now with these values

$$z\frac{du}{dz} = \frac{Uz + z^2\Theta_1(U, V, z)}{Kz + z\Phi(U, V, z)} = \frac{U + z\Theta_1(U, V, z)}{K + \Phi(U, V, z)},$$

$$z\frac{dv}{dz} = \frac{Vz + z^2\Theta_2(U, V, z)}{Kz + z\Phi(U, V, z)} = \frac{V + z\Theta_2(U, V, z)}{K + \Phi(U, V, z)},$$

where Θ_1 and Θ_2 are regular functions of their arguments, which do not necessarily vanish when $U=0$, $V=0$, $z=0$; Φ is a regular function of its arguments, which does vanish when $U=0$, $V=0$, $z=0$; and K is not zero. Hence

$$z^2\frac{dU}{dz} + zU = c'z + a'z\frac{du}{dz} + b'z\frac{dv}{dz}$$

$$= c'z + \frac{a'U + b'V + z\{a'\Theta_1(U, V, z) + b'\Theta_2(U, V, z)\}}{K + \Phi(U, V, z)},$$

and therefore $\quad z^2\dfrac{dU}{dz} = \dfrac{a'}{K}U + \dfrac{b}{K}V + c_1'z + \text{higher powers};$

and similarly, $\quad z^2\dfrac{dV}{dz} = \dfrac{a''}{K}U + \dfrac{b''}{K}V + c_1''z + \text{higher powers};$

the value of K being

$$\begin{vmatrix} a', & b', & c' \\ a'', & b'', & c'' \\ a, & b, & c \end{vmatrix} \div \begin{vmatrix} a', & b' \\ a'', & b'' \end{vmatrix}.$$

The equations are now in their reduced typical forms.

Ex. 3. Obtain typical reduced forms for the sets of equations : —

$$\left.\begin{aligned} \frac{du}{dz} &= \frac{av^2 + bz^2}{au + \beta v + \gamma z} \\ \frac{dv}{dz} &= \frac{a'u^2 + b'z^2}{au + \beta v + \gamma z} \end{aligned}\right\};$$

$$\left.\begin{aligned} \frac{du}{dz} &= \frac{au^3 + a'v^3 + \kappa_3 z^3}{cu + c'v + \kappa_1 z} \\ \frac{dv}{dz} &= \frac{bu^2 + b'v^2 + \kappa_2 z^2}{cu + c'v + \kappa_1 z} \end{aligned}\right\};$$

$$\left.\begin{aligned} z\frac{du}{dz} &= \frac{(a, b, c, f, g, h\,\backslash\!\!\backslash u, v, z)^2}{(a, \beta, \gamma\,\backslash\!\!\backslash u, v, z)} \\ z\frac{dv}{dz} &= \frac{(a', b', c', f', g', h'\,\backslash\!\!\backslash u, v, z)^2}{(a, \beta, \gamma\,\backslash\!\!\backslash u, v, z)} \end{aligned}\right\};$$

$$\left.\begin{aligned} z\frac{du}{dz} &= \frac{au^\alpha + a'v^\alpha + \kappa_2 z}{cu^\gamma + c'v^\gamma + \lambda z} \\ z\frac{dv}{dz} &= \frac{bu^\beta + b'v^\beta + \kappa_1 z}{cu^\gamma + c'v^\gamma + \lambda z} \end{aligned}\right\};$$

$$\left.\begin{aligned} z\frac{du}{dz} &= \frac{au + bv + cz}{a'u + b'v + c'z} \\ z\frac{dv}{dz} &= \frac{au + \beta v + \gamma z}{a'u + \beta'v + \gamma'z} \end{aligned}\right\}.$$

Ex. 4. As an indication of another method of proceeding, consider the equations of the first example in the specially simple case

$$\frac{du}{dz} = \frac{a'u + b'v + c'z}{au + bv + cz}, \quad \frac{dv}{dz} = \frac{a''u + b''v + c''z}{au + bv + cz}.$$

Introduce a new variable ξ, such that

$$\frac{du}{d\xi} = a'u + b'v + c'z,$$

$$\frac{dv}{d\xi} = a''u + b''v + c''z,$$

$$\frac{dz}{d\xi} = au + bv + cz.$$

Let w_1, w_2, w_3 be the roots of the cubic

$$\begin{vmatrix} a' - w, & b' , & c' \\ a'' , & b'' - w, & c'' \\ a , & b , & c - w \end{vmatrix} = 0,$$

which is the cubic that occurs in the earlier reduction. Then the most general solution of the equations is

$$u = A_1 e^{w_1 \xi} + A_2 e^{w_2 \xi} + A_3 e^{w_3 \xi},$$

$$v = B_1 e^{w_1 \xi} + B_2 e^{w_2 \xi} + B_3 e^{w_3 \xi},$$

$$z = C_1 e^{w_1 \xi} + C_2 e^{w_2 \xi} + C_3 e^{w_3 \xi},$$

where the ratios of quantities A_i, B_i, C_i associated with w_i are determined by any two of the equations

$$\left. \begin{array}{l} (a' - w_i) A_i + b' B_i + c' C_i = 0 \\ a'' A_i + (b'' - w_i) B_i + c'' C_i = 0 \\ a A_i + b B_i + (c - w_i) C_i = 0 \end{array} \right\}.$$

Now in place of ξ introduce a variable η, such that

$$\frac{d\eta}{\eta} = w_1 d\xi,$$

so that the form of the equations is

$$\left. \begin{array}{l} w_1 \eta \dfrac{du}{d\eta} = a'u + b'v + c'z \\[2mm] w_1 \eta \dfrac{dv}{d\eta} = a''u + b''v + c''z \\[2mm] w_1 \eta \dfrac{dz}{d\eta} = au + bv + cz \end{array} \right\} ;$$

then each of the quantities u, v, z is expressible in the form

$$A + B\eta^\lambda + C\eta^\mu,$$

where

$$\lambda = \frac{w_2}{w_1}, \quad \mu = \frac{w_3}{w_1}.$$

This method can be applied also when the expressions for $\dfrac{du}{dz}$ and $\dfrac{dv}{dz}$ have the fuller generality as in Ex. 1 : and then η is a parametric quantity in terms of which u, v, z are expressible. The most useful application arises when u, v, z, η, λ, μ all are real ; the differential equations are then equations of skew curves in ordinary space of three dimensions. This application will, however, not be developed here, largely for the reasons that led to the omission of the corresponding application in connection with a single equation*.

Ex. 5.   It might prove of some interest to examine in detail one special instance of the general case considered in this chapter, viz. that in which the equation is one of the second order, and of such a form as to give $\dfrac{d^2y}{dx^2}$ explicitly in terms of x, y, p. Taking y and p as dependent variables : assigning the condition that y (though not necessarily p) vanishes with x and, as an alternative, that y and p vanish with x; it would be necessary to obtain the various reduced forms for the equation

$$\frac{dp}{dx} = f(x, y, p),$$

for values of the variables in the vicinity of every non-ordinary combination of $f(x, y, p)$.

See, in this connection, Chapter xiv.

---

* See Ex. 1, § 60 ; and the footnote at the beginning of Chapter viii.

CHAPTER XII.

The Integrals of the Reduced Forms of a System of Equations, chiefly of Two Dependent Variables*.

164. Before proceeding to the establishment of the integrals of the typical reduced forms of the equations, an indication of some of their properties can be obtained from the discussion of a simple case which can be solved in finite terms. Take the equations

$$z \frac{dU}{dz} = \alpha_1 U + \beta_1 V + \gamma_1 z^m$$
$$z \frac{dV}{dz} = \alpha_2 U + \beta_2 V + \gamma_2 z^n$$,

where m and n are positive integers. Let $z = e^\zeta$, and denote $\frac{d}{d\zeta}$ by D; the equations are

$$(D - \alpha_1) U - \beta_1 V = \gamma_1 e^{m\zeta}$$
$$- \alpha_2 U + (D - \beta_2) V = \gamma_2 e^{n\zeta}$$,

and therefore

$$\{(D - \alpha_1)(D - \beta_2) - \alpha_2 \beta_1\} U = \gamma_1 (m - \beta_2) e^{m\zeta} + \gamma_2 \beta_1 e^{n\zeta};$$

and similarly

$$\{(D - \alpha_1)(D - \beta_2) - \alpha_2 \beta_1\} V = \gamma_1 \alpha_2 e^{m\zeta} + \gamma_2 (n - \alpha_1) e^{n\zeta}.$$

---

* In order to avoid prolixity of formulæ, the investigations are confined almost entirely to the case when the system of ordinary equations involves only two dependent variables: the appropriate generalisation to the case of n dependent variables is stated in § 187 without proof, because the proof is sufficiently indicated for the case $n = 2$. Some references are given in the course of the chapter: a more general reference is due to Königsberger's treatise (cited p. 8), ch. v, where a number of memoirs are quoted in the preface; and to part of a memoir by Goursat, *Amer. Journ. Math.*, t. xi (1889), pp. 329–372. See also a memoir by the author in the Stokes Jubilee volume (1899) of the *Camb. Phil. Trans.*, t. xviii, pp. 35–90.

Let ξ_1 and ξ_2 be the roots of the quadratic

$$(\xi - \alpha_1)(\xi - \beta_2) - \alpha_2\beta_1 = 0;$$

then

$$
\left.
\begin{aligned}
U &= \frac{\gamma_1(m - \beta_2)}{(m - \alpha_1)(m - \beta_2) - \alpha_2\beta_1}\, e^{m\zeta} + \frac{\gamma_2\beta_1}{(n - \alpha_1)(n - \beta_2) - \alpha_2\beta_1}\, e^{n\zeta} \\
&\qquad\qquad + Pe^{\xi_1\zeta} + Qe^{\xi_2\zeta} \\
V &= \frac{\gamma_1\alpha_2}{(m - \alpha_1)(m - \beta_2) - \alpha_2\beta_1}\, e^{m\zeta} + \frac{\gamma_2(n - \alpha_1)}{(n - \alpha_1)(n - \beta_2) - \alpha_2\beta_1}\, e^{n\zeta} \\
&\qquad\qquad + \frac{\xi_1 - \alpha_1}{\beta_1}\, Pe^{\xi_1\zeta} + \frac{\xi_2 - \alpha_1}{\beta_1}\, Qe^{\xi_2\zeta}
\end{aligned}
\right\},
$$

which, on the substitution of z for e^ζ, give the solution of the original equations. The quantities P and Q are arbitrary constants.

Hence also the solution of the equations

$$
\left.
\begin{aligned}
z\frac{dU}{dz} &= \alpha_1 U + \beta_1 V + \Sigma\epsilon_s z^s \\
z\frac{dV}{dz} &= \alpha_2 U + \beta_2 V + \Sigma\epsilon_s' z^s
\end{aligned}
\right\}
$$

is given by

$$
\left.
\begin{aligned}
U &= Pz^{\xi_1} + Qz^{\xi_2} + \Sigma\frac{(s - \beta_2)\epsilon_s + \beta_1\epsilon_s'}{(s - \alpha_1)(s - \beta_2) - \alpha_2\beta_1}\, z^s \\
V &= \frac{(\xi_1 - \alpha_1)}{\beta_1}\, Pz^{\xi_1} + \frac{(\xi_2 - \alpha_1)}{\beta_1}\, Qz^{\xi_2} + \Sigma\frac{\alpha_2\epsilon_s + (s - \alpha_1)\epsilon_s'}{(s - \alpha_1)(s - \beta_2) - \alpha_2\beta_1}\, z^s
\end{aligned}
\right\},
$$

where ξ_1 and ξ_2 are the roots of the quadratic

$$(\xi - \alpha_1)(\xi - \beta_2) - \alpha_2\beta_1 = 0,$$

and P, Q are arbitrary constants.

The solution thus obtained is the complete solution: it is satisfactory and definite, provided

(i) neither root of the quadratic is a positive integer; and

(ii) the roots of the quadratic are unequal.

It is clear that the power-series in the expressions for U and V converge when the power-series in the expressions for $z\dfrac{dU}{dz}$ and $z\dfrac{dV}{dz}$ converge.

The solution obtained ceases to be satisfactory, as regards form and completeness, if the conditions specified are not satisfied. Corresponding to this alternative, there are four subsidiary cases, viz.

I.   When $\xi_1 = \kappa$, an integer, and ξ_2 is not an integer:

II. When $\xi_1 = \kappa$, an integer, and $\xi_2 = \lambda$, another integer:

III. When $\xi_1 = \xi_2$, the common value θ not being an integer:

IV. When $\xi_1 = \xi_2$, the common value ϑ being an integer:

the integers in each case being positive.

Case I. It is clear that the terms in U and V which involve z^{ξ_2} will be unaffected by the hypothesis, as will also all the terms involving powers of z in the two summations other than the κth power. We have, on taking $s = \kappa + \Delta$ (Δ being an infinitesimal real quantity),

$$z^s = z^\kappa (1 + \Delta \log z + \ldots),$$

$$(s - \alpha_1)(s - \beta_2) - \alpha_2 \beta_1 = (s - \kappa)(s - \xi_2) = \Delta (\kappa - \xi_2 + \Delta),$$

so that the term in U in the summation gives rise (i) to a term in z^κ with a coefficient which, though infinite in the limit when $\Delta = 0$, can be absorbed in the coefficient of $P z^{\xi_1}$, that is, of $P z^\kappa$: and likewise for the term in V; (ii) to a term

$$\frac{(\kappa - \beta_2) \epsilon_\kappa + \beta_1 \epsilon_\kappa'}{\kappa - \xi_2} z^\kappa \log z$$

in U, and to a term

$$\frac{\alpha_2 \epsilon_\kappa + (\kappa - \alpha_1) \epsilon_\kappa'}{\kappa - \xi_2} z^\kappa \log z$$

in V; and (iii) to terms which vanish in the limit when $\Delta = 0$. To determine more precisely the relation of the arbitrary coefficients of z^κ in U and V respectively, we take

$$\left. \begin{aligned}
U &= Q z^{\xi_2} + \Sigma' \frac{(s - \beta_2) \epsilon_s + \beta_1 \epsilon_s'}{(s - \alpha_1)(s - \beta_2) - \alpha_2 \beta_1} z^s \\
&\quad + \frac{(\kappa - \beta_2) \epsilon_\kappa + \beta_1 \epsilon_\kappa'}{\kappa - \xi_2} z^\kappa \log z + P_1 z^\kappa \\
V &= \frac{\xi_2 - \alpha_1}{\beta_1} Q z^{\xi_2} + \Sigma' \frac{\alpha_2 \epsilon_s + (s - \alpha_1) \epsilon_s'}{(s - \alpha_1)(s - \beta_2) - \alpha_2 \beta_1} z^s \\
&\quad + \frac{\alpha_2 \epsilon_\kappa + (\kappa - \alpha_1) \epsilon_\kappa'}{\kappa - \xi_2} z^\kappa \log z + P_2 z^\kappa
\end{aligned} \right\},$$

where the term corresponding to $s = \kappa$ is omitted from the summation Σ'. Substituting these values, we find that both equations are identically satisfied if

$$(\kappa - \alpha_1)\left(P_1 - \frac{\epsilon_\kappa}{\kappa - \xi_2}\right) = \beta_1\left(P_2 - \frac{\epsilon_\kappa'}{\kappa - \xi_2}\right),$$

$$\alpha_2\left(P_1 - \frac{\epsilon_\kappa}{\kappa - \xi_2}\right) = \beta_2\left(P_2 - \frac{\epsilon_\kappa'}{\kappa - \xi_2}\right).$$

These equations are consistent with one another; and therefore we may take

$$P_1 = \frac{\epsilon_\kappa}{\kappa - \xi_2} + \beta_1 S \left.\begin{matrix} \\ \\ \end{matrix}\right\}$$

$$P_2 = \frac{\epsilon_\kappa'}{\kappa - \xi_2} + (\kappa - \alpha_1) S$$

where $\dot{S}$ is an arbitrary constant. When these values of P_1 and P_2 are substituted, the resulting expressions give the values of the integrals.

Case II. It is clear that the terms in U and V, which arise through the summation for various values of s, are unaffected by the hypothesis except the two which involve the κth power and the λth power of z respectively. Proceeding as in the last case, we find

$$U = \Sigma'' \frac{(s - \beta_2)\epsilon_s + \beta_1\epsilon_s'}{(s - \alpha_1)(s - \beta_2) - \alpha_2\beta_1} + P_1 z^\kappa + Q_1 z^\lambda$$
$$+ \left[\{(\kappa - \beta_2)\epsilon_\kappa + \beta_1\epsilon_\kappa'\} z^\kappa - \{(\lambda - \beta_2)\epsilon_\lambda + \beta_1\epsilon_\lambda'\} z^\lambda\right]\frac{\log z}{\kappa - \lambda}$$

$$V = \Sigma'' \frac{\alpha_2\epsilon_s + (s - \alpha_1)\epsilon_s'}{(s - \alpha_1)(s - \beta_2) - \alpha_2\beta_1} + P_2 z^\kappa + Q_2 z^\lambda$$
$$+ \left[\{\alpha_2\epsilon_\kappa + (\kappa - \alpha_1)\epsilon_\kappa'\} z^\kappa - \{\alpha_2\epsilon_\lambda + (\lambda - \alpha_1)\epsilon_\lambda'\} z^\lambda\right]\frac{\log z}{\kappa - \lambda}$$

where the summation in Σ'' extends to all values of s except only κ and λ; where P_1 and P_2, Q_1 and Q_2, are connected by the respective relations

$$P_1 = \frac{\epsilon_\kappa}{\kappa - \lambda} + \beta_1 S \left.\begin{matrix} \\ \\ \end{matrix}\right\} \qquad Q_1 = \frac{\epsilon_\lambda}{\lambda - \kappa} + \beta_1 T \left.\begin{matrix} \\ \\ \end{matrix}\right\}$$

$$P_2 = \frac{\epsilon_\kappa'}{\kappa - \lambda} + (\kappa - \alpha_1) S \qquad Q_2 = \frac{\epsilon_\lambda'}{\lambda - \kappa} + (\lambda - \alpha_1) T$$

and S and T are arbitrary constants.

Case III. This is the customary case that occurs in connection with the complementary function in the solution of a

linear equation when the roots of the critical equation are equal.
The solution is easily proved to be

$$
\begin{aligned}
U &= \Sigma \frac{(s - \beta_2)\,\epsilon_s + \beta_1 \epsilon_s'}{(s - \alpha_1)(s - \beta_2) - \alpha_2\beta_1} z^s + A\beta_1 z^\theta \log z + \frac{\beta_1(R - A)}{\theta - \alpha_1} z^\theta \\
V &= \Sigma \frac{\alpha_2 \epsilon_s + (s - \alpha_1)\,\epsilon_s'}{(s - \alpha_1)(s - \beta_2) - \alpha_2\beta_1} z^s + A(\theta - \alpha_1) z^\theta \log z + R z^\theta
\end{aligned} \Bigg\} ,
$$

where A and R are arbitrary constants, θ is the (repeated) root of
the quadratic

$$
(\theta - \alpha_1)(\theta - \beta_2) - \alpha_2\beta_1 = 0,
$$

the summations in U and V are for all values of s, and θ is not
an integer.

Case IV. This case differs from the last in the fact that θ is
an integer $\Im$, so that the form ceases to be satisfactory for the
term in the summation for which $s = \Im$. To obtain a suggestion
as to the form, take

$$
s = \Im + \Delta,
$$

where Δ ultimately will be made zero; then

$$
\begin{aligned}
\frac{z^s}{(s - \alpha_1)(s - \beta_2) - \alpha_2\beta_1} &= \frac{z^s}{(s - \Im)^2} \\
&= \frac{1}{\Delta^2} z^\Im \left\{ 1 + \Delta \log z + \frac{\Delta^2}{2} (\log z)^2 + \ldots \right\}.
\end{aligned}
$$

Consequently we assume

$$
\begin{aligned}
U &= \Sigma' \frac{(s - \beta_2)\,\epsilon_s + \beta_1 \epsilon_s'}{(s - \alpha_1)(s - \beta_2) - \alpha_2\beta_1} z^s + \tfrac{1}{2}\{(\Im - \beta_2)\,\epsilon_\Im + \beta_1\epsilon_\Im'\} z^\Im (\log z)^2 \\
&\qquad + H_1 z^\Im \log z + K_1 z^\Im \\
V &= \Sigma' \frac{\alpha_2 \epsilon_s + (s - \alpha_1)\,\epsilon_s'}{(s - \alpha_1)(s - \beta_2) - \alpha_2\beta_1} z^s + \tfrac{1}{2}\{\alpha_2\epsilon_\Im + (\Im - \alpha_1)\,\epsilon_\Im'\} z^\Im (\log z)^2 \\
&\qquad + H_2 z^\Im \log z + K_2 z^\Im
\end{aligned} \Bigg\} ,
$$

where the summation in Σ' is for all values of s except only $s = \Im$.
In order that the equations may be identically satisfied, we must
have

$$
\begin{aligned}
H_1 - \epsilon_\Im &= \beta_1 A \\
H_2 - \epsilon_\Im' &= (\Im - \alpha_1) A \\
K_1 &= \frac{\beta_1}{\Im - \alpha_1} (K_2 - A)
\end{aligned} \Bigg\} ,
$$

where A and K_2 are arbitrary constants. When these values of H_1, H_2, K_1 are inserted in the expressions for U and V, the resulting forms are the integrals for the case when the quadratic has the integer ϑ for a repeated root.

Note. It may be pointed out that, if the method of § 146 had been adopted so as to introduce a new variable ξ, the equations would have been

$$\frac{dU}{d\xi} = \alpha_1 U + \beta_1 V + \gamma_1 z + \dots,$$

$$\frac{dV}{d\xi} = \alpha_2 U + \beta_2 V + \gamma_2 z + \dots,$$

$$\frac{dz}{d\xi} = z;$$

the critical cubic is

$$\begin{vmatrix} \alpha_1 - \omega, & \beta_1, & \gamma_1 \\ \alpha_2, & \beta_2 - \omega, & \gamma_2 \\ 0, & 0, & 1 - \omega \end{vmatrix} = 0,$$

the roots of which are $\omega = 1$ and the roots of the quadratic

$$(\alpha_1 - \omega)(\beta_2 - \omega) - \alpha_2 \beta_1 = 0,$$

which is the critical quadratic of the preceding method.

165. When it is necessary to consider the character of integrals of the equations

$$\left. \begin{array}{l} t \dfrac{dU}{dt} = \alpha_1 U + \beta_1 V + \gamma_1 t + \dots = \theta_1(U, V, t) \\[2mm] t \dfrac{dV}{dt} = \alpha_2 U + \beta_2 V + \gamma_2 t + \dots = \theta_2(U, V, t) \end{array} \right\},$$

which vanish when $t = 0$, the preceding example gives some indications of results to be obtained. The equation

$$(\xi - \alpha_1)(\xi - \beta_2) - \alpha_2 \beta_1 = 0,$$

which is of importance, will be called the *critical quadratic*. The following theorems are suggested (though of course not with adequate justification of the conditions):

(i) If the roots ξ_1 and ξ_2 of the critical quadratic be neither of them a positive integer, then integrals U and V of the equations exist, which vanish at $t = 0$ and are regular functions of t.

(This is suggested by taking $P = 0$, $Q = 0$ in the solution of the unconditioned form of the example, and by taking $A = 0$, $R = 0$ in Case III.)

(ii) If the real part of one of the roots, say ξ_1, be positive, neither ξ_1 nor ξ_2 being a positive integer, then integrals U and V of the equations may exist, which vanish at $t = 0$ and, though not regular functions of t, are regular functions of t and t^{ξ_1}.

Similarly if the real part of ξ_2 be positive, though neither ξ_1 nor ξ_2 is a positive integer, integrals may exist, which are regular functions of t and t^{ξ_2} and vanish with t.

And if the real parts of both ξ_1 and ξ_2 be positive, though neither ξ_1 nor ξ_2 is a positive integer, integrals may exist, which are regular functions of t, t^{ξ_1}, t^{ξ_2} and vanish with t.

(Manifestly limitations must be introduced, so as to take account of the possibility that ξ_1 and ξ_2 may be commensurable positive quantities in the respective cases, and of the possibility that $\xi_1 - \xi_2$ may be an integer, including zero.)

(iii) If the real part of each of the roots ξ_1 and ξ_2 be negative, then the regular integrals are the only integrals of the equation which exist, subject to the condition of vanishing with t.

(iv) If one root (but not the other root) of the quadratic be an integer, and if its value be κ, then no regular integrals of the equations exist vanishing with t, unless the coefficients in $\theta_1(U, V, t)$ and $\theta_2(U, V, t)$ satisfy certain conditions. If however these conditions are satisfied, then there can be an infinitude of regular integrals vanishing with t.

If either (or both) of the roots of the quadratic be a positive integer, then integrals of the equation may exist, which vanish with t and are regular functions of t and $t \log t$.

To the consideration of these theorems we now proceed*, taking account of the successive different general cases that can arise owing to the various possibilities of the form of the roots of

---

* The first of them was enunciated by Picard, *Comptes Rendus*, t. LXXXVII (1878), pp. 430—432, 743—745; it is developed, further than his earliest enunciation, in his *Cours d'Analyse*, t. III, ch. I. See also Goursat, *Amer. Journ. Math.*, t. XI (1889), p. 335.

The third was enunciated by Goursat in the memoir just quoted (p. 342); and a

the critical quadratic. For this purpose, the method given by Jordan, for the corresponding case of a single equation and used in § 71, is adapted to the system of two equations. The various cases are :—

I. The quadratic has unequal roots :—

  (a) neither root being a positive integer :

  (b) one root being a positive integer, the other not :

  (c) both roots being positive integers.

II. The quadratic has equal roots :—

  (a) the (repeated) root not being a positive integer :

  (b) the (repeated) root being a positive integer.

It should be added that a further assumption will be made for the present purpose, viz. that the critical quadratic has not a zero-root. As a matter of fact, the existence of a zero-root would imply (as for a single equation of the first order) that the reduced form of the system belongs (§§ 159, 160) to a type different from that here considered.

166. It is convenient to transform the variables. When the roots of the critical quadratic

$$(\xi - \alpha_1)\,(\xi - \beta_2) - \alpha_2\beta_1 = 0$$

very special case of one of the portions of the second is remarked by him in the form

$$\left.\begin{aligned} z\frac{du}{dz} &= \frac{1}{2}u \\ z\frac{dv}{dz} &= -v + \frac{5}{2}u^3 \end{aligned}\right\},$$

which are satisfied by $u = z^{\frac{1}{2}}$, $v = z^{\frac{3}{2}}$.

A considerable portion of Chapter v of Königsberger's treatise, already frequently cited, is devoted to the corresponding discussion for n equations. Some difficulties, as regards adequacy of proof of the theorems, independently of the general statement, prevent me from thinking the investigation entirely satisfactory. An example, illustrating the general result for $n=2$, and differing from Königsberger's implied form of integral, is given subsequently (§ 176, Ex. 2).

Some papers by Horn, *Crelle*, t. cxvi (1896), pp. 265—306, *ib.* t. cxvii (1897), pp. 104—128, 254—266, may also be consulted; further references will be found in them.

are unequal, say ξ_1 and ξ_2, we introduce new variables u and v, such that

$$u = \lambda U + \mu V, \quad v = \lambda' U + \mu' V,$$

where $(\alpha_1 - \xi_1) \lambda + \alpha_2 \mu = 0\big\}$ $(\alpha_1 - \xi_2) \lambda' + \alpha_2 \mu' = 0\big\}$;
        $\beta_1 \lambda + (\beta_2 - \xi_1) \mu = 0\big\}$, $\beta_1 \lambda' + (\beta_2 - \xi_2) \mu' = 0\big\}$

the ratios $\lambda : \mu$ and $\lambda' : \mu'$ are unequal, and consequently the new variables u, v are distinct. The equations become

$$t \frac{du}{dt} = \xi_1 u + \phi_1 (u, v, t)\Bigg\}$$
$$t \frac{dv}{dt} = \xi_2 v + \phi_2 (u, v, t)\Bigg\} ,$$

where ϕ_1, ϕ_2 are regular functions of their arguments and vanish with them; except for a term in t, they have all their terms of the second or higher orders in u, v, t combined.

When the roots of the critical quadratic are equal, having a value ξ, say, we introduce a new variable u such that

$$u = \lambda U + \mu V,$$

where $(\alpha_1 - \xi) \lambda + \alpha_2 \mu = 0, \quad \beta_1 \lambda + (\beta_2 - \xi) \mu = 0.$
Then we have

$$t \frac{du}{dt} = \xi u + \phi_1 (u, v, t),$$

$$t \frac{dV}{dt} = \frac{\alpha_2}{\lambda} u + \xi V + \phi_2 (u, V, t)$$

$$= \kappa u + \xi V + \phi_2 (u, V, t),$$

say.

It therefore appears that the equations, corresponding to the cases I (a), I (b), I (c), are

$$t \frac{du}{dt} = \xi_1 u + \phi_1 (u, v, t)\Bigg\}$$
$$t \frac{dv}{dt} = \xi_2 v + \phi_2 (u, v, t)\Bigg\} ,$$

where ξ_1 and ξ_2 are unequal to one another: and that the equations, corresponding to the cases II (a), II (b), are

$$t \frac{du}{dt} = \xi u + \phi_1 (u, v, t)\Bigg\}$$
$$t \frac{dv}{dt} = \kappa u + \xi v + \phi_2 (u, v, t)\Bigg\} .$$

In both forms, the functions ϕ_1 and ϕ_2 are regular functions of their arguments and vanish with them; and the only term of the first order in ϕ_1 and ϕ_2 is possibly a term in t. For both forms, the initial conditions are that $u = 0$, $v = 0$, when $t = 0$.

For brevity, integrals, which are regular functions of t, will be called *regular integrals*: and integrals, which are not regular functions of t, but are regular functions of quantities that themselves are not regular in t, will be called *non-regular integrals*. The results are obtained for the transformed equations in u and v; since U and V are linear homogeneous combinations of u and v, the results apply to the original equations. We shall first discuss the regular integrals for all the cases in turn; afterwards, the non-regular integrals.

REGULAR INTEGRALS.

CASE I (a): *the critical quadratic has unequal roots, neither being a positive integer.*

167. If the equations

$$t\frac{du}{dt} = \xi_1 u + \phi_1(u, v, t), \qquad t\frac{dv}{dt} = \xi_2 v + \phi_2(u, v, t),$$

possess regular integrals vanishing with t, these integrals must have the form

$$u = \sum_{n=1}^{\infty} a_n t^n, \qquad v = \sum_{n=1}^{\infty} b_n t^n.$$

That they may have significance, the power-series must converge; that they may be solutions, they must satisfy the equations identically.

Accordingly, substituting the expressions and comparing coefficients of t^n, we have

$$(n - \xi_1) a_n = f_n, \qquad (n - \xi_2) b_n = g_n,$$

where f_n and g_n are the coefficients of t^n in ϕ_1 and ϕ_2 respectively after the expressions for u and v are substituted. From the forms of ϕ_1 and ϕ_2, it is clear that f_n and g_n are linear combinations of the coefficients of ϕ_1 and ϕ_2, that they are rational integral combinations of the coefficients a_1, a_2, ..., b_1, b_2, ..., and that they contain no coefficient a after a_{n-1} and no coefficient b after b_{n-1} in the respective sets. Since neither ξ_1 nor ξ_2 is a positive integer, the equations can be solved in succession for increasing values of n, so as to determine formal expressions for all the coefficients. In particular, a_n and b_n are obtained each of them as sums of quotients; the numerators of these quotients are rational integral functions of the coefficients in ϕ_1 and ϕ_2, and the denominators are products of powers of the quantities

$$1 - \xi_1, \quad 2 - \xi_1, \quad \ldots, \quad n - 1 - \xi_1, \quad n - \xi_1,$$

$$1 - \xi_2, \quad 2 - \xi_2, \quad \ldots, \quad n - 1 - \xi_2, \quad n - \xi_2.$$

To discuss the convergence of the power-series, we introduce an associated set of dominant equations. The functions ϕ_1 and ϕ_2 are regular in the vicinity of $u = 0$, $v = 0$, $t = 0$; let their domain of existence include a region $|t| \leqslant r$, $|u| \leqslant \rho'$, $|v| \leqslant \rho''$: of the two quantities ρ' and ρ'', let ρ denote the smaller, so that ϕ_1 and ϕ_2 are certainly regular in a region $|t| \leqslant r$, $|u| \leqslant \rho$, $|v| \leqslant \rho$. Within that region, let M' denote the greatest value of $|\phi_1|$, and M'' the greatest value of $|\phi_2|$: of the two quantities M' and M'', let M denote the larger, so that

$$|\phi_1| \leqslant M, \quad |\phi_2| \leqslant M,$$

within the region specified, and M is a finite magnitude. Then if f_{ijk} and g_{ijk} are the coefficients of $u^i v^j t^k$ in ϕ_1 and ϕ_2 respectively, it is known* that

$$|f_{ijk}| \leqslant \frac{M}{\rho^{i+j} r^k}, \quad |g_{ijk}| \leqslant \frac{M}{\rho^{i+j} r^k}.$$

Further, no one of the quantities $m - \xi_1$, $m - \xi_2$ vanishes for integer values of m; there is therefore a least (and non-zero) value of $|m - \xi_1|$, $|m - \xi_2|$, for the various values of m; let it be denoted by ϵ.

Now consider the equations

$$\left.\begin{aligned}
\epsilon X &= \frac{M}{r} t + \underset{i}{\Sigma}\underset{j}{\Sigma}\underset{k}{\Sigma} \frac{M}{\rho^{i+j} r^k} X^i Y^j t^k \\
\epsilon Y &= \frac{M}{r} t + \underset{i}{\Sigma}\underset{j}{\Sigma}\underset{k}{\Sigma} \frac{M}{\rho^{i+j} r^k} X^i Y^j t^k
\end{aligned}\right\},$$

* *Th. Fns.*, § 22.

where the triple summation is for integer values of i, j, k such that $i + j + k \geqslant 2$. Clearly $X = Y$; and each of them is given by

$$\epsilon X = \frac{M}{r} t + \underset{i\ j\ k}{\Sigma \Sigma \Sigma} \frac{M}{\rho^{i+j} r^k} X^{i+j} t^k$$

$$= \frac{M}{\left(1 - \dfrac{t}{r}\right)\left(1 - \dfrac{X}{\rho}\right)^2} - 2\frac{M}{\rho} X - M,$$

and therefore

$$X \left(\epsilon + 2\frac{M}{\rho}\right)\left(1 - \frac{X}{\rho}\right)^2 = \frac{M}{1 - \dfrac{t}{r}} - M\left(1 - \frac{X}{\rho}\right)^2.$$

In this cubic equation, the term independent of X vanishes when $t = 0$; and the term involving the first power does not vanish, because ϵ is not zero. Hence when $t = 0$, the cubic equation has one root and only one root which vanishes. It therefore follows, from the continuity of roots of an algebraical equation, that the cubic has one root, which vanishes with t, and which is a regular function of t for values of $|t|$ less than the least modulus of a root of the discriminant, that is, for a finite range. To obtain the expansion of this root as a regular function, it is sufficient to determine the coefficients in the power-series

$$X = A_1 t + A_2 t^2 + \ldots + A_n t^n + \ldots,$$

so that the equation

$$\epsilon X = \frac{M}{r} t + \underset{i\ j\ k}{\Sigma \Sigma \Sigma} \frac{M}{\rho^{i+j} r^k} X^{i+j} t^k$$

is identically satisfied; because the root which vanishes with t is the only root of the cubic of that type, and the series for X is known to converge within the finite range indicated. Clearly we have

$$A_n = \frac{F_n}{\epsilon},$$

where F_n is the coefficient of t^n on the right-hand side of the equation for ϵX. This relation can be used for successive values of n, so as to give values of $A_1, \ldots, A_{n-1}$. When these values are substituted in F_n, the ultimate formal expression obtained for F_n is the quotient, by a power of ϵ, of an expression which is rational and integral in the coefficients $\dfrac{M}{\rho^{i+j} r^k}$.

Comparing the quantities $|f_n|$ and F_n, we note that a quantity greater than $|f_n|$ is obtained when in its numerator every term is replaced by its modulus; that this greater quantity is further increased when the modulus of the coefficient of $u^i v^j t^k$ in ϕ_1 or in ϕ_2 is replaced by $\dfrac{M}{\rho^{i+j}\gamma^k}$; and that this increased quantity is still further appreciated when every factor of the type $|m - \xi|$ in the denominator is replaced by ϵ. But, on comparing the two coefficients a_n and A_n, it is clear that these three changes turn f_n into F_n; accordingly

$$|f_n| < F_n.$$

Similarly for g_n and F_n, so that

$$|g_n| < F_n.$$

Also

$$|n - \xi_1| \geqslant \epsilon, \quad |n - \xi_2| \geqslant \epsilon;$$

hence

$$|a_n| < A_n, \quad |b_n| < A_n.$$

The series

$$A_1 t + A_2 t^2 + A_3 t^3 + \dots$$

converges within a finite region round $t = 0$; therefore also the series

$$a_1 t + a_2 t^2 + a_3 t^3 + \dots,$$
$$b_1 t + b_2 t^2 + b_3 t^3 + \dots,$$

converge within that region.

Hence *the differential equations possess regular integrals which vanish with t.* It is not difficult to prove, as in § 12 or in § 13, that *they are the only regular integrals which vanish with t.*

CASE I (b): *the critical quadratic has unequal roots, one of which is a positive integer and the other of which is not a positive integer.*

168.   The equations may be taken in the form

$$\left. \begin{aligned} t\,\frac{du}{dt} &= mu + \alpha t + \theta\,(u, v, t) \\ t\,\frac{dv}{dt} &= \xi v + \beta t + \phi\,(u, v, t) \end{aligned} \right\},$$

where m is a positive integer, ξ is not a positive integer, θ and ϕ are regular functions of their arguments, which vanish with u, v, t, and contain no terms of dimensions lower than 2.

If regular solutions exist, which vanish with t, we can take

$$u = t(\lambda + u_1), \quad v = t(\mu + v_1),$$

choosing the constants λ and μ so that u_1 and v_1 vanish with t. Then t^2 is a factor of θ and ϕ after this substitution is made, say

$$\theta(u, v, t) = t^2\theta_1'(u_1, v_1, t), \quad \phi(u, v, t) = t^2\phi_1'(u_1, v_1, t);$$

but θ_1' and ϕ_1' do not necessarily vanish when t, u_1, v_1 vanish. The equations for the new variables are

$$\left. \begin{aligned}
t\frac{du_1}{dt} &= (m-1)\lambda + \alpha + (m-1)u_1 + t\theta_1'(u_1, v_1, t) \\
t\frac{dv_1}{dt} &= (\xi-1)\mu + \beta + (\xi-1)v_1 + t\phi_1'(u_1, v_1, t)
\end{aligned} \right\}.$$

Now as u_1, v_1 are regular functions of t, the expressions on the left-hand side vanish when $t = 0$; hence

$$(m-1)\lambda + \alpha = 0, \quad (\xi-1)\mu + \beta = 0.$$

If $\theta_1'(0, 0, 0) = \alpha_1$, $\phi_1'(0, 0, 0) = \beta_1$, the equations are

$$\left. \begin{aligned}
t\frac{du_1}{dt} &= (m-1)u_1 + \alpha_1 t + t\theta_1(u_1, v_1, t) \\
t\frac{dv_1}{dt} &= (\xi-1)v_1 + \beta_1 t + t\phi_1(u_1, v_1, t)
\end{aligned} \right\},$$

where θ_1 and ϕ_1 are regular functions of their arguments, which vanish when $u_1 = 0$, $v_1 = 0$, $t = 0$. The equations are, in form, the same as before, except that the coefficients of the first power of the dependent variables on the right-hand side have been reduced by unity; and the relation between the two sets of dependent variables is

$$u = t\left(-\frac{\alpha}{m-1} + u_1\right), \quad v = t\left(-\frac{\beta}{\xi-1} + v_1\right).$$

It is manifest that the equations in u_1 and v_1 can be subjected to a similar transformation with a corresponding result; and that, as m is a positive integer while ξ is not, the process can be carried out $m-1$ times, but not more. Denoting the dependent variables by u', v' after all these transformations have been effected, we have equations in the form

$$t\frac{du'}{dt} = u' + at + h(u', v', t) \left.\vphantom{\begin{array}{c}a\\b\end{array}}\right\}$$
$$t\frac{dv'}{dt} = \kappa v' + bt + k(u', v', t) \left.\vphantom{\begin{array}{c}a\\b\end{array}}\right\},$$

where κ, $= \xi - m + 1$, is not a positive integer; h, k are regular functions of their arguments, which vanish with t and contain no terms of order less than 2. The relation between the variables u, v and u', v' is of the form

$$u = P_{m-1} + t^{m-1}u', \quad v = Q_{m-1} + t^{m-1}v',$$

where P_{m-1} and Q_{m-1} are algebraical polynomials of degree $m-1$ vanishing with t; and $u' = 0$, $v' = 0$ when $t = 0$. The coefficients a and b are rational integral functions of the original coefficients.

The equations can possess regular integrals only if a is zero. For regular integrals must be of the form

$$u' = p_1 t + p_2 t^2 + \dots, \quad v' = q_1 t + q_2 t^2 + \dots;$$

substituting these, remembering that h and k are then of the second order at least in t, and equating coefficients of t in the first of the equations, we must have

$$p_1 = p_1 + a,$$

which is possible, for non-infinite values of p_1, only if a is zero.

Suppose now that a is zero. Since u' and v' (if they exist as regular functions of t) vanish with t, we can assume

$$u' = t\eta_1, \quad v' = t\eta_2,$$

the sole transferred condition being that η_1 and η_2 are regular functions of t, which now need not necessarily vanish with t. We have

$$t^2\frac{d\eta_1}{dt} = h(t\eta_1, t\eta_2, t) = t^2 H(\eta_1, \eta_2, t),$$

$$t^2\frac{d\eta_2}{dt} = (\kappa - 1)t\eta_2 + bt + k(t\eta_1, t\eta_2, t)$$
$$= (\kappa - 1)t\eta_2 + bt + t^2 K(\eta_1, \eta_2, t),$$

where H and K are regular functions of their arguments. The second equation shews that, when $t = 0$, then $(\kappa - 1)\,\eta_2 + b = 0$; accordingly taking

$$\eta_2 = \frac{b}{1 - \kappa} + \zeta_2,$$

we have ζ_2 vanishing when $t = 0$. As regards η_1, there is, as yet, no restriction upon its value when $t = 0$; denoting it by A, we take

$$\eta_1 = A + \zeta_1,$$

where ζ_1 vanishes when $t = 0$. Both ζ_1 and ζ_2 are regular functions of t. When the values are substituted, A remains undetermined by the equations; and therefore an arbitrary (finite) value can be assigned to A. The equations for ζ_1 and ζ_2 now are

$$\left. \begin{aligned}
t\,\frac{d\zeta_1}{dt} &= tH\left(A + \zeta_1,\; \frac{b}{1 - \kappa} + \zeta_2,\; t\right) \\
t\,\frac{d\zeta_2}{dt} &= (\kappa - 1)\,\zeta_2 + tK\left(A + \zeta_1,\; \frac{b}{1 - \kappa} + \zeta_2,\; t\right)
\end{aligned} \right\},$$

with the condition that ζ_1 and ζ_2 must be regular functions of t vanishing with t.

Let them, if they exist, be denoted by

$$\zeta_1 = \sum_{n=1} a_n t^n, \quad \zeta_2 = \sum_{n=1} b_n t^n;$$

substituting in the equations which must be satisfied identically, and equating coefficients, we have relations

$$n a_n = f_n, \quad (n - \kappa + 1)\,b_n = g_n,$$

similar to those in the Case I (a).

These equations are treated in the same way as in the Case I (a). Since κ is not a positive integer, no one of the coefficients of b_n vanishes; and thence it is easy to see that the whole of the treatment in I (a) subsequent to the corresponding stage can, with only slight changes in the analysis, be applied to the present case. It leads to the result that the power-series for ζ_1 and ζ_2 converge for a finite region round $t = 0$; and from the form of the equations for ζ_1 and ζ_2, it is clear that the coefficients in the power-series will involve the arbitrary constant A.

Hence it follows that, *unless the condition represented by* $a = 0$ *be satisfied, the equations do not possess regular integrals vanishing with* $t = 0$. *If that condition be satisfied, the equations possess regular integrals vanishing with* $t = 0$ *and involving an arbitrary constant: in other words, they possess a single infinitude of regular integrals vanishing with* $t = 0$.

The condition, represented by $a = 0$, can be obtained from the original equations

$$t \frac{du}{dt} = mu + \alpha t + \theta (u, v, t) \Bigg\}$$
$$t \frac{dv}{dt} = \xi u + \beta t + \phi (u, v, t) \Bigg\},$$

as follows. Let

$$u = \sum_{p=1}^{m-1} f_p t^p + t^m U,$$

$$v = \sum_{p=1}^{m-1} g_p t^p + t^m V;$$

substitute in the equations, and determine (by comparison of the coefficients) the values of $f_1, \ldots, f_{m-1}, g_1, \ldots, g_{m-1}$. Then the condition is that the coefficient of t^m in

$$\alpha t + \theta \left(\sum_{p=1}^{m-1} f_p t^p, \sum_{p=1}^{m-1} g_p t^p, t \right)$$

shall be zero. This statement can be easily verified.

Ex. To obtain the regular integrals of

$$t \frac{dt_1}{dt} = t_1 + \theta_1 (t_1, t_2, t) \Bigg\}$$
$$t \frac{dt_2}{dt} = \tfrac{1}{2} t_2 + \tfrac{1}{2} \theta t + \theta_2 (t_1, t_2, t) \Bigg\},$$

where
$$\theta_1 = (a, b, c, f, g, h \rangle t_1, t_2, t)^2, \qquad \theta_2 = (a', b', c', f', g', h' \rangle t_1, t_2, t)^2,$$

the integrals being required to vanish with t, let

$$t_1 = p_1 t + p_2 t^2 + \ldots,$$

$$t_2 = q_1 t + q_2 t^2 + \ldots;$$

then

$$(n - 1) p_n = \text{coefficient of } t^n \text{ in } \theta_1,$$

$$(n - \tfrac{1}{2}) q_n = \ldots\ldots\ldots\ldots\ldots\ldots\ldots \tfrac{1}{2} \theta t + \theta_2.$$

Then p_1 is undetermined by these equations: let its value be A, where A is arbitrary: also

$$\tfrac{1}{2} q_1 = \tfrac{1}{2} \theta,$$

so that $q_1 = \theta$.

Next, we have

$$p_2 = (a,\, b,\, c,\, f,\, g,\, h\,\rrbracket A,\, \theta,\, 1)^2,$$

$$\tfrac{3}{2} q_2 = (a',\, b',\, c',\, f',\, g',\, h'\,\rrbracket A,\, \theta,\, 1)^2.$$

Next, we have

$$2p_3 = \text{coefficient of } t^3 \text{ in } (a,\, b,\, c,\, f,\, g,\, h\,\rrbracket At + p_2 t^2,\; \theta t + q_2 t^2,\; t)^2$$

$$= \left(p_2 \frac{\partial}{\partial A} + q_2 \frac{\partial}{\partial \theta}\right)(a,\, b,\, c,\, f,\, g,\, h\,\rrbracket A,\, \theta,\, 1)^2$$

$$= p_2 \frac{\partial p_2}{\partial A} + q_2 \frac{\partial p_2}{\partial \theta},$$

$$\tfrac{5}{2} q_3 = p_2 \frac{\partial q_2}{\partial A} + q_2 \frac{\partial q_2}{\partial \theta};$$

and so on.

CASE I (c): *the critical quadratic has unequal roots, both of which are positive integers.*

169.   Let m and n be the two unequal roots, of which m is the smaller, so that the equations may be taken in the form

$$t\,\frac{du}{dt} = mu + \alpha' t + \theta\,(u,\, v,\, t) \Big\rbrace$$
$$t\,\frac{dv}{dt} = nv + \beta' t + \phi\,(u,\, v,\, t) \Big\rbrace .$$

These equations can be transformed, as in the Case I (b), by $m-1$ substitutions in turn; and ultimately they acquire the form

$$t\,\frac{du'}{dt} = u' + at + h\,(u',\, v',\, t) \Big\rbrace$$
$$t\,\frac{dv'}{dt} = \kappa v' + bt + k\,(u',\, v',\, t) \Big\rbrace ,$$

where $\kappa, = n - m + 1$, is a positive integer greater than unity, u' and v' are regular functions of t vanishing when $t = 0$, and the functions h, k have the same signification as in Case I (b).

If the equations possess regular solutions, the latter must be of the form

$$u' = \sum_{l=1} p_l t^l, \quad v' = \sum_{l=1} q_l t^l\,;$$

substituting these values and equating coefficients, we have

$$p_1 = p_1 + a, \quad q_1 = \kappa q_1 + b,$$

$$(l-1)\,p_l = \text{coefficient of } t^l \text{ in } h\,(u',\, v',\, t),$$

$$(l-\kappa)\,q_l = \dots\dots\dots\dots\dots\; t^l \text{ in } k\,(u',\, v',\, t).$$

It is clear that, if a is different from zero, the equation cannot be satisfied; and therefore, as one condition for the possession of regular integrals, a must be zero. Assuming this satisfied, we see that p_1 is left undetermined: let a value α, provisionally arbitrary, be assigned to it.

Solving now the remaining equations for the values $l = 1, 2, \ldots,$ $\kappa - 1$ in successive sets, each set being associated with one value of l, we have the values of $p_1, \ldots, p_{\kappa-1}, q_1, \ldots, q_{\kappa-1}$; all these in general involve α. In order that q_κ may have a finite value, so that $(l - \kappa) q_l$ vanishes for $l = \kappa$, we must have the coefficient of t^κ in $k(u', v', t)$ zero. When this is the case, the value of q_κ is undetermined; let a value β, provisionally arbitrary, be assigned to it. For the remaining values of l, the equations determine formal expressions for the remaining coefficients, involving α and β: and no further formal conditions need be imposed. When the values of $p_1, \ldots, p_{\kappa-1}, q_1, \ldots, q_{\kappa-1}$ are inserted in $k(u', v', t)$, the coefficient of t^κ in that quantity may be an identical zero; in that case, the functions u', v' involve two arbitrary constants α and β, so that, if the functions actually exist, there is a double infinitude of regular solutions vanishing with t. Or the coefficient may be zero, only if some relation among the constants of the original equations be satisfied; if the relation is not satisfied, there are no regular integrals of the original equations vanishing with t: if the relation is satisfied, there is a double infinitude of regular integrals. Or the coefficient may be zero, only if some relation among the constants of the original equations and α be satisfied; this relation is to be regarded as determining α, and then for each value of α so determined, there is a single infinitude of regular solutions vanishing with t.

These results are stated on the assumption that the power-series, as obtained with the coefficients p and q, converge: the assumption can be justified as follows. Let

$$A_{\kappa-1} = p_1 t + p_2 t^2 + \ldots + p_{\kappa-1} t^{\kappa-1},$$
$$B_{\kappa-1} = q_1 t + q_2 t^2 + \ldots + q_{\kappa-1} t^{\kappa-1},$$

the coefficients p and q being known; if functions u' and v' exist of the specified form, we can assume

$$u' = A_{\kappa-1} + t^{\kappa-1} U',$$
$$v' = B_{\kappa-1} + t^{\kappa-1} V',$$

where U' and V' are regular functions of t that vanish with t. Thus, assuming $a = 0$, we have

$$t\frac{dA_{\kappa-1}}{dt} + t^\kappa\frac{dU'}{dt} + (\kappa - 1)t^{\kappa-1}U'$$

$$= A_{\kappa-1} + t^{\kappa-1}U' + h\,(A_{\kappa-1} + t^{\kappa-1}U',\ B_{\kappa-1} + t^{\kappa-1}V',\ t).$$

Now the quantity

$$t\frac{dA_{\kappa-1}}{dt} - A_{\kappa-1}$$

is equal to the aggregate of the terms involving t, t^2, ..., $t^{\kappa-1}$ in

$$h\,(A_{\kappa-1},\ B_{\kappa-1},\ t).$$

Also in $h\,(u', v', t)$, there are no terms of dimensions lower than 2; so that, in

$$h\,(A_{\kappa-1} + t^{\kappa-1}U',\ B_{\kappa-1} + t^{\kappa-1}V',\ t) - h\,(A_{\kappa-1},\ B_{\kappa-1},\ t),$$

the coefficients of $t^{\kappa-1}U'$, $t^{\kappa-1}V'$ are of dimension at least unity, and therefore this expression may be taken as equal to

$$t^{\kappa-1}H_1\,(U',\ V',\ t),$$

where H_1 is a regular function of its arguments, which vanishes with them and contains no terms of dimension lower than two. Also let the terms in $h\,(A_{\kappa-1}, B_{\kappa-1}, t)$, of order higher than $\kappa - 1$, be

$$c_\kappa t^\kappa + c_{\kappa+1}t^{\kappa+1} + \ldots ;$$

then

$$t^\kappa\frac{dU'}{dt} + (\kappa - 2)t^{\kappa-1}U' = c_\kappa t^\kappa + c_{\kappa+1}t^{\kappa+1} + \ldots + t^{\kappa-1}H_1\,(U',\ V',\ t),$$

and therefore

$$t\frac{dU'}{dt} = (2 - \kappa)\,U' + c_\kappa t + H\,(U',\ V',\ t),$$

on absorbing the other powers of t into H_1 and denoting by H the new function which has the same character as H_1. Similarly

$$t\frac{dV'}{dt} = V' + b_\kappa t + K\,(U',\ V',\ t),$$

where the terms in $k\,(A_{\kappa-1}, B_{\kappa-1}, t)$, of order higher than $\kappa - 1$, are

$$b_\kappa t^\kappa + \ldots,$$

and K is a function of the same character as H.

As κ is a positive integer > 1, $2 - \kappa$ is not a positive integer $\geqslant 1$. Thus the coefficient of U' is not a positive integer, while the coefficient of V' is unity; and thus the two equations for U'

and V' are a particular instance of the general form discussed in I (b). There is no regular integral vanishing with t, unless $b_\kappa = 0$; the significance of this condition, either as an identity, or as a relation among the constants of the original equations, or as an equation determining α, has already been discussed. Assuming the condition $b_\kappa = 0$ satisfied, it is known from the preceding result that the equations in U' and V' possess regular integrals, which vanish with t and involve an arbitrary constant that does not appear in the differential equations. The inferences stated earlier are therefore established.

It appears, from the investigation, that two conditions must be satisfied to allow the possession of regular integrals: one of them is a relation among the constants of the equation represented by $a = 0$: the other of them is the relation represented by $b_\kappa = 0$. To obtain them directly from the original forms, we can proceed as follows. Let

$$u = \sum_{l=1} p_l t^l, \quad v = \sum_{l=1} q_l t^l,$$

be substituted in the original equations: and determine p_1, ..., p_{m-1}, q_1, ..., q_{m-1}. The first condition is that the coefficient of t^m in

$$\theta \left(\sum_{l=1}^{m-1} p_l t^l, \ \sum_{l=1}^{m-1} q_l t^l, \ t \right)$$

shall vanish. Take $p_m = \alpha$; and then from the original equations determine p_{m+1}, ..., p_{n-1}, q_m, ..., q_{n-1}. The second condition is that the coefficient of t^n in

$$\phi \left(\sum_{l=1}^{n-1} p_l t^l, \ \sum_{l=1}^{n-1} q_l t^l, \ t \right)$$

shall vanish. It is not difficult to verify these statements.

Summarising the results, it appears that, *unless one condition be satisfied, the equations possess no regular integrals vanishing with t. When the condition is satisfied, another relation must be satisfied. If this relation determines a parameter, the equations possess a single infinitude of regular integrals; if it involves only the constants in the differential equations, then, when it is not satisfied, there are no regular integrals vanishing with t: and, when it is satisfied, there is a double infinitude of such integrals.*

Ex. Consider the equations

$$t\frac{dt_1}{dt}=t_1+(a,\,b,\,c,\,f,\,g,\,h\tilde{X}t_1,\,t_2,\,t)^2+(a,\,\beta,\,\gamma,\,\delta\tilde{X}t_1,\,t_2)^3,$$

$$t\frac{dt_2}{dt}=3t_2-2\theta t+(a',\,b',\,c',\,f',\,g',\,h'\tilde{X}t_1,\,t_2,\,t)^2+(a',\,\beta',\,\gamma',\,\delta'\tilde{X}t_1,\,t_2)^3.$$

If they possess regular solutions that vanish with t, the expressions for the solutions must be of the form

$$t_1=At+p_2t^2+p_3t^3+\dots,$$
$$t_2=q_1t+q_2t^2+q_3t^3+\dots,$$

where A is initially an arbitrary quantity. Substituting, the first equation becomes

$$p_2t^2+2p_3t^3+\dots=t^2(a,\,b,\,c,\,f,\,g,\,h\tilde{X}A,\,q_1,\,1)^2$$

$$+t^3\left\{\left(p_2\frac{\partial}{\partial A}+q_2\frac{\partial}{\partial q_1}\right)(a,\,b,\,c,\,f,\,g,\,h\tilde{X}A,\,q_1,\,1)^2+(a,\,\beta,\,\gamma,\,\delta\tilde{X}A,\,q_1)^3\right\}+\dots,$$

and the second

$$-2q_1t-q_2t^2-0.q_3t^3+\dots=-2\theta t+t^2(a',\,b',\,c',\,f',\,g',\,h'\tilde{X}A,\,q_1,\,1)^2$$

$$+t^3\left\{\left(p_2\frac{\partial}{\partial A}+q_2\frac{\partial}{\partial q_1}\right)(a',\,b',\,c',\,f',\,g',\,h'\tilde{X}A,\,q_1,\,1)^2+(a',\,\beta',\,\gamma',\,\delta'\tilde{X}A,\,q_1)^3\right\}+\dots.$$

Hence

$$q_1=\theta\,;$$
$$p_2=(a,\,b,\,c,\,f,\,g,\,h\tilde{X}A,\,\theta,\,1)^2,$$
$$-q_2=(a',\,b',\,c',\,f',\,g',\,h'\tilde{X}A,\,\theta,\,1)^2\,;$$
$$2p_3=p_2\frac{\partial p_2}{\partial A}+q_2\frac{\partial p_2}{\partial\theta}+(a,\,\beta,\,\gamma,\,\delta\tilde{X}A,\,\theta)^3\,;$$

and in order that q_3 may be finite, so that the coefficient of t^3 must vanish on both sides of the second equation, we must have

$$p_2\frac{\partial q_2}{\partial A}+q_2\frac{\partial q_2}{\partial\theta}-(a',\,\beta',\,\gamma',\,\delta'\tilde{X}A,\,\theta)^3=0.$$

Since p_2 and q_2 are quadratic in A, this is a cubic equation in general; and therefore, in general, there are three values of A. But for special relations among the coefficients, the degree of the equation in A may be less than 3; for instance, if

$$2a'(a-h')-a'=0,$$

the term in A^3 disappears, and so for the other terms.

If the equation is evanescent, then A remains arbitrary; but if it involves A, then A is thereby a determinate constant. In either case, q_3 is not determinate; we can assign an arbitrary value B to q_3. The remaining coefficients $p_4,\,q_4$; $p_5,\,q_5$; ... involve B. In the former case, there is a double infinitude of solutions; they involve the arbitrary constants A and B, and are regular functions of t which vanish with t. In the latter case, there is a single infinitude of such solutions: they involve the arbitrary constant B.

If the equation does not involve A and is not evanescent, then no regular solutions exist which vanish with t.

CASE II (a): *the critical quadratic has equal roots,*
not a positive integer.

170. The equations are

$$t \frac{du}{dt} = \xi u + \phi_1 (u, v, t) \left.\begin{array}{l} \\ \\ \\ \\ \end{array}\right\},$$
$$t \frac{dv}{dt} = \kappa u + \xi v + \phi_2 (u, v, t)$$

where ξ is not a positive integer; the functions ϕ_1 and ϕ_2 are regular and (with the possible exception of a term in t) contain no terms of order lower than 2. If they possess regular integrals vanishing with t, these must have the forms

$$u = \sum_{n=1} p_n t^n, \quad v = \sum_{n=1} q_n t^n.$$

Substituting these expressions and equating coefficients, we find

$$(n - \xi) p_n = f_n \left.\begin{array}{l} \\ \\ \end{array}\right\},$$
$$(n - \xi) q_n = g_n + \kappa p_n$$

where f_n and g_n are the coefficients of t^n in ϕ_1 and ϕ_2 respectively, when the series for u and v are substituted. It is clear that f_n and g_n are linear in the coefficients of ϕ_1 and ϕ_2, that they are polynomials in $p_1, p_2, ..., q_1, q_2, ...$, and that they contain no coefficient p or q in the succession later than p_{n-1} and q_{n-1}. As ξ is not an integer, the foregoing equations, taken for successive values of n, determine formal expressions for the whole set of coefficients p and q; in particular, p_n and q_n are obtained as sums of quotients, the numerators of which are integral functions of the coefficients in ϕ_1 and ϕ_2, and the denominators of which are products of powers of the quantities

$$1 - \xi, \quad 2 - \xi, ..., \quad n - \xi.$$

To discuss the convergence of the power-series for u and v with these coefficients, we introduce an associated set of dominant equations. Let a common region of existence of ϕ_1 and ϕ_2 be determined by the range

$$|u| \leqslant \rho, \quad |v| \leqslant \sigma, \quad |t| \leqslant r;$$

within this region let the greatest value of $|\phi_1|$ be M, and that of $|\phi_2|$ be N, so that within the region

$$|\phi_1| \leqslant M, \quad |\phi_2| \leqslant N,$$

M and N denoting finite magnitudes. Also, let ϵ denote the smallest value of $|m - \xi|$ for values of the integer m; and let $|\kappa| = c$. Then we consider the dominant equations given by

$$\epsilon X = \frac{M}{r} t + \Sigma\Sigma\Sigma \; \frac{M}{\rho^i \sigma^j r^k} X^i Y^j t^k,$$

$$\epsilon Y = cX + \frac{N}{r} t + \Sigma\Sigma\Sigma \; \frac{N}{\rho^i \sigma^j r^k} X^i Y^j t^k,$$

where the summations on the right-hand side are for integer values of i, j, k such that $i + j + k \geqslant 2$.

The general course of the argument is similar to that in I (a). In the first place, X and Y can be determined by the simultaneous equations

$$\left.\begin{aligned}
\epsilon X &= \frac{M}{\left(1 - \dfrac{t}{r}\right)\left(1 - \dfrac{X}{\rho}\right)\left(1 - \dfrac{Y}{\sigma}\right)} - M - \frac{M}{\rho} X - \frac{M}{\sigma} Y \\
\epsilon Y &= cX + \frac{N}{\left(1 - \dfrac{t}{r}\right)\left(1 - \dfrac{X}{\rho}\right)\left(1 - \dfrac{Y}{\sigma}\right)} - N - \frac{N}{\rho} X - \frac{N}{\sigma} Y
\end{aligned}\right\}.$$

From these we have

$$N\epsilon X = M(\epsilon Y - cX),$$

so that

$$Y = X\left(\frac{N}{M} + \frac{c}{\epsilon}\right);$$

when this value is substituted for Y in either equation, say in the first, we have

$$\left(\epsilon + \frac{M}{\rho} + \frac{N}{\sigma} + \frac{cM}{\sigma\epsilon}\right) X \left(1 - \frac{X}{\rho}\right)\left\{1 - \frac{X}{\sigma}\left(\frac{N}{M} + \frac{c}{\epsilon}\right)\right\}$$

$$= \frac{M}{1 - \dfrac{t}{r}} - M\left(1 - \frac{X}{\rho}\right)\left\{1 - \frac{X}{\sigma}\left(\frac{N}{M} + \frac{c}{\epsilon}\right)\right\},$$

a cubic equation in X. The term independent of X vanishes when $t = 0$; and the term involving the first power of X does not vanish when $t = 0$, because ϵ is not zero. Hence the cubic has one (and only one) root vanishing when $t = 0$.

It follows, as before, that this root of the cubic can be expressed as a regular function of t in the form of a power-series, which converges absolutely for values of $|t|$ less than the least modulus of a root of the discriminant, that is, for a finite range. When the power-series for X has been obtained, the power-series for Y is given by

$$Y = X \left(\frac{N}{M} + \frac{c}{\epsilon} \right),$$

say

$$X = P_1 t + P_2 t^2 + \dots + P_n t^n + \dots \\ Y = Q_1 t + Q_2 t^2 + \dots + Q_n t^n + \dots \Big\} .$$

In the second place it can, as before, be shewn that the analysis, effective for the determination of p_n and q_n in connection with the original equations, is effective for the determination of P_n and Q_n in connection with the dominant equations by merely making literal changes, and that these literal changes are such as to give

$$|p_n| < P_n, \quad |q_n| < Q_n.$$

for all values of n.   It therefore follows that the series

$$p_1 t + p_2 t^2 + p_3 t^3 + \dots,$$

$$q_1 t + q_2 t^2 + q_3 t^3 + \dots,$$

converge within a finite region round $t = 0$. Consequently *the equations possess regular integrals vanishing with* t: and it is not difficult to prove that *these regular integrals are unique as regular integrals with the assigned conditions.*

CASE II (*b*): *when the critical quadratic has equal roots, the repeated root being a positive integer.*

171.   The equations are

$$t \frac{du}{dt} = mu + \alpha t + \theta (u, v, t) \\ t \frac{dv}{dt} = \kappa u + mv + \beta t + \phi (u, v, t) \Bigg\} ,$$

where m is a positive integer. The functions θ and ϕ are regular; they vanish with u, v, t, and contain no terms of dimensions lower than 2.

We transform the equations as in I (*b*) by successive substitutions, each of which leads to new equations of a similar form with a diminution by one unit in the coefficients of u and of v after each operation. We take

$$u = t (\lambda + u_1), \quad v = t (\mu + v_1),$$

choosing λ and μ so that u_1 and v_1 vanish with t: then u_1 and v_1 are regular functions of t, if the equations possess regular integrals. To secure this form of transformation, we must have

$$(m-1)\lambda + \alpha = 0,$$

$$\kappa\lambda + (m-1)\mu + \beta = 0,$$

so that

$$\lambda = -\frac{\alpha}{m-1}, \qquad \mu = \frac{\kappa\alpha}{(m-1)^2} - \frac{\beta}{m-1};$$

and the new equations are

$$t\frac{du_1}{dt} = (m-1)u_1 + \alpha't + \theta_1(u_1, v_1, t) \left.\right\}.$$
$$t\frac{dv_1}{dt} = \kappa u_1 + (m-1)v_1 + \beta't + \phi_1(u_1, v_1, t) \left.\right\}$$

A similar transformation can be effected upon this pair, with a similar result; and the process can be carried out $m-1$ times in all, leading to equations

$$t\frac{du'}{dt} = u' + at + h(u', v', t) \left.\right\},$$
$$t\frac{dv'}{dt} = \kappa u' + v' + bt + k(u', v', t) \left.\right\}$$

where h, k are regular functions, which vanish with u', v', t, and contain no terms of dimensions lower than 2; also u', v' are to vanish with t.

There are two sub-cases to be considered, according as κ is zero or κ is different from zero.

First, let κ be 0; so that the equations are

$$t\frac{du'}{dt} = u' + at + h(u', v', t) \left.\right\}.$$
$$t\frac{dv'}{dt} = v' + bt + k(u', v', t) \left.\right\}$$

It is easy to see, by substituting expressions of the form

$$u' = p_1 t + p_2 t^2 + \dots, \quad v' = q_1 t + q_2 t^2 + \dots,$$

that the equations cannot possess regular integrals vanishing with t unless

$$a = 0, \quad b = 0.$$

Assume, therefore, that $a = 0$, $b = 0$. If the equations then possess regular integrals vanishing with t, we can take

$$u' = tU', \quad v' = tV',$$

where now the only transferred condition to be imposed upon U' and V' is that they are to be regular functions of t. Substituting these values, we find

$$t^2 \frac{dU'}{dt} = h(tU', tV', t) = t^2 H(U', V', t),$$

$$t^2 \frac{dV'}{dt} = k(tU', tV', t) = t^2 K(U', V', t),$$

so that

$$\frac{dU'}{dt} = H(U', V', t), \quad \frac{dV'}{dt} = K(U', V', t),$$

where H and K are regular functions of their arguments. To these equations, Cauchy's general existence-theorem can be applied; it shews that they possess integrals, which are regular functions of t and assume assigned (arbitrary) values when $t = 0$. Accordingly, *the equations in u' and v', in the case when the conditions $a = 0$, $b = 0$ are satisfied and when the constant κ is zero, possess a double infinitude of regular integrals which vanish when $t = 0$.*

Secondly, let κ be different from zero. If regular integrals exist, they are expressible in the form

$$u' = a_1 t + a_2 t^2 + \ldots, \quad v' = b_1 t + b_2 t^2 + \ldots;$$

substituting these, and taking account of the first power of t on the two sides of both equations, we have

$$a_1 = a_1 + a, \quad b_1 = \kappa a_1 + b_1 + b.$$

Hence we must have $a = 0$; then b_1 is undetermined, and

$$a_1 = -\frac{b}{\kappa},$$

a finite quantity because κ is not zero.

Assuming that the condition $a = 0$ is satisfied, and assigning an arbitrary value A to b_1, let

$$u' = t\left(-\frac{b}{\kappa} + \eta_1\right), \quad v' = t(A + \eta_2),$$

so that η_1 and η_2 are to be regular functions of t vanishing with t; the equations for η_1 and η_2 are

$$t^2 \frac{d\eta_1}{dt} = h\left(-t\frac{b}{\kappa} + t\eta_1, \quad tA + t\eta_2, \quad t\right)$$

$$= t^2 H\left(\eta_1, \eta_2, t\right),$$

$$t^2 \frac{d\eta_2}{dt} = \kappa t\eta_1 + k\left(-t\frac{b}{\kappa} + t\eta_1, \quad tA + t\eta_2, \quad t\right)$$

$$= \kappa t\eta_1 + t^2 K\left(\eta_1, \eta_2, t\right),$$

that is, they are

$$\left. \begin{aligned} t\frac{d\eta_1}{dt} &= tH\left(\eta_1, \eta_2, t\right) \\ t\frac{d\eta_2}{dt} &= \kappa\eta_1 + tK\left(\eta_1, \eta_2, t\right) \end{aligned} \right\},$$

where H, K are regular functions of their arguments and involve the arbitrary constant A.

These equations are now the same as in the Case II (a) when ξ is made zero. Accordingly, all the analysis of that earlier discussion applies, when in it ϵ is taken equal to unity. The equations in η_1 and η_2 possess regular integrals vanishing with t, and their expression involves A, the arbitrary constant; and therefore *the original equations in u and v possess no regular integrals vanishing with t, unless the condition represented by $a = 0$ be satisfied; but if that condition be satisfied, they possess a simple infinitude of regular integrals vanishing with t.*

The conditions represented by $a = 0$ and $b = 0$ in the sub-case when κ is zero, and the condition represented by $a = 0$ in the sub-case when κ is different from zero, can be expressed as before. For the former sub-case, we determine coefficients a and b, so that

$$\left. \begin{aligned} u &= a_1 t + \ldots + a_{m-1} t^{m-1} + \ldots \\ v &= b_1 t + \ldots + b_{m-1} t^{m-1} + \ldots \end{aligned} \right\}$$

satisfy the equations

$$\left. \begin{aligned} t\frac{du}{dt} &= mu + \alpha t + \theta\left(u, v, t\right) \\ t\frac{dv}{dt} &= mv + \beta t + \phi\left(u, v, t\right) \end{aligned} \right\};$$

the conditions are that the coefficient of t^m in

$$\alpha t + \theta \left(\sum_{l=1}^{m-1} a_l t^l, \quad \sum_{l=1}^{m-1} b_l t^l, \quad t \right),$$

and the same coefficient in

$$\beta t + \phi \left(\sum_{l=1}^{m-1} a_l t^l, \quad \sum_{l=1}^{m-1} b_l t^l, \quad t \right),$$

shall vanish. For the latter sub-case, we determine the $2(m-1)$ coefficients in u and v so that the equations

$$t \frac{du}{dt} = mu + \alpha t + \theta (u, v, t) \Bigg\}$$
$$t \frac{dv}{dt} = \kappa u + mv + \beta t + \phi (u, v, t) \Bigg\}$$

are satisfied; and the single condition is that the coefficient of t^m in

$$\alpha t + \theta \left(\sum_{l=1}^{m-1} a_l t^l, \quad \sum_{l=1}^{m-1} b_l t^l, \quad t \right)$$

shall vanish.

This completes the discussion of the regular integrals vanishing with t, with the respective results as enunciated in the various cases.

Note. Now that the possession of regular integrals, either unconditionally or conditionally, is established in the several cases, it is unnecessary to transform a system

$$t \frac{dU}{dt} = a_1 U + \beta_1 V + \gamma_1 t + \vartheta_1 (U, V, t) = \theta_1 (U, V, t) \Bigg\}$$
$$t \frac{dV}{dt} = a_2 U + \beta_2 V + \gamma_2 t + \vartheta_2 (U, V, t) = \theta_2 (U, V, t) \Bigg\}$$

in the manner adopted in the general investigation. Because the regular integrals vanish with t, they have the forms

$$U = k_1 t + k_2 t^2 + \ldots + k_n t^n + \ldots \Bigg\}$$
$$V = l_1 t + l_2 t^2 + \ldots + l_n t^n + \ldots \Bigg\};$$

when the coefficients are obtained by means of the condition that U and V satisfy the equations identically, the resulting power-series are known to converge. Now noting that ϑ_1 and ϑ_2 are regular functions of their arguments, which contain no terms of dimensions lower than two, and denoting by η_n and ζ_n the coefficients of t^n in ϑ_1 and ϑ_2 respectively after substitution is made for U and V, we have (for $n > 1$)

$$(n - a_1) k_n - \beta_1 l_n = \eta_n \Bigg\}$$
$$- a_2 k_n + (n - \beta_2) l_n = \zeta_n \Bigg\},$$

where η_n and ζ_n involve no coefficients k and l except $k_1, ..., k_{n-1}, l_1, ..., l_{n-1}$. To determine k_1 and l_1, we have

$$(1-a_1)k_1 - \beta_1 l_1 = \gamma_1 \atop -a_2 k_1 + (1-\beta_2)l_2 = \gamma_2 \Bigg\},$$

so that

$$k_1 = \frac{(1-\beta_2)\gamma_1 + \beta_1\gamma_2}{Q(1)}, \quad l_1 = \frac{a_2\gamma_1 + (1-a_1)\gamma_2}{Q(1)},$$

where $Q(n)$ denotes

$$(n-a_1)(n-\beta_2) - a_2\beta_1.$$

When these values of k_1 and l_1 are substituted in η_2 and ζ_2, the values of k_2 and l_2 are

$$k_2 = \frac{(2-\beta_2)\eta_2 + \beta_1\zeta_2}{Q(2)}, \quad l_2 = \frac{a_2\eta_2 + (2-a_1)\zeta_2}{Q(2)};$$

and so on, for the coefficients in succession.

Non-regular Integrals.

172. It has been seen that, either in general or subject to certain conditions, the equations

$$t\frac{dU}{dt} = a_1 U + \beta_1 V + \gamma_1 t + ... = \theta_1(U, V, z) \atop t\frac{dV}{dt} = a_2 U + \beta_2 V + \gamma_2 t + ... = \theta_2(U, V, z) \Bigg\}$$

possess regular integrals which vanish with t: and these are unique as regular integrals. Denoting them by u_1, v_1, let

$$U = u_1 + x, \qquad V = v_1 + y;$$

so that, if functions x and y exist, different from constant zero, they are non-regular functions of t; and they must vanish with t because U, u_1, V, v_1 all vanish with t. Then

$$t\frac{dx}{dt} = \theta_1(u_1 + x,\ v_1 + y,\ t) - \theta_1(u_1,\ v_1,\ t) \atop = \sum_{n=1}^{\infty} \frac{1}{n!}\left(x\frac{\partial}{\partial u_1} + y\frac{\partial}{\partial v_1}\right)^n \theta_1(u_1,\ v_1,\ t) \atop t\frac{dy}{dt} = \sum_{n=1}^{\infty} \frac{1}{n!}\left(x\frac{\partial}{\partial u_1} + y\frac{\partial}{\partial v_1}\right)^n \theta_2(u_1,\ v_1,\ t) \Bigg\}$$

are equations to determine x and y. On the right-hand sides there are no terms involving t alone; the only terms of the first order are $a_1 x + \beta_1 y$, $a_2 x + \beta_2 y$ respectively; and the coefficients of the other powers of x and y are functions of t and of u_1, v_1, that is, after substitution of the values of u_1, v_1, these coefficients are regular functions of t. Hence we may take the equations in the form

$$t\frac{dx}{dt} = \alpha_1 x + \beta_1 y + \vartheta_1(x, y, t)$$
$$t\frac{dy}{dt} = \alpha_2 x + \beta_2 y + \vartheta_2(x, y, t)$$

where ϑ_1 and ϑ_2 are regular functions of x, y, t, which vanish when $x = 0$, $y = 0$, and contain no terms of dimensions lower than 2 in x, y, and t. The dependent variables x and y, if they exist as other than zero constants (which manifestly satisfy the equations), are to be non-regular functions of t which vanish when $t = 0$.

It is convenient to transform the equations by linear changes of the dependent variables, as was done in the discussion of regular integrals: the new forms depending upon the roots of the critical quadratic

$$(\xi - \alpha_1)(\xi - \beta_2) - \alpha_2\beta_1 = 0.$$

When the roots of the quadratic are unequal, say ξ_1 and ξ_2, we take new variables

$$t_1 = \lambda x + \mu y, \qquad t_2 = \lambda' x + \mu' y,$$

where

$$(\alpha_1 - \xi_1)\lambda + \alpha_2\mu = 0 \atop \beta_1\lambda + (\beta_2 - \xi_1)\mu = 0 \bigg\} , \qquad (\alpha_1 - \xi_2)\lambda' + \alpha_2\mu' = 0 \atop \beta_1\lambda' + (\beta_2 - \xi_2)\mu' = 0 \bigg\} ;$$

the equations become

$$t\frac{dt_1}{dt} = \xi_1 t_1 + \phi_1(t_1, t_2, t)$$
$$t\frac{dt_2}{dt} = \xi_2 t_2 + \phi_2(t_1, t_2, t)$$

where the regular functions ϕ_1 and ϕ_2 vanish when $t_1 = 0$, $t_2 = 0$, and contain no terms in t_1, t_2, t of dimensions lower than 2.

When the roots of the quadratic are equal, the common value being ξ, the corresponding forms are

$$t\frac{dt_1}{dt} = \xi t_1 + \phi_1(t_1, t_2, t)$$
$$t\frac{dt_2}{dt} = \kappa t_1 + \xi t_2 + \phi_2(t_1, t_2, t)$$

with the same characteristic properties of the functions ϕ_1 and ϕ_2 as in the former case; here $t_2 = y$ and $t_1 = \lambda x + \mu y$, where

$$(\alpha_1 - \xi)\lambda + \alpha_2\mu = 0, \qquad \beta_1\lambda + (\beta_2 - \xi)\mu = 0,$$

and the constant κ is given by $\kappa\lambda = \alpha_2$.

We proceed to deal with the various alternative cases, as for the regular integrals. For those instances of the original equations, which do not possess regular integrals because the appropriate condition is not satisfied, it will be necessary to return to those original equations for the discussion of the non-regular integrals.

Some indication of the character of the solutions may be derived from the consideration of two simple examples, one of each form.

A simple example of the case, when the roots of the critical quadratic are unequal, is

$$\left. \begin{array}{l} t\dfrac{dt_1}{dt} = \lambda t_1 + a t t_2 \\[2mm] t\dfrac{dt_2}{dt} = \mu t_2 + b t t_1 \end{array} \right\};$$

integrals (if they exist) are required which vanish when $t = 0$. The solution of these equations, which are linear, can be made to depend upon that of a linear equation of the second order having $t = 0$ for a singularity: it appears that the integrals are normal in the vicinity of $t = 0$. Their full expression is

$$t_1 = A t^\lambda \left\{ 1 + \frac{abt^2}{2(1+\rho)} + \frac{(abt^2)^2}{2.4(1+\rho)(3+\rho)} + \dots \right\}$$
$$+ \frac{a}{1-\rho} B t^{\mu+1} \left\{ 1 + \frac{abt^2}{2(3-\rho)} + \frac{(abt^2)^2}{2.4(3-\rho)(5-\rho)} + \dots \right\},$$

$$t_2 = \frac{b}{1+\rho} A t^{\lambda+1} \left\{ 1 + \frac{abt^2}{2(3+\rho)} + \frac{(abt^2)^2}{2.4(3+\rho)(5+\rho)} + \dots \right\}$$
$$+ B t^\mu \left\{ 1 + \frac{abt^2}{2(1-\rho)} + \frac{(abt^2)^2}{2.4(1-\rho)(3-\rho)} + \dots \right\},$$

where $\rho = \lambda - \mu$: in order that the solution may be satisfactory, it is manifest that ρ may not be an integer, positive or negative. For the present purpose, the general integrals must be chosen so that they vanish with t; and consequently the most important terms in the immediate vicinity of $t = 0$ are

$$\left. \begin{array}{l} t_1 = A t^\lambda + \dfrac{a}{1-\rho} B t^{\mu+1} \\[2mm] t_2 = \dfrac{b}{1+\rho} A t^{\lambda+1} + B t^\mu \end{array} \right\},$$

the quantities A and B being arbitrary.

If the real part of λ and the real part of μ be both positive, then, when the variable t approaches its origin, not making an infinite number of circuits round that origin, t_1 and t_2 ultimately vanish when $t=0$; that is, as λ and μ are not integers, there is a double infinitude of non-regular integrals vanishing with t.

If the real part of λ be positive and the real part of μ be negative, then, when t tends to zero as before, t_2 can tend to zero only if B be zero : and if $B=0$, then t_1 and t_2 ultimately vanish when $t=0$; that is, there is a single infinitude of non-regular integrals vanishing with t.

Similarly, if the real part of λ be negative and the real part of μ be positive, there is a single infinitude of non-regular integrals vanishing with t.

If both the real part of λ and the real part of μ be negative, then t_1 and t_2 vanish with t only if $A=0$, $B=0$: that is, non-regular integrals vanishing with t do not then exist. This last result is in accordance with Goursat's result already quoted (§ 165, note).

It will be noticed that the parts depending upon t^λ alone, when they exist, are of the form

$$t_1 = t^\lambda \rho_1, \qquad t_2 = t^\lambda \rho_2,$$

where ρ_1 is an arbitrary finite quantity, and ρ_2 is zero, when $t=0$; and that the parts depending upon t^μ alone, when they exist, are of the form

$$t_1 = t^\mu \sigma_1, \qquad t_2 = t^\mu \sigma_2,$$

where σ_2 is another arbitrary finite quantity, and σ_1 is zero, when $t=0$. These particular results are general and, in this form, can be established by an appropriate modification of Goursat's argument. They are included in the more general theorems that will be considered immediately.

A simple example of the case, when the roots of the critical quadratic are equal, is

$$\left. \begin{array}{l} t\dfrac{dt_1}{dt} = \lambda t_1 + a t t_2 \\[2mm] t\dfrac{dt_2}{dt} = \kappa t_1 + \lambda t_2 \end{array} \right\} \ ;$$

integrals (if they exist) are required which vanish when $t=0$. The solution of these equations can, as for the preceding example, be made to depend upon the solution of a linear equation of the second order, having $t=0$ for a singularity ; and their expressions can be obtained in the form

$$t_1 = A a t^{\lambda+1}(1+\tfrac{1}{2}a\kappa t + \ldots) + Ba\,\{a\kappa t^{\lambda+1}(1+\tfrac{1}{2}a\kappa t + \ldots)\log t + (1 - \tfrac{3}{4}a^2\kappa^2 t^2 - \ldots)\,t^\lambda\},$$

$$t_2 = A t^\lambda (1 + a\kappa t + \ldots) + B\,\{a\kappa t^\lambda(1 + a\kappa t + \ldots)\log t + (a\kappa - a^2\kappa^2 t - \ldots)\,t^\lambda\}.$$

When the real part of λ is positive, these integrals vanish with t ; and there is a double infinitude of them. When the real part of λ is negative, then it is necessary that A and B both vanish : that is, the integrals do not exist if they are to vanish with t.

When B is zero, then the integrals become of the form

$$t_1 = t^\lambda \rho_1, \qquad t_2 = t^\lambda \rho_2,$$

where ρ_2 is an arbitrary finite quantity and ρ_1 is zero, when $t = 0$. This result is general. There is no corresponding simple inference from the parts that depend solely upon B : the complication is caused by the term κt_1 in the second equation.

The special results here obtained are included in the theorems relating to the equations in their general form : they suggest that integrals exist which are regular functions of t, t^λ, and $t^\lambda \log t$, when the real part of λ is positive.

CASE I (a): *the critical quadratic has unequal roots, neither of them being a positive integer.*

173. It has been proved that the original equations in this case possess regular integrals vanishing with t : and therefore, in order to consider the non-regular integrals (if any) that vanish with t, we transform the equations as in § 172, and we study the derived system

$$\left. \begin{aligned} t \frac{dt_1}{dt} &= \xi_1 t_1 + \phi_1 (t_1, t_2, t) \\ t \frac{dt_2}{dt} &= \xi_2 t_2 + \phi_2 (t_1, t_2, t) \end{aligned} \right\},$$

where ϕ_1 and ϕ_2 are regular functions of their arguments, which vanish when $t_1 = 0$, $t_2 = 0$, and contain no terms of dimensions less than two in t_1, t_2, t. The integrals t_1 and t_2 are to be non-regular functions of t, and they are required to vanish with t.

The main theorem is as follows :—

When the roots of the critical quadratic ξ_1 and ξ_2 have their real parts positive, and are such that no one of the quantities

$$(\lambda - 1) \xi_1 + \mu \xi_2 + \nu, \quad \lambda \xi_1 + (\mu - 1) \xi_2 + \nu,$$

vanishes for positive integer values of λ, μ, ν such that $\lambda + \mu + \nu \geqslant 2$, then the equations possess a double infinitude of non-regular integrals, which vanish * *with t, these integrals being regular functions of t, t^{ξ_1}, t^{ξ_2}.*

Immediate corollaries, when once this theorem is established, are as follows :—

If the real part of ξ_1 be positive and that of ξ_2 be negative, there is only a single infinitude of non-regular integrals vanishing with t : they are regular functions of t and t^{ξ_1}.

---

* It is unnecessary here to discuss the path of approach of t to its origin, in order to secure that t^{ξ_1} and t^{ξ_2} vanish at the origin; for it practically would be a repetition of the corresponding discussion in § 65.

Likewise, if the real part of ξ_2 be positive and that of ξ_1 be negative, there is only a single infinitude of non-regular integrals vanishing with t: they are regular functions of t and t^{ξ_2}.

If the real part both of ξ_1 and of ξ_2 be negative, there are no non-regular integrals of the equations that vanish with t.

These results will be found sufficiently obvious after the establishment of the main theorem.

174. In discussing the equations, it will be convenient to replace t^{ξ_1} and t^{ξ_2} by new variables, say

$$t^{\xi_1} = z_1, \quad t^{\xi_2} = z_2,$$

so that, by the general theorem, regular functions of z_1, z_2, t are to be established as solutions of the equations. Accordingly, regarding t_1 and t_2 as functions of these three arguments, assume

$$\left. \begin{array}{l} t_1 = \Sigma\Sigma\Sigma a_{mnp}\, z_1{}^m z_2{}^n t^p \\ t_2 = \Sigma\Sigma\Sigma b_{mnp}\, z_1{}^m z_2{}^n t^p \end{array} \right\} ,$$

where the summation is for all positive (and zero) values of the integers m, n, p, with the conventions

$$a_{000} = 0, \quad b_{000} = 0.$$

Moreover

$$t\,\frac{d}{dt} = t\,\frac{\partial}{\partial t} + \xi_1 z_1\,\frac{\partial}{\partial z_1} + \xi_2 z_2\,\frac{\partial}{\partial z_2}\,.$$

Hence the differential equations are

$$\left. \begin{array}{l} t\,\dfrac{\partial t_1}{\partial t} + \xi_1 z_1\,\dfrac{\partial t_1}{\partial z_1} + \xi_2 z_2\,\dfrac{\partial t_1}{\partial z_2} = \xi_1 t_1 + \phi_1\,(t_1,\ t_2,\ t) \\[2ex] t\,\dfrac{\partial t_2}{\partial t} + \xi_1 z_1\,\dfrac{\partial t_2}{\partial z_1} + \xi_2 z_2\,\dfrac{\partial t_2}{\partial z_2} = \xi_2 t_2 + \phi_2\,(t_1,\ t_2,\ t) \end{array} \right\}.$$

Substituting the assumed values of t_1 and t_2, and afterwards equating coefficients of $z_1{}^m z_2{}^n t^p$, we have

$$\left. \begin{array}{l} \{(m-1)\,\xi_1 + n\xi_2 + p\}\, a_{mnp} = \alpha'_{mnp} \\ \{m\xi_1 + (n-1)\,\xi_2 + p\}\, b_{mnp} = \beta'_{mnp} \end{array} \right\} :$$

where α'_{mnp} is a rational function of the coefficients in ϕ_1, of those coefficients $a_{m'n'p'}$ in t_1 for which

$$m' \leqslant m, \quad n' \leqslant n, \quad p' \leqslant p, \quad m' + n' + p' < m + n + p,$$

and of the coefficients $b_{m'n'p'}$ in t_2 with the same restrictions: and likewise for β'_{mnp} in relation to ϕ_2.

As there is no term in $\phi_1(t_1, t_2, t)$ of dimension unity in t, t_1, t_2, there can be no term of dimension unity in z_1, z_2, t after substitution of the values of t_1 and t_2: hence

$$\{(m-1)\xi_1 + n\xi_2 + p\}\, a_{mnp} = 0,$$

when $m + n + p = 1$. Accordingly

$$a_{010} = 0, \quad a_{001} = 0 ;$$

but there is no limitation upon a_{100}, so that it can be taken arbitrarily: we assume

$$a_{100} = A.$$

For similar reasons

$$\{m\xi_1 + (n-1)\xi_2 + p\}\, b_{mnp} = 0,$$

when $m + n + p = 1$; and we infer that

$$b_{100} = 0, \quad b_{001} = 0, \quad b_{010} = B,$$

where B is arbitrary.

Suppose now that no one of the quantities

$$(m-1)\xi_1 + n\xi_2 + p, \quad m\xi_1 + (n-1)\xi_2 + p,$$

for positive integer values of m, n, p such that

$$m + n + p \geqslant 2,$$

vanishes. Then when the equations

$$\left.\begin{array}{l} \{(m-1)\xi_1 + n\xi_2 + p\}\, a_{mnp} = \alpha'_{mnp} \\ \{m\xi_1 + (n-1)\xi_2 + p\}\, b_{mnp} = \beta'_{mnp} \end{array}\right\}$$

are solved in groups for the same value of $m + n + p$, and in successive groups for increasing values of $m + n + p$ beginning with 2, they lead to results of the form

$$a_{mnp} = \alpha_{mnp}, \quad b_{mnp} = \beta_{mnp},$$

where $\alpha_{mnp}, \beta_{mnp}$ are rational integral functions of the coefficients that occur in ϕ_1 and ϕ_2, these functions being divided by a product of factors of the forms

$$(m-1)\xi_1 + n\xi_2 + p, \quad m\xi_1 + (n-1)\xi_2 + p, \text{ for } m + n + p \geqslant 2.$$

It has been seen that $a_{001} = 0$, $b_{001} = 0$: we easily see that $a_{00p} = 0$, $b_{00p} = 0$, for all values of p. For every term in $\phi_1(t_1, t_2, t)$ and every term in $\phi_2(t_1, t_2, t)$ involve t_1, or t_2, or both: and the equations for a_{00p}, b_{00p} are

$$(p - \xi_1)\, a_{00p} = A_{00p}, \quad (p - \xi_2)\, b_{00p} = B_{00p},$$

where A_{oop}, B_{oop} are integral functions of the coefficients in ϕ_1 and ϕ_2, and of coefficients $a_{oop'}$, $b_{oop'}$ such that $p' < p$, these integral functions being divided by factors of the form $p' - \xi_1$, $p' - \xi_2$. No term occurs either in A_{oop}, B_{oop} independent of $a_{oop'}$, $b_{oop'}$, because there is no term in ϕ_1 or in ϕ_2 independent of t_1 and t_2. Hence if all the coefficients $a_{oop'}$, $b_{oop'}$ vanish when $p' < p$, then a_{oop}, b_{oop} also vanish. But $a_{001} = 0$, $b_{001} = 0$: hence $a_{002} = 0$, $b_{002} = 0$: and so on with the whole series.

Consequently in the expressions for t_1 and t_2, there occur no terms that involve t alone without either z_1, or z_2, or z_1 and z_2: which is therefore one general characteristic of the non-regular integrals if they exist.

From t_1 and t_2, let all the terms which do not involve z_2 be gathered together. By what has just been proved, there are no terms which involve t alone: hence the aggregates of the selected terms contain z_1 as a factor, and the aggregates of the remainders contain z_2 as a factor, so that we can write

$$t_1 = z_1\rho + z_2\Theta_1,$$

$$t_2 = z_1\tau + z_2\Theta_2,$$

where ρ and τ are regular functions of t and z_1, which will be proved to be such that $\rho = A$, $\tau = 0$, when $t = 0$, A being an arbitrary constant: and Θ_1, Θ_2 are regular functions of t, z_1, z_2, which will be proved to be such that $\Theta_1 = 0$, $\Theta_2 = B$, when $t = 0$, B being an arbitrary constant.

The first stage of the proof will establish the existence of the parts $z_1\rho$, $z_1\tau$: the second stage will establish the existence of the parts $z_2\Theta_1$, $z_2\Theta_2$. It may be added that, had it been deemed desirable, a selection from t_1 and t_2 of all the terms which do not involve z_1 might first have been made: the forms of t_1 and t_2 would then have been

$$t_1 = z_2\rho_1 + z_1\Psi_1, \quad t_2 = z_2\tau_1 + z_1\Psi_2,$$

where $\rho_1 = 0$, $\tau_1 = B$, when $t = 0$, and ρ_1, τ_1 are regular functions of t and z_2: also $\Psi_1 = A$, $\Psi_2 = 0$, when $t = 0$, and Ψ_1, Ψ_2 are regular functions of t, z_1, z_2. Further, it will be seen from the forms of the functions that ρ, τ, Ψ_1, Ψ_2 all vanish when $A = 0$: and that Θ_1, Θ_2, ρ_1, τ_1 all vanish when $B = 0$.

175. It is clear that, if the equations under consideration possess integrals of the form

$$t_1 = \rho z_1, \quad t_2 = \tau z_1,$$

where ρ and τ are to be regular functions of z and z_1, then, taking account of the forms of ϕ_1 and ϕ_2, the quantities ρ and τ must satisfy the equations

$$\left.\begin{aligned}
t\frac{\partial \rho}{\partial t} + \xi_1 z_1 \frac{\partial \rho}{\partial z_1} &= t\frac{d\rho}{dt} = \psi_1\,(\rho,\,\tau,\,z_1,\,t) \\[2mm]
t\frac{\partial \tau}{\partial t} + \xi_1 z_1 \frac{\partial \tau}{\partial z_1} &= t\frac{d\tau}{dt} = (\xi_2 - \xi_1)\,\tau + \psi_2\,(\rho,\,\tau,\,z_1,\,t)
\end{aligned}\right\}.$$

The functions ψ_1, ψ_2 are regular in their arguments: both of them vanish when $\rho = 0$, $\tau = 0$: in each of them, every term, which is of dimensions λ in ρ and τ combined, possesses a factor $z_1{}^{\lambda-1}$: and no term is of dimensions less than two in ρ, τ, t combined. Because ρ and τ are to be regular functions of t and z_1, they will be expressible in the forms

$$\rho = \Sigma\Sigma k_{mn} z_1{}^m t^n, \quad \tau = \Sigma\Sigma l_{mn} z_1{}^m t^n\,;$$

substituting these values and equating coefficients on the two sides of both equations, we find

$$\left.\begin{aligned}
(n + m\xi_1)\,k_{mn} &= k'_{mn} \\
\{n + (m+1)\,\xi_1 - \xi_2\}\,l_{mn} &= l'_{mn}
\end{aligned}\right\},$$

where k'_{mn} and l'_{mn} are linear in the coefficients of ψ_1 and ψ_2 respectively, and are rational integral functions of those coefficients $k_{m'n'}$, $l_{m'n'}$ in ρ and τ, for which $m' \leqslant m$, $n' \leqslant n$, $m' + n' < m + n$.

From the forms of the functions ψ_1 and ψ_2, we have $k'_{00} = 0$, $l'_{00} = 0$. Hence when $m = 0$, $n = 0$, the first of the coefficient-equations leaves k_{00} undetermined: we therefore make it an arbitrary (finite) quantity A: the second of the coefficient-equations gives $l_{00} = 0$, for ξ_1 and ξ_2 are unequal.

Since no one of the quantities

$$(m - 1)\,\xi_1 + n\xi_2 + p, \quad m\xi_1 + (n - 1)\,\xi_2 + p$$

vanishes for integer values of m, n, p such that $m + n + p \geqslant 2$, it follows that no one of the quantities

$$n + m\xi_1, \quad n + (m+1)\,\xi_1 - \xi_2$$

vanishes for integer values of m and n such that $m + n \geqslant 1$. Hence when the coefficient-equations for k and l are solved in groups for the same value of $m + n$, and in successive groups for increasing values of $m + n$ beginning with 1, they lead to results of the form

$$k_{mn} = \gamma_{mn}, \quad l_{mn} = \lambda_{mn},$$

where the quantities γ and λ are integral functions of the coefficients that occur in ψ_1 and ψ_2, each divided by a product of factors of the forms

$$n + m\xi_1, \quad n + (m + 1)\xi_1 - \xi_2.$$

Moreover each of the coefficients k and l, thus determined, contains A as a factor.

It now is necessary to prove that the series for ρ and τ, the formal expressions of which have been deduced, are converging series. For this purpose, we construct dominant equations as follows.

Let a region of common existence of the functions ψ_1 and ψ_2 be defined by the ranges $|t| \leqslant r$, $|z_1| \leqslant r_1$, $|\rho| \leqslant \alpha$, $|\tau| \leqslant \beta$: so that ψ_1 and ψ_2 are regular functions of their arguments within these ranges. In this region, let M_1 be the greatest value of $|\psi_1|$, and M_2 the greatest value of $|\psi_2|$: let M denote the greater of the two quantities M_1 and M_2. Further, since the quantities $n + m\xi_1$, $n + (m + 1)\xi_1 - \xi_2$, do not vanish for integer values of m and n such that $m + n \geqslant 1$, there must be a least value for the moduli of the quantities for the various combinations of m and n; let this value be η, so that

$$|n + m\xi_1| \leqslant \eta, \quad |n + (m + 1)\xi_1 - \xi_2| \leqslant \eta,$$

in all instances. Also let $|A| = A'$. Then the dominant equations are chosen to be

$$\eta(P - A') = \eta T$$

$$= \frac{1}{z_1} \left\{ \frac{M}{\left(1 - \dfrac{t}{r}\right)\left(1 - \dfrac{z_1 P}{r_1 \alpha}\right)\left(1 - \dfrac{z_1 T}{r_1 \beta}\right)} - \frac{M}{1 - \dfrac{t}{r}} - M\frac{z_1 P}{r_1 \alpha} - M\frac{z_1 T}{r_1 \beta} \right\}.$$

Clearly $P - A' = T$: their common values are given as the roots of the cubic equation

$$\left\{ \left(\eta + \frac{M}{r_1 \beta} + \frac{M}{r_1 \alpha}\right) T + \frac{MA'}{r_1 \alpha} \right\} \left(1 - \frac{z_1 T}{r_1 \beta}\right) \left(1 - \frac{z_1 A'}{r_1 \alpha} - \frac{z_1 T}{r_1 \alpha}\right)$$

$$= \frac{M}{1 - \dfrac{t}{r}} \left\{ \frac{A'}{r_1 \alpha} + T\left(\frac{1}{r_1 \alpha} + \frac{1}{r_1 \beta} - \frac{A' z_1}{r_1^2 \alpha \beta}\right) - T^2 \frac{z_1}{r_1^2 \alpha \beta} \right\}.$$

When $t = 0$ and $z_1 = 0$, the term in this equation independent of T vanishes: but the term in the first power of T does not vanish because η is not zero. Hence there is one root, and only one root, of the cubic equation which vanishes when $t = 0$ and $z_1 = 0$; it is a regular function of t and z_1 in the immediate vicinity of $t = 0$, over a region which is not infinitesimal. Actually solving the equation for this root, we find

$$T = \frac{MA'}{\eta r_1 \alpha} \left(\frac{t}{r} + \frac{A'}{r_1 \alpha} z_1 \right) + \text{higher powers of } t \text{ and } z_1;$$

and then

$$P = A' + \frac{MA'}{\eta r_1 \alpha} \left(\frac{t}{r} + \frac{A'}{r_1 \alpha} z_1 \right) + \text{higher powers of } t \text{ and } z_1.$$

Now knowing that such a solution of the dominant equations exists, we can obtain its formal expression otherwise. Let

$$\begin{aligned} P &= A' + \Sigma\Sigma z_1{}^m t^n \Gamma_{mn} \\ T &= \Sigma\Sigma z_1{}^m t^n \Gamma_{mn} \end{aligned} \Bigg\};$$

substitute these values in the dominant equations, expand their right-hand sides in the form of regular series, and equate coefficients of $z_1{}^m t^n$ on the two sides. We find

$$\Gamma_{mn} = K'_{mn},$$

say. Instead of actually evaluating K'_{mn}, the analysis used to determine γ_{mn} can be adopted. To this end, construct the value of $|\gamma_{mn}|$ and, in its expression, effect the following changes in succession :—

i. Replace every modulus of a sum by the sum of the moduli of its terms:

ii. Replace each denominator-factor

$$|n + m\xi_1| \text{ and } |n + (m+1)\xi_1 - \xi_2| \text{ by } \eta:$$

iii. Replace the coefficients of $\rho^{m_1} \tau^{n_1} z_1{}^{p_1} t^{q_1}$ in ϕ_1 and ϕ_2 by $M \div \alpha^{m_1} \beta^{n_1} r_1{}^{p_1} r^{q_1}$, for all values of m_1, n_1, p_1, q_1:

iv. Replace $|A|$ by A'.

The final expression, so modified, is K'_{mn}. But the effect, upon the initial expression for $|\gamma_{mn}|$, of each of these changes is to appreciate the value: hence, taking the cumulative result, we have

$$|\gamma_{mn}| < \Gamma_{mn}.$$

Similarly

$$|\lambda_{mn}| < \Gamma_{mn}.$$

But the series

$$A' + \Sigma\Sigma z_1^m t^n \Gamma_{mn}$$

converges for a finite region round the origin $t = 0$; hence the series

$$\left. \begin{array}{l} \rho = A + \Sigma\Sigma\gamma_{mn} z_1^m t^n \\ \tau = \Sigma\Sigma\lambda_{mn} z_1^m t^n \end{array} \right\}$$

likewise converge: that is to say, the formal expressions ρ and τ have significance, being regular functions of z_1 and t. The equations accordingly have integrals

$$t_1 = \rho z_1, \quad t_2 = \tau z_1,$$

of the characteristics indicated.

This completes the first stage of the proof.

176. For the second stage, let

$$t_1 = \rho z_1 + T_1, \quad t_2 = \tau z_1 + T_2;$$

the equations for T_1 and T_2 are

$$\begin{aligned} t\frac{dT_1}{dt} &= \xi_1 T_1 + \phi_1(\rho z_1 + T_1, \tau z_1 + T_2, t) - \phi_1(\rho z_1, \tau z_1, t) \\ &= \xi_1 T_1 + \psi_1(T_1, T_2, z_1, t) \\ t\frac{dT_2}{dt} &= \xi_2 T_2 + \psi_2(T_1, T_2, z_1, t) \end{aligned} \Bigg\} \,,$$

after substitution for ρ and τ. Here ψ_1 and ψ_2 are regular functions of their arguments and vanish when $T_1 = 0$, $T_2 = 0$; they contain no terms of aggregate dimensions lower than 2 in T_1, T_2, z_1, t. In accordance with the statement in § 174, it has to be proved that these equations possess solutions of the form

$$T_1 = z_2\Theta_1, \quad T_2 = z_2\Theta_2,$$

where Θ_1 and Θ_2 are regular functions of t, z_1, z_2: it will appear that $\Theta_2 = B$ (an arbitrary constant) and $\Theta_1 = 0$, when $t = 0$. Substituting these values for T_1 and T_2, we find

$$\left. \begin{array}{l} t\dfrac{\partial\Theta_1}{\partial t} + \xi_1 z_1\dfrac{\partial\Theta_1}{\partial z_1} + \xi_2 z_2\dfrac{\partial\Theta_1}{\partial z_2} + (\xi_2 - \xi_1)\Theta_1 = f_1(\Theta_1, \Theta_2, z_1, z_2, t) \\[2ex] t\dfrac{\partial\Theta_2}{\partial t} + \xi_1 z_1\dfrac{\partial\Theta_2}{\partial z_1} + \xi_2 z_2\dfrac{\partial\Theta_2}{\partial z_2} = f_2(\Theta_1, \Theta_2, z_1, z_2, t) \end{array} \right\} \,;$$

the functions f_1 and f_2 are regular in their arguments, every term involves Θ_1 or Θ_2 or both, and a term involving Θ_1 and Θ_2 in the form $\Theta_1^\lambda\Theta_2^\mu$ has also a factor $z_2^{\lambda+\mu-1}$.

If quantities Θ_1, Θ_2 exist, which are regular functions of t, z_1, z_2, and satisfy these equations, the substitution of expressions of the form

$$\Theta_1 = \Sigma\Sigma\Sigma p_{lmn} z_1{}^l z_2{}^m t^n, \qquad \Theta_2 = \Sigma\Sigma\Sigma q_{lmn} z_1{}^l z_2{}^m t^n,$$

in these equations must lead to identities. Accordingly, equating coefficients of $z_1{}^l z_2{}^m t^n$ on the two sides of both equations, we have

$$\{n + (l-1)\,\xi_1 + (m+1)\,\xi_2\}\,p_{lmn} = \pi'_{lmn},$$

$$(n + l\xi_1 + m\xi_2)\,q_{lmn} = \kappa'_{lmn},$$

where π'_{lmn}, κ'_{lmn} are linear functions of the coefficients in f_1 and f_2, and are integral functions of the coefficients $p_{l'm'n'}$ and $q_{l'm'n'}$, such that

$$l' \leqslant l, \quad m' \leqslant m, \quad n' \leqslant n, \quad l' + m' + n' < l + m + n.$$

Owing to the forms of f_1 and f_2, we have

$$\pi'_{000} = 0, \quad \kappa'_{000} = 0.$$

Hence $p_{000} = 0$, and q_{000} is left undetermined; we take

$$q_{000} = B,$$

where B is an arbitrary constant. Moreover, no one of the quantities

$$n + (l-1)\,\xi_1 + m\xi_2, \quad n + l\xi_1 + (m-1)\,\xi_2,$$

vanishes for values of l, m, n, such that $n + l + m \geqslant 2$; hence in the equations for p_{lmn}, q_{lmn}, no one of the coefficients of p_{lmn}, q_{lmn} vanishes when $n+l+m \geqslant 1$. The equations can therefore be solved for all the coefficients p and q after p_{000}, q_{000}. They are most conveniently solved in groups for the same value of $n + l + m$, and in succeeding groups for increasing values of $n + l + m$, beginning with 1; the results are

$$p_{lmn} = \pi_{lmn}, \quad q_{lmn} = \kappa_{lmn},$$

where π_{lmn}, κ_{lmn} are sums of integral functions of the coefficients in f_1 and f_2, each divided by products of factors of the types

$$n + (l-1)\,\xi_1 + (m+1)\,\xi_2, \quad n + l\xi_1 + m\xi_2.$$

Expressions are thus obtained as formal solutions of the equations: it is necessary to establish the convergence of the infinite series. As before, we construct dominant equations for this purpose, as follows.

Let a common region of existence of the functions f_1 and f_2, which are regular in their arguments, be defined by the ranges

$$|t| \leqslant r, \quad |z_1| \leqslant \rho_1, \quad |z_2| \leqslant \rho_2, \quad |\Theta_1| \leqslant \sigma_1, \quad |\Theta_2| \leqslant \sigma_2:$$

and within this region, let N denote the maximum value of $|f_1|$ and $|f_2|$, so that N is a finite quantity. Also let η denote the least among the values of

$$|n + (l-1)\xi_1 + (m+1)\xi_2|, \quad |n + l\xi_1 + m\xi_2|,$$

for the various combinations of the integers l, m, n such that $l + m + n \geqslant 1$; and let $|B| = B'$. Then the dominant equations to be considered are

$$\eta \Phi_1 = \eta (\Phi_2 - B')$$

$$= \frac{1}{z_2} \left\{ \frac{N}{\left(1 - \dfrac{t}{r}\right)\left(1 - \dfrac{z_1}{\rho_1}\right)\left(1 - \dfrac{z_2 \Phi_1}{\rho_2 \sigma_1}\right)\left(1 - \dfrac{z_2 \Phi_2}{\rho_2 \sigma_2}\right)} \right.$$

$$\left. - \frac{N}{\left(1 - \dfrac{t}{r}\right)\left(1 - \dfrac{z_1}{\rho_1}\right)} - \frac{N z_2 \Phi_1}{\rho_2 \sigma_1} - \frac{N z_2 \Phi_2}{\rho_2 \sigma_2} \right\}.$$

The common values of Φ_1 and $\Phi_2 - B'$ are determined as roots of the cubic equation

$$\left\{ \Phi_1 \left(\eta + \frac{N}{\rho_2 \sigma_1} + \frac{N}{\rho_2 \sigma_2} \right) + \frac{NB'}{\rho_2 \sigma_2} \right\} \left(1 - \frac{z_2 \Phi_1}{\rho_2 \sigma_1}\right) \left(1 - \frac{z_2 B'}{\rho_2 \sigma_2} - \frac{z_2 \Phi_1}{\rho_2 \sigma_2}\right)$$

$$= \frac{N}{\left(1 - \dfrac{t}{r}\right)\left(1 - \dfrac{z_1}{\rho_1}\right)} \left\{ \frac{B'}{\rho_2 \sigma_2} + \Phi_1 \left(\frac{1}{\rho_2 \sigma_1} + \frac{1}{\rho_2 \sigma_2} - \frac{B' z_2}{\rho_2{}^2 \sigma_1 \sigma_2} \right) - \Phi_1{}^2 \frac{z_2}{\rho_2{}^2 \sigma_1 \sigma_2} \right\}.$$

When $t = 0$, $z_1 = 0$, $z_2 = 0$, the term in this equation independent of Φ_1 vanishes: but the term in the first power of Φ_1 does not then vanish, because η is different from zero. Hence there is one root, and only one root, of the cubic which vanishes when $t = 0$, $z_1 = 0$, $z_2 = 0$: and it is a regular function of t, z_1, z_2, in the immediate vicinity* of $t = 0$. Actually solving the equation for this root, we find

$$\Phi_1 = \frac{NB'}{\eta \rho_2 \sigma_2} \left(\frac{t}{r} + \frac{z_1}{\rho_1} + \frac{z_2 B'}{\rho_2 \sigma_2} \right) + \text{terms of higher orders};$$

and then we have

$$\Phi_2 = B' + \frac{NB'}{\eta \rho_2 \sigma_2} \left(\frac{t}{r} + \frac{z_1}{\rho_1} + \frac{z_2 B'}{\rho_2 \sigma_2} \right) + \text{terms of higher orders.}$$

* It remains a regular function, so long as $|t|$ is less than the least of the moduli of the roots of the discriminant of the cubic.

As in the preceding stage of proof of the main theorem, we can obtain the expression of these particular quantities Φ_1 and Φ_2 otherwise. Knowing that Φ_1 and $\Phi_2 - B'$, equal to one another, are regular functions of t, z_1, z_2, let

$$\Phi_1 = \Phi_2 - B' = \Sigma\Sigma\Sigma P_{lmn} z_1{}^l z_2{}^m t^n;$$

substitute in the dominant equations, expand the right-hand side in the form of regular series, and equate the coefficients of $z_1{}^l z_2{}^m t^n$ on the two sides. We find

$$P_{lmn} = \Pi_{lmn}.$$

But instead of actually deriving Π_{lmn} from the equations so obtained, we can utilise the analysis that leads to the quantities π_{lmn}, κ_{lmn}, as follows. Construct $|\pi_{lmn}|$ and, in its analytical expression, effect the following changes in succession:—

i. Replace every modulus of a sum by the sum of the moduli of the terms:

ii. Replace each denominator-factor

$$|n + (l-1)\xi_1 + (m+1)\xi_2| \text{ and } |n + l\xi_1 + m\xi_2| \text{ by } \eta:$$

iii. Replace the coefficient of $\Theta_1{}^{m_1}\Theta_2{}^{m_2} z_1{}^{n_1} z_2{}^{n_2} t^p$ in f_1 and f_2 by $N \div \sigma_1{}^{m_1}\sigma_2{}^{m_2}\rho_1{}^{n_1}\rho_2{}^{n_2} r^p$, for all values of m_1, m_2, n_1, n_2, p:

iv. Replace $|B|$ by B'.

The final expression, after all these modifications have been made, is Π_{lmn}. But the effect, upon the initial expression for $|\pi_{lmn}|$, of each of the modifications is to appreciate the value; hence taking the cumulative effect, we have

$$|\pi_{lmn}| < \Pi_{lmn}.$$

Similarly
$$|\kappa_{lmn}| < \Pi_{lmn}.$$

Now the series for Φ_2, when P_{lmn} is replaced by Π_{lmn}, converges for a finite region round the origin; hence the series

$$\begin{aligned}
\Theta_1 &= \quad \Sigma\Sigma\Sigma\pi_{lmn} z_1{}^l z_2{}^m t^n \\
\Theta_2 &= B + \Sigma\Sigma\Sigma\kappa_{lmn} z_1{}^l z_2{}^m t^n
\end{aligned}\Big\}$$

also converge for that region. Consequently the modified equations have integrals of the character

$$T_1 = z_2\Theta_1, \quad T_2 = z_2\Theta_2:$$

and therefore the original equations have integrals

$$t_1 = \rho z_1 + z_2 \Theta_1, \quad t_2 = \tau z_1 + z_2 \Theta_2,$$

where ρ and τ are regular functions of t and z_1: and Θ_1, Θ_2 are regular functions of

$$t, \, z_1, \, z_2.$$

This completes the proof of the main theorem with the specified conditions.

Ex. 1. Consider the equations

$$z\frac{dx}{dz} = x + y - x^2 \left.\begin{array}{c}\\\\\end{array}\right\}\, ,$$
$$z\frac{dy}{dz} = Q_1 x + Q_2 y - xy + zQ_3$$

where Q_1, Q_2, Q_3 are regular functions of z which do not vanish with z, their expansions being

$$Q_i = q_i + zr_i + z^2 s_i + \dots,$$

for $i = 1, 2, 3$.

The critical quadratic is

$$(\xi - 1)(\xi - q_2) - q_1 = 0.$$

Let the roots of this quadratic be ξ_1 and ξ_2: and assume that neither ξ_1 nor ξ_2 is an integer, that $\xi_1 - \xi_2$ is not zero nor an integer, and that the real parts of ξ_1 and ξ_2 are positive. The further conditions in the theorem are that

$$(\lambda - 1)\xi_1 + \mu\xi_2 + \nu, \quad \lambda\xi_1 + (\mu - 1)\xi_2 + \nu,$$

do not vanish for positive integer values of λ, μ, ν, such that

$$\lambda + \mu + \nu \geqslant 2;$$

these also will be assumed to be satisfied. (The final result shews that, for this particular set of equations, they cease to provide conditions additional to those specified: though they are necessary for the general case.)

It happens that, owing to the form of the equations, the integrals can be obtained so as to shew explicitly the double infinitude of non-regular solutions. Take

$$x = \frac{z}{u}\frac{du}{dz};$$

then the first equation gives

$$y = z\frac{dx}{dz} - x + x^2$$
$$= \frac{z^2}{u}\frac{d^2u}{dz^2}.$$

When these values are substituted in the second equation and the result is reduced, it is found that u is to be determined so as to satisfy

$$\frac{d^3u}{dz^3} + \frac{2 - Q_2}{z}\frac{d^2u}{dz^2} - \frac{Q_1}{z^2}\frac{du}{dz} - \frac{Q_3}{z^2}u = 0,$$

a linear equation.

What is desired is the expression of the values of x and y in the vicinity of $z=0$; accordingly, the value of u in that vicinity must be obtained. The fundamental determining equation for $z=0$ is

$$m(m-1)(m-2)+(2-q_2)m(m-1)-q_1m=0,$$

of which one root is $m=0$, and the other two roots are given by

$$m^2-(1+q_2)m+q_2-q_1=0:$$

that is, the three roots are 0, ξ_1, ξ_2. Having regard to the limitations which have been imposed upon ξ_1 and ξ_2, we infer from the ordinary theory of linear differential equations that three linearly independent solutions of the equation in u are provided by

$$G_0, \quad z^{\xi_1}G_1, \quad z^{\xi_2}G_2,$$

where G_0, G_1, G_2 are regular functions of z, which do not vanish when $z=0$ and can be constructed in the usual manner; and consequently that its complete solution is given by

$$u=AG_0+Bz^{\xi_1}G_1+Cz^{\xi_2}G_2,$$

where A, B, C are arbitrary constants. Hence x and y are

$$\frac{Az\dfrac{dG_0}{dz}+Bz^{\xi_1}\left(\xi_1G_1+z\dfrac{dG_1}{dz}\right)+Cz^{\xi_2}\left(\xi_2G_2+z\dfrac{dG_2}{dz}\right)}{AG_0+Bz^{\xi_1}G_1+Cz^{\xi_2}G_2},$$

$$\frac{Az^2\dfrac{d^2G_0}{dz^2}+Bz^{\xi_1}\left\{(\xi_1{}^2-\xi_1)G_1+2\xi_1z\dfrac{dG_1}{dz}+z^2\dfrac{d^2G_1}{dz^2}\right\}+Cz^{\xi_2}\left\{(\xi_2{}^2-\xi_2)G_2+2\xi_2z\dfrac{dG_2}{dz}+z^2\dfrac{d^2G_2}{dz^2}\right\}}{AG_0+Bz^{\xi_1}G_1+Cz^{\xi_2}G_2},$$

respectively. These integrals clearly vanish when $z=0$: and they constitute a double infinitude of solutions, for they contain the two ratios of the arbitrary constants.

When $B=0$, $C=0$, the solutions are regular functions; with the assumed properties of ξ_1 and ξ_2, they are the only regular functions when A is not zero.

When $A=0$, $C=0$, the expressions become regular functions, and they are solutions of the equations; but they do not vanish when $z=0$, and therefore they do not satisfy the conditions. Similarly for the solutions that arise from the assumptions $A=0$, $B=0$.

The solutions, corresponding to the values ρz_1 and τz_1 in the text for the modified variables in § 173, are easily seen to be given by

$$\left.\begin{aligned}
x&=\frac{z\dfrac{dG_0}{dz}+Kz_1\left(\xi_1G_1+z\dfrac{dG_1}{dz}\right)}{G_0+Kz_1G_1}-\frac{z\dfrac{dG_0}{dz}}{G_0}\\[2ex]
&=z_1K\frac{\xi_1G_0G_1+z\left(G_0\dfrac{dG_1}{dz}-G_1\dfrac{dG_0}{dz}\right)}{G_0(G_0+Kz_1G_1)}\\[2ex]
y&=z_1K\frac{(\xi_1{}^2-\xi_1)G_0G_1+2\xi_1zG_0\dfrac{dG_1}{dz}+z^2\left(G_0\dfrac{d^2G_1}{dz^2}-G_1\dfrac{d^2G_0}{dz^2}\right)}{G_0(G_0+Kz_1G_1)}
\end{aligned}\right\},$$

where $K=\dfrac{B}{A}$.

Similarly, for the solutions corresponding to the values $z_2\Theta_1$ and $z_2\Theta_2$ of the last set of dependent variables in § 173, we have

$$z_2\Theta_1 = \frac{z\dfrac{dG_0}{dz} + Kz_1\left(\xi_1 G_1 + z\dfrac{dG_1}{dz}\right) + Lz_2\left(\xi_2 G_2 + z\dfrac{dG_2}{dz}\right)}{G_0 + Kz_1 G_1 + Lz_2 G_2}$$

$$-\frac{z\dfrac{dG_0}{dz} + Kz_1\left(\xi_1 G_1 + z\dfrac{dG_1}{dz}\right)}{G_0 + Kz_1 G_1}$$

$$z_2\Theta_2 = \frac{z^2\dfrac{d^2G_0}{dz^2} + Kz_1\left\{(\xi_1^2-\xi_1)G_1 + 2\xi_1 z\dfrac{dG_1}{dz} + z^2\dfrac{d^2G_1}{dz^2}\right\} + Lz_2\left\{(\xi_2^2-\xi_2)G_2 + 2\xi_2 z\dfrac{dG_2}{dz} + z^2\dfrac{d^2G_2}{dz^2}\right\}}{G_0 + Kz_1 G_1 + Lz_2 G_2}$$

$$-\frac{z^2\dfrac{d^2G_0}{dz^2} + Kz_1\left\{(\xi_1^2-\xi_1)G_1 + 2\xi_1 z\dfrac{dG_1}{dz} + z^2\dfrac{d^2G_1}{dz^2}\right\}}{G_0 + Kz_1 G_1}$$

where $K = \dfrac{B}{A}$, $L = \dfrac{C}{A}$.

The quantities t_1 and t_2 are

$$t_1 = \frac{z\dfrac{dG_0}{dz} + Kz_1\left(\xi_1 G_1 + z\dfrac{dG_1}{dz}\right) + Lz_2\left(\xi_2 G_2 + z\dfrac{dG_2}{dz}\right)}{G_0 + Kz_1 G_1 + Lz_2 G_2} - \frac{z\dfrac{dG_0}{dz}}{G_0}$$

$$t_2 = \frac{z^2\dfrac{d^2G_0}{dz^2} + Kz_1\left\{(\xi_1^2-\xi_1)G_1 + 2\xi_1 z\dfrac{dG_1}{dz} + z^2\dfrac{d^2G_1}{dz^2}\right\} + Lz_2\left\{(\xi_2^2-\xi_2)G_2 + 2\xi_2 z\dfrac{dG_2}{dz} + z^2\dfrac{d^2G_2}{dz^2}\right\}}{G_0 + Kz_1 G_1 + Lz_2 G_2}$$

$$-\frac{z^2\dfrac{d^2G_2}{dz^2}}{G_0}$$

being of the forms indicated.

Ex. 2. Consider the equations

$$z\frac{dx}{dz} = \lambda_1 x + azx + bzy + cy^2$$
$$z\frac{dy}{dz} = \lambda_2 y + azx + \beta zy + \gamma x^2$$

where λ_1 and λ_2 are subject to the same restrictions as ξ_1 and ξ_2 in the text.

Proceeding, as in § 174, to obtain the expansions of x and y as regular functions of z, ζ_1, ζ_2, where

$$\zeta_1 = z^{\lambda_1}, \quad \zeta_2 = z^{\lambda_2},$$

we find

$$x = A\zeta_1 + aAz\zeta_1 + \frac{bB}{1+\lambda_2-\lambda_1}z\zeta_2 + \frac{cB^2}{2\lambda_2-\lambda_1}\zeta_2^2 + \ldots,$$

$$y = B\zeta_2 + \frac{aA}{1+\lambda_1-\lambda_2}z\zeta_1 + \beta Bz\zeta_2 + \frac{\gamma A^2}{2\lambda_1-\lambda_2}\zeta_1^2 + \ldots,$$

where the unexpressed terms are of higher dimensions in z, ζ_1, ζ_2 than two; and A, B are arbitrary.

Taking the parts which contain ζ_1 as a factor, we have

$$x_1 = \zeta_1 \rho = \zeta_1 (A + aAz + \text{terms of second and higher orders}),$$

$$y_1 = \zeta_1 \tau = \zeta_1 \left(\frac{aA}{1 + \lambda_1 - \lambda_2} z + \frac{\gamma A^2}{2\lambda_1 - \lambda_2} \zeta_1 + \ldots \right),$$

shewing that $\rho = A$, $\tau = 0$ when $z = 0$. Similarly for the parts which contain ζ_2 alone.

The only way to secure a solution, which makes ζ_1 a factor of x, is to take $B = 0$; and then ζ_2 cannot be a factor of y. Similarly, the only way to secure a solution, which makes ζ_2 a factor of y, is to take $A = 0$; and then ζ_1 cannot be a factor of x.

This is the example referred to in the note (p. 47 of this volume) to § 165. Königsberger's investigation seems to imply that the non-regular integrals of

$$x \frac{dY_1}{dx} = \lambda_1 Y_1 + [x, Y_1, Y_2] \Bigg\} ,$$

$$x \frac{dY_2}{dx} = \lambda_2 Y_2 + [x, Y_1, Y_2] \Bigg\} ,$$

when the real parts of λ_1 and λ_2 are positive, are

$$Y_1 = x^{\lambda_1} \Sigma \, c x^{\mu_1 + \lambda_1 \nu_{11} + \lambda_2 \nu_{12}} \Bigg\} ,$$

$$Y_2 = x^{\lambda_2} \Sigma \, c' x^{\mu_2 + \lambda_1 \nu_{21} + \lambda_2 \nu_{22}} \Bigg\} ,$$

so that x^{λ_1} is a factor of Y_1 and x^{λ_2} a factor of Y_2.

CASE I (b): *one root of the critical quadratic is a positive integer, the other is not a positive integer.*

177. Let the integer root be denoted by m, the non-integer root by ξ; the equations can be taken in the form

$$t \frac{du}{dt} = mu + \alpha t + \theta (u, v, t) \Bigg\rangle ,$$

$$t \frac{dv}{dt} = \xi v + \beta t + \phi (u, v, t) \Bigg\rangle ,$$

where θ and ϕ are regular functions of their arguments, which vanish with u, v, t and contain no terms of dimensions lower than two. The same transformations as were used in § 168, viz.

$$u = t \left(-\frac{\alpha}{m-1} + u_1 \right), \qquad v = t \left(-\frac{\beta}{\xi - 1} + v_1 \right),$$

can be applied $m - 1$ times in succession : and ultimately we have equations

$$t\frac{dt_1}{dt} = \left. \begin{array}{l} t_1 + at + f_1(t_1,\ t_2,\ t) \\[2mm] t\frac{dt_2}{dt} = \kappa t_2 + bt + f_2(t_1,\ t_2,\ t) \end{array} \right\},$$

where $\kappa,\ = \xi - m + 1$, is not a positive integer, the functions f_1 and f_2 are regular functions of their arguments of the same type as θ and ϕ above, and the integrals t_1 and t_2 are to vanish with t.

It has been proved that there are no regular integrals of the equation vanishing with t unless a is zero : and that, if $a = 0$, there exists a simple infinitude of regular integrals satisfying the equations. We proceed, not in the first place to the complete theorem but only to a partial theorem, by shewing that *when a is not zero, there exists a simple infinitude of non-regular integrals vanishing with t, these integrals being regular functions of t and t* log *t : and when a is zero, these non-regular integrals do not exist.*

To establish this result, we proceed from equations

$$t\frac{dx}{dt} = \left. \begin{array}{l} \sigma x + at + \theta_1(x,\ y,\ t) \\[2mm] t\frac{dy}{dt} = \kappa y + bt + \theta_2(x,\ y,\ t) \end{array} \right\},$$

where σ is taken to be a real positive quantity, a little less than 1 initially and equal to 1 ultimately : and, as the explicit forms of θ_1 and θ_2 are required, we suppose

$$\theta_1(x,\ y,\ t) = \Sigma\Sigma\Sigma a_{ijp}x^i y^j t^p,$$
$$(i + j + p \geqslant 2).$$
$$\theta_2(x,\ y,\ t) = \Sigma\Sigma\Sigma b_{ijp}x^i y^j t^p,$$

With these equations, we associate a set of dominant equations. Let

$$|a_{ijp}| = A_{ijp},\quad |b_{ijp}| = B_{ijp},\quad |a| = A ;$$

then the dominant equations are

$$t\frac{dX}{dt} - \sigma X + At = \left. \begin{array}{l} \Theta_1(X,\ Y,\ t) \\[2mm] t\frac{dY}{dt} - \kappa Y \pm Bt = \Theta_2(X,\ Y,\ t) \end{array} \right\},$$

where

$$\Theta_1(X,\ Y,\ t) = \Sigma\Sigma\Sigma A_{ijp} X^i Y^j t^p \Big\}$$
$$\Theta_2(X,\ Y,\ t) = \Sigma\Sigma\Sigma B_{ijp} X^i Y^j t^p \Big\}.$$

If κ be real, not being a positive integer, we choose that sign for the term $\pm Bt$, which makes

$$\frac{B}{\kappa - 1}$$

a positive quantity; if κ be complex, we choose a term $+ Bt$, such that

$$\frac{B}{\kappa - 1}$$

is a real positive quantity, and $|B| \geqslant |b|$.

By the theorem of § 175, we know that solutions exist, which vanish with t and are expressible as regular functions of t and t^σ. Let a new variable θ be introduced, defined by the equation

$$t^\sigma - t = (1 - \sigma)\,\theta\,;$$

and, in the solutions indicated, replace t^σ by $t + (1 - \sigma)\theta$; they then become regular functions of t and θ, expressed as converging power-series. To obtain their coefficients in this form directly, let

$$X = \Sigma\Sigma a_{mn}\,\theta^m t^n,$$
$$Y = \Sigma\Sigma b_{mn}\,\theta^m t^n,$$

where $a_{00} = 0$, $b_{00} = 0$; then since

$$t\,\frac{d\theta}{dt} = \sigma\theta - t,$$

we have

$$t\,\frac{dX}{dt} = \Sigma\Sigma a_{mn}\left\{n\theta^m t^n + m\theta^{m-1} t^n\,(\sigma\theta - t)\right\}$$
$$= \Sigma\Sigma \left\{(n + \sigma m)\,\theta^m t^n - m\theta^{m-1} t^{n+1}\right\} a_{mn},$$

and

$$t\,\frac{dY}{dt} = \Sigma\Sigma \left\{(n + \sigma m)\,\theta^m t^n - m\theta^{m-1} t^{n+1}\right\} b_{mn}.$$

Substituting in the differential equations and comparing coefficients, we have

$$(n + \sigma m - \sigma)\,a_{mn} - (m + 1)\,a_{m+1,\,n-1} = H_{m,\,n} \Big\}$$
$$(n + \sigma m - \kappa)\,b_{mn} - (m + 1)\,b_{m+1,\,n-1} = K_{m,\,n} \Big\},$$

where $H_{m,n}$ and $K_{m,n}$ are sums of terms of the form

$$A_{ijp} a_{m_1 n_1} \dots a_{m_i n_i} b_{m_1' n_1'} \dots b_{m_j' n_j'},$$

and similarly for $K_{m,n}$, such that

$$\left. \begin{array}{l} i+j+p \geqslant 2 \\ m_1 + \dots + m_i + m_1' + \dots + m_j' = m \\ p + n_1 + \dots + n_i + n_1' + \dots + n_j' = n \end{array} \right\}.$$

As regards the initial coefficients, we have the following expressions.

For $m + n = 0$, so that $m = 0$, $n = 0$; then

$$a_{00} = 0, \quad b_{00} = 0.$$

For $m + n = 1$, so that $m = 1$, $n = 0$: and $m = 0$, $n = 1$; then

$$0 \cdot a_{10} = 0, \quad (\sigma - \kappa) b_{10} = 0;$$

$$(1 - \sigma) a_{01} - a_{10} = -A, \quad (1 - \kappa) b_{01} - b_{10} = \mp B;$$

so that

$$a_{10} = (1 - \sigma) a_{01} + A, \quad (1 - \kappa) b_{01} = \mp B, \quad b_{10} = 0;$$

thus a_{01} is undetermined and therefore can be taken arbitrarily, say $= C$, where C is positive. Thus a_{01}, a_{10}, b_{01} are positive.

For $m + n = 2$, so that $m = 2$, $n = 0$: $m = 1$, $n = 1$: and $m = 0$, $n = 2$; then

$$\left. \begin{array}{l} \sigma a_{20} = A_{200} a^2_{10} \\ (2\sigma - \kappa) b_{20} = B_{200} a^2_{10} \end{array} \right\},$$

$$\left. \begin{array}{l} a_{11} - 2a_{20} = 2A_{200} a_{10} a_{01} + A_{110} a_{10} b_{01} + A_{101} a_{10} \\ (1 + \sigma - \kappa) b_{11} - 2b_{20} = 2B_{200} a_{10} a_{01} + B_{110} a_{10} b_{01} + B_{101} a_{10} \end{array} \right\},$$

$$\left. \begin{array}{l} (2 - \sigma) a_{02} - a_{11} = A_{200} a^2_{01} + A_{110} a_{01} b_{01} + A_{020} b^2_{01} + A_{101} a_{01} + A_{011} b_{01} + A_{002} \\ (2 - \kappa) b_{02} - b_{11} = B_{200} a^2_{01} + B_{110} a_{01} b_{01} + B_{020} b^2_{01} + B_{101} a_{01} + B_{011} b_{01} + B_{002} \end{array} \right\}.$$

And so on, taking in succession the groups of terms for increasing values of $m + n$, and taking, in each group, the equations for increasing values of n beginning with zero. The result is to give

$$a_{mn} = \theta_{mn}, \quad b_{mn} = \phi_{mn},$$

where θ_{mn} and ϕ_{mn} are sums of a number of terms; each term is a quotient, the numerator being a positive integral function of the coefficients of θ_1 and θ_2 and containing $a_{10}{}^m$ as a factor, and the denominator being a product of quantities of the form

$$n + \sigma m - \sigma, \quad n + \sigma m - \kappa.$$

It can be proved, by an argument precisely similar to that in § 71, that the number of quantities entering into the denominator product for each of the terms in θ_{mn} and ϕ_{mn} is

$$\leqslant m + 2n - 1.$$

On account of the theorem of § 175, establishing the existence of the integrals as regular functions of t and t^σ, it follows that the series

$$\Sigma\Sigma a_{mn}\theta^m t^n, \quad \Sigma\Sigma b_{mn}\theta^m t^n$$

converge.

Now proceed to the limit in which σ increases to, and ultimately acquires, the value unity; then θ becomes $-t \log t$, the differential equations become

$$\left. \begin{aligned} t\,\frac{dX}{dt} - X + At &= \Theta_1\,(X,\,Y,\,t) \\[2mm] t\,\frac{dY}{dt} - \kappa Y \pm Bt &= \Theta_2\,(X,\,Y,\,t) \end{aligned} \right\},$$

and the integrals change to

$$\Sigma\Sigma a'_{mn}\theta^m t^n, \quad \Sigma\Sigma b'_{mn}\theta^m t^n,$$

where a'_{mn} and b'_{mn} are the values of a_{mn} and b_{mn} when σ is replaced by 1.

In θ_{mn}, let T be any one of the terms, and let T' be the value of T when σ is replaced by 1. As regards the numerator in T, it is the sum of a series of positive quantities: and it is unaffected by the change of σ, except that a_{10} is replaced by A, that is, by a diminished quantity; hence the numerator of T' is less than that of T. As regards the numerical denominator, each factor $n + \sigma m - \sigma$ is replaced by $n + m - 1$, which is a greater quantity than the factor it replaces, unless m vanishes; but when $m = 0$, then

$$\frac{n-\sigma}{n-1} \leqslant 2 - \sigma,$$

because then $n \geqslant 2$. Also every factor $n + \sigma m - \kappa$ is replaced by $n + m - \kappa$; the imaginary portions (if any) of these two are the same, but the real part of the new factor is greater than that of the old except when $m = 0$, and then they are the same. The number of factors in the denominator is not greater than $m + 2n - 1$: hence

$$\Pi \left| \frac{n + \sigma m - \sigma}{n + m - 1} \cdot \frac{n + \sigma m - \kappa}{n + m - \kappa} \right| \leqslant (2 - \sigma)^{m + 2n - 1}$$

$$\leqslant (2 - \sigma)^{2m + 2n}.$$

The changes made have diminished the numerator of T; thus

$$\left| \frac{T'}{T} \right| < \Pi \left| \frac{n + \sigma m - \sigma}{n + m - 1} \cdot \frac{n + \sigma m - \kappa}{n + m - \kappa} \right|$$

$$< (2 - \sigma)^{2m + 2n}.$$

Remembering that θ_{mn} is a sum of terms T and bearing in mind the character of T, we have

$$\left| \frac{a'_{mn}}{a_{mn}} \right| < (2 - \sigma)^{2m + 2n}.$$

Similarly

$$\left| \frac{b'_{mn}}{b_{mn}} \right| < (2 - \sigma)^{2m + 2n}.$$

Now the series

$$\Sigma\Sigma a_{mn} \theta^m t^n, \quad \Sigma\Sigma b_{mn} \theta^m t^n$$

converge for a finite region round the origin. Let this be defined by $|t| \leqslant r$, $|\theta| \leqslant s$; and let M_1, M_2 be the respective maximum values of the moduli of the series within the region. Then

$$a_{mn} < \frac{M_1}{s^m r^n}, \quad b_{mn} < \frac{M_2}{s^m r^n};$$

and therefore

$$|a'_{mn}| < \frac{M_1}{\left\{ \frac{s}{(2-a)^2} \right\}^m \left\{ \frac{r}{(2-a)^2} \right\}^n},$$

$$|b'_{mn}| < \frac{M_2}{\left\{ \frac{s}{(2-a)^2} \right\}^m \left\{ \frac{r}{(2-a)^2} \right\}^n}.$$

Consequently the series

$$\Sigma\Sigma a'_{mn} \theta^m t^n, \quad \Sigma\Sigma b'_{mn} \theta^m t^n$$

converge for a finite region round $t = 0$.

If the original equations

$$t \frac{dx}{dt} = x + at + \theta_1(x, y, t) \left.\vphantom{\frac{dx}{dt}}\right\}$$
$$t \frac{dy}{dt} = \kappa y + bt + \theta_2(x, y, t) \left.\vphantom{\frac{dy}{dt}}\right\}$$

possess integrals, which vanish with t and are regular functions of t and $t \log t$, these integrals may be assumed to be

$$x = \Sigma\Sigma f_{mn} \theta^m t^n \left.\vphantom{\Sigma}\right\} :$$
$$y = \Sigma\Sigma g_{mn} \theta^m t^n \left.\vphantom{\Sigma}\right\}$$

when substituted, they must satisfy the equations identically. Choose f_{01} so that

$$|f_{01}| = C,$$

where C is the arbitrary constant in the integrals of the preceding equations.

When the relations that arise from the comparison of the coefficients are solved so as to give f_{mn}, g_{mn}, it is easy to see that the same results are obtained as would be given by changing, in a'_{mn} and b'_{mn}, A into $-a$, B into $\mp b$, A_{ijp} into a_{ijp}, and B_{ijp} into b_{ijp}, for all values of i, j, p. Bearing in mind that

$$|a| = A, \quad |b| \leqslant |B|, \quad |a_{ijp}| = A_{ijp}, \quad |b_{ijp}| = B_{ijp},$$

it is manifest that the real positive quantities $|a'_{mn}|$ and $|b'_{mn}|$ are superior limits for $|f_{mn}|$ and $|g_{mn}|$; that is,

$$|f_{mn}| < |a'_{mn}|, \quad |g_{mn}| < |b'_{mn}|.$$

But the series

$$\Sigma\Sigma a'_{mn} \theta^m t^n, \quad \Sigma\Sigma b'_{mn} \theta^m t^n$$

converge : hence the series

$$\Sigma\Sigma f_{mn} \theta^m t^n, \quad \Sigma\Sigma g_{mn} \theta^m t^n$$

also converge, and the equations accordingly possess integrals as stated in the theorem.

Note. If a is zero, then $a'_{10} = 0$; $a'_{20} = 0$, $a'_{11} = 0$; and it is immediately obvious that

$$a'_{mn} = 0,$$

for all values of $m > 0$ and all values of n. Similarly

$$b'_{mn} = 0,$$

for the same combinations of m and n. In this case, θ disappears entirely from the expressions

$$\Sigma\Sigma f_{mn}\theta^m t^n, \quad \Sigma\Sigma g_{mn}\theta^m t^n;$$

so that the integrals become regular functions of t, which are known to be solutions of the equations when $a = 0$.

178. The main theorems as to the equations

$$t\frac{dt_1}{dt} = t_1 + at + f_1(t_1,\ t_2,\ t)\left.\right]$$
$$t\frac{dt_2}{dt} = \kappa t_2 + bt + f_2(t_1,\ t_2,\ t)\left.\right]$$

so far as concerns the non-regular solutions, are :—

When a is not zero, so that the equations do not possess any regular solutions that vanish with t, they possess non-regular solutions that vanish with t. If κ have its real part positive, not itself being a positive integer, there is a double infinitude of such solutions; they are regular functions of t, t^κ and $t \log t$. If κ have its real part negative, there is only a single infinitude of such solutions; they are regular functions of t and $t \log t$.

When a is zero, so that the equations possess a single infinitude of regular solutions vanishing with t, then if κ have its real part positive, not itself being a positive integer, there is a single infinitude of non-regular solutions vanishing with t which are regular functions of t and t^κ; but if κ have its real part negative, the equations possess no non-regular solutions vanishing with t.

These theorems can be established by analysis and a course of argument similar to those which have been adopted, wholly or partially, in preceding cases. The actual expressions for the integrals, when a is not zero, are

$$t_1 = a\theta + At + \Sigma\Sigma\Sigma g_{lmn}\,\zeta^l\theta^m t^n\left.\right]$$
$$t_2 = \frac{b}{1-\kappa}t + B\zeta + \Sigma\Sigma\Sigma h_{lmn}\,\zeta^l\theta^m t^n\left.\right]$$

where the summation is for values of l, m, n, such that $l + m + n \geqslant 2$, the coefficients A and B are arbitrary, ζ denotes t^κ and θ denotes $t \log t$.

When a is zero, all the coefficients g_{lmn}, h_{lmn} for values of $m > 0$ vanish; so that θ disappears from the expressions for t_1

and t_2. The resulting expressions then can be resolved each into the sum of two functions: one a regular function of t which involves A, the other a regular function of t and ζ which involves B and vanishes when $B = 0$.

It may be noted that a slight degeneration occurs in the solutions, when κ is the reciprocal of a positive integer; a regular function of t and t^κ is then merely a regular function of t^κ.

When the equations in their first transformed expression are

$$\left. \begin{aligned} t\,\frac{du}{dt} &= mu + at + \theta\,(u,\ v,\ t) \\[2mm] t\,\frac{dv}{dt} &= \xi v + \beta t + \phi\,(u,\ v,\ t) \end{aligned} \right\},$$

the general results are the same as above; the value of κ is $\xi - m + 1$, and the critical condition, which is represented by $a = 0$, is stated at the end of § 168.

CASE I (c) : *the roots of the critical quadratic are unequal, and both are positive integers.*

179. Denoting the roots by m and n, of which m may be taken as the smaller integer, the equations can be transformed so as to become

$$\left. \begin{aligned} t\,\frac{du}{dt} &= mu + at + \theta\,(u,\ v,\ t) \\[2mm] t\,\frac{dv}{dt} &= nv + \beta t + \phi\,(u,\ v,\ t) \end{aligned} \right\}.$$

They can be modified by substitutions similar to those adopted in the preceding case; such substitutions can be applied $m - 1$ times in succession, leading to the forms

$$\left. \begin{aligned} t\,\frac{dt_1}{dt} &= t_1 + at + f_1\,(t_1,\ t_2,\ t) \\[2mm] t\,\frac{dt_2}{dt} &= \kappa t_2 + bt + f_2\,(t_1,\ t_2,\ t) \end{aligned} \right\},$$

where $\kappa,\ = n - m + 1$, is a positive integer greater than 1, the integrals t_1 and t_2 are to vanish with t, and the functions $f_1,\ f_2$ are regular functions, which vanish with their arguments and contain no terms of dimensions lower than two in $t_1,\ t_2,\ t$ combined.

It has already been proved (§ 169) that the equations possess no regular integrals vanishing with t, unless two relations among the constants be satisfied; one of them is represented by $a = 0$, the other by (say) $C = 0$, where C is a definite combination of a, b, and the constant coefficients in f_1 and f_2. The theorem as regards the non-regular integrals is:

The equations in general possess a double infinitude of non-regular integrals which vanish with t; they are regular functions of t, and $t \log t$. If both of the conditions represented by $a = 0$, $C = 0$ are satisfied, the equations possess no non-regular integrals vanishing with t: they are known to possess a double infinitude of regular integrals which vanish with t.

The method of establishing this theorem is similar to that for the case when κ is unity, so that the critical quadratic has a repeated root. As that case will be discussed later in full detail, we shall not here reproduce the analysis and the argument, which follow closely the corresponding analysis and argument in that later discussion.

The conditions for the equations

$$t \frac{du}{dt} = mu + \alpha t + \theta (u, v, t) \left.\right\}$$
$$t \frac{dv}{dt} = nv + \beta t + \phi (u, v, t) \left.\right\}$$,

represented for the modified forms by $a = 0$, $C = 0$, have already been given (§ 169).

CASE II (a): *the critical quadratic has equal roots,
not a positive integer.*

180. It has been proved that, in this case, the original equations possess regular integrals vanishing with t: and therefore, in order to consider the non-regular integrals (if any) that vanish with t, we transform the equations as in § 172, and we study the derived system

$$t \frac{dt_1}{dt} = \xi t_1 + \phi_1 (t_1, t_2, t) \left.\right\}$$
$$t \frac{dt_2}{dt} = \theta t_1 + \xi t_2 + \phi_2 (t_1, t_2, t) \left.\right\}$$,

where ϕ_1 and ϕ_2 are regular functions of their arguments; they vanish when $t_1 = 0$, $t_2 = 0$, and contain no terms of dimensions less than 2 in t_1, t_2, t combined. The integrals t_1 and t_2 are to be non-regular functions of t, required to vanish with t.

The non-regular integrals are given by the theorem:

When the repeated root ξ of the critical quadratic has its real part positive, and is not itself a positive integer, there is a double infinitude of non-regular integrals vanishing with t, these integrals being regular functions of t, t^ξ, $t^\xi \log t$.

When the theorem is established, there is an immediate corollary:

If the real part of the repeated root ξ of the critical quadratic be negative, then the equations do not possess non-regular integrals vanishing with t; the regular integrals possessed by the original system of equations are the only integrals that vanish with t.

The forms of the theorem and the corollary are indicated, by proceeding towards the limit of the theorems for the case of I (a) when the roots of the critical quadratic are equal to one another. If $\xi_2 = \xi_1 + \delta$, where δ is infinitesimal, then

$$t^{\xi_2} = t^{\xi_1} (1 + \delta \log t + \dots),$$

so that a function of t, t^{ξ_1}, t^{ξ_2} becomes a function of t, t^{ξ_1}, $t^{\xi_1} \log t$; but further investigation is needed in order to shew that, in passing to the limit, the functions under consideration continue to exist. Instead of adopting this method of proof, we proceed independently.

It is convenient to take

$$\zeta = t^\xi, \quad -\eta = t^\xi \log t.$$

If therefore integrals of the character indicated in the theorem exist, they can be expressed in the forms

$$\left. \begin{array}{l} t_1 = \Sigma\Sigma\Sigma \, a_{lmn} \zeta^l \eta^m t^n \\ t_2 = \Sigma\Sigma\Sigma \, b_{lmn} \zeta^l \eta^m t^n \end{array} \right\};$$

and these values must, when substituted, satisfy the differential equations identically. Now

$$t \frac{d\zeta}{dt} = \xi\zeta, \quad t \frac{d\eta}{dt} = \xi\eta - \zeta,$$

so that

$$t \frac{d}{dt} (\zeta^l \eta^m t^n) = (n + l\xi + m\xi) \zeta^l \eta^m t^n - m\zeta^{l+1} \eta^{m-1} t^n.$$

Hence equating coefficients of $\zeta^l \eta^m t^n$ on the two sides of both equations, after substitution of the assumed values of t_1 and t_2, we have

$$\left.\begin{aligned} \{n+(l+m-1)\,\xi\}\,a_{lmn} - (m+1)\,a_{l-1,\,m+1,\,n} &= \alpha'_{lmn} \\ \{n+(l+m-1)\,\xi\}\,b_{lmn} - (m+1)\,b_{l-1,\,m+1,\,n} &= \theta a_{lmn} + \beta'_{lmn} \end{aligned}\right\},$$

where α'_{lmn}, β'_{lmn}, being the coefficients of $\zeta^l \eta^m t^n$ in ϕ_1 and ϕ_2 respectively, are linear functions of the constants in ϕ_1 and ϕ_2, and are integral functions of the coefficients $a_{l'm'n'}$, $b_{l'm'n'}$, such that $l' \leqslant l$, $m' \leqslant m$, $n' \leqslant n$, $l' + m' + n' < l + m + n$.

Assuming that the real part of ξ is positive, but that ξ is not itself a positive integer, we see that no one of the quantities $n+(l+m-1)\,\xi$ can vanish if $l + m + n \geqslant 2$.

If $l = m = n = 0$, then $\alpha'_{lmn} = 0$, $\beta'_{lmn} = 0$; hence

$$a_{000} = 0,$$

$$-\xi b_{000} = \theta a_{000} = 0.$$

For values such that $l + m + n = 1$, we have

$$0 \cdot a_{010} = 0, \text{ that is, } a_{010} = K,$$

$$a_{001} = 0,$$

$$0 \cdot a_{100} - a_{010} = 0, \text{ that is, } a_{100} = A, \text{ and } K = 0;$$

$$0 \cdot b_{010} = \theta \cdot a_{010} = \theta K = 0, \text{ that is, } b_{010} = B;$$

$$b_{001} = \theta \cdot a_{001} = 0,$$

$$0 \cdot b_{100} - b_{010} = \theta \cdot a_{100} = \theta A.$$

The last equation shews that

$$\theta A + B = 0,$$

which determines L; and then b_{010}, b_{100} are arbitrary; that is, we have

$$a_{010} = 0, \quad a_{001} = 0, \quad a_{100} = A;$$

$$b_{010} = B, \quad b_{001} = 0, \quad b_{100} = C.$$

To obtain the terms of dimension two in ζ, η, t in t_1 and t_2, we require the explicit expressions of ϕ_1 and ϕ_2: let them be

$$\phi_1 = a t t_1 + b t t_2 + c t_1^2 + e t_1 t_2 + k t_2^2 + \dots,$$

$$\phi_2 = \alpha t t_1 + \beta t t_2 + \gamma t_1^2 + \epsilon t_1 t_2 + \kappa t_2^2 + \dots.$$

The terms in t_1 and t_2 of dimension one, obtained as above, are

$$t_1 = A\zeta, \quad t_2 = C\zeta + B\eta,$$

so that, as far as terms of dimension two in ϕ_1 and ϕ_2 after substitution, we have

$$\phi_1 = (cA^2 + eAC + kC^2)\,\zeta^2 + (eAB + 2kBC)\,\zeta\eta + kB^2\eta^2 \\ + (aA + bC)\,\zeta t + bB\eta t,$$

$$\phi_2 = (\gamma A^2 + \epsilon AC + \kappa C^2)\,\zeta^2 + (\epsilon AB + 2\kappa BC)\,\zeta\eta + \kappa B^2\eta^2 \\ + (\alpha A + \beta C)\,\zeta t + \beta B\eta t.$$

Accordingly, for $l + m + n = 2$, we have

$$\xi a_{020} = kB^2, \quad a_{011} = bB, \quad (2 - \xi)\,a_{002} = 0,$$

$$a_{101} - a_{011} = aA + bC, \quad \xi a_{110} - 2a_{020} = eAB + 2kBC,$$

$$\xi a_{200} - a_{110} = cA^2 + eAC + kC^2;$$

$$\xi b_{020} = \kappa B^2 + \theta a_{020}, \quad b_{011} = \beta B + \theta a_{011}, \quad (2 - \xi)\,b_{002} = \theta a_{002},$$

$$b_{101} - b_{011} = \alpha A + \beta C + \theta a_{101}, \quad \xi b_{110} - 2b_{020} = \epsilon AB + 2\kappa BC + \theta a_{110},$$

$$\xi b_{200} - b_{110} = \gamma A^2 + \epsilon AC + \kappa C^2 + \theta a_{200}.$$

These relations, taken in succession, determine the values of the coefficients a_{lmn}, b_{lmn}, such that $l + m + n = 2$; when the values are substituted, we obtain the terms in t_1 and t_2, which are of dimensions two in the arguments ζ, η, t. And so on, for successive groups of terms.

The equations, when solved in groups for the same value of $l + m + n$ beginning with a zero value of l, and solved in successive groups for increasing values of $l + m + n$, give values of a_{lmn}, b_{lmn}, which are sums of integral functions of the literal coefficients of ϕ_1 and ϕ_2, and of the arbitrary coefficients B and C, each such integral function being divided by a product of factors of the form $n + (l + m - 1)\,\xi$. Let the values thus obtained be

$$a_{lmn} = \alpha_{lmn}, \quad b_{lmn} = \beta_{lmn}.$$

As in § 174 for the former case, it can be proved that

$$a_{00p} = 0, \quad b_{00p} = 0,$$

for all positive integer values of p, so that there are no terms in t_1

or in t_2, which involve t alone; every term involves either ζ or η or both ζ and η.

To establish the convergence of the series thus obtained, we proceed in two stages as in the corresponding question (§§ 175, 176), when the roots of the critical quadratic are unequal.

Extract from t_1 and t_2 all the terms which are free from η; as each of them involves ζ, their aggregate can be taken in the respective forms $\zeta\rho$, $\zeta\tau$. The remaining terms then have η for a factor, so that we may write

$$t_1 = \zeta\rho + \eta\Theta_1,$$
$$t_2 = \zeta\tau + \eta\Theta_2.$$

It will be proved, first, that solutions of the form

$$t_1 = \zeta\rho, \quad t_2 = \zeta\tau$$

exist, where ρ and τ are regular functions of t and ζ, ρ vanishing at $t = 0$ and τ having an arbitrary value there: so that the functions involve one arbitrary constant, and there consequently is a simple infinitude of such solutions.

Then substituting

$$t_1 = \zeta\rho + \eta\Theta_1, \quad t_2 = \zeta\tau + \eta\Theta_2,$$

it will be proved that functions Θ_1 and Θ_2 exist, which are regular in their arguments ζ, η, t, and involve an arbitrary constant C; Θ_1 vanishes at $t = 0$, and Θ_2 acquires the value C there. Thus for an assigned value of B, these will represent another (and an independent) simple infinitude of integrals.

In each stage, the details of the analysis follow the detailed analysis of the former case somewhat closely: it therefore will be abbreviated for the present purpose.

181.   Substituting $t_1 = \zeta\rho$, $t_2 = \zeta\tau$ in the equations for t_1 and t_2, we find ρ and τ determined by

$$\left. \begin{aligned} t\frac{d\rho}{dt} &= \psi_1\left(\rho,\ \tau,\ \zeta,\ t\right) \\ t\frac{d\tau}{dt} &= \theta\rho + \psi_2(\rho,\ \tau,\ \zeta,\ t) \end{aligned} \right\},$$

where the general character of ψ_1 and ψ_2 is as before. If these are satisfied by regular functions of t and ζ, their expressions

$$\rho = \Sigma\Sigma k_{mn}\zeta^m t^n,$$
$$\tau = \Sigma\Sigma j_{mn}\zeta^m t^n,$$

when substituted in the above equations, must satisfy them identically. Accordingly, comparing coefficients of $\zeta^m t^n$ on the two sides of both equations, we have

$$(n + m\xi)\,k_{mn} = K'_{mn},$$
$$(n + m\xi)\,j_{mn} = J'_{mn} + \theta k_{mn},$$

where K'_{mn}, J'_{mn} are linear in the literal coefficients of ρ and τ, and are integral functions of $k_{m'n'}$, $j_{m'n'}$, such that $m' \leqslant m$, $n' \leqslant n$, $m' + n' < m + n$. Also, from the form of ψ_1 and ψ_2, $K'_{00} = 0$, $J'_{00} = 0$; hence we have

$$k_{00} = 0.$$

But j_{00} is undetermined, and it can therefore be taken arbitrarily: let its value be B, where B is any arbitrary constant.

When the equations for k_{mn} and j_{mn} are solved, in groups for the same value of $m + n$ and in succeeding groups for increasing values of $m + n$, they lead to results of the form

$$k_{mn} = \kappa_{mn}, \quad j_{mn} = \iota_{mn},$$

where κ_{mn}, ι_{mn} are sums of integral functions of the coefficients in ψ_1 and ψ_2, divided by products of factors of the form $n + m\xi$.

The dominant functions are constructed as before. Let ϵ denote the least value of $|n + m\xi|$ for integer values of m and n, so that ϵ is a finite (non-vanishing) quantity; and let $|\theta| = \Theta$, $|C| = C'$. Also, let a common region of existence for the functions ψ_1 and ψ_2 be given by the ranges $|t| \leqslant r$, $|\zeta| \leqslant r_1$, $|\rho| \leqslant h$, $|\tau| \leqslant k$; within this region, let M be the greatest of the values of $|\psi_1|$ and of $|\psi_2|$. Consider functions P and T, defined by the equations

$$\epsilon P = \frac{1}{\zeta}\left\{\frac{M}{\left(1 - \dfrac{t}{r}\right)\left(1 - \dfrac{P\zeta}{hr_1}\right)\left(1 - \dfrac{T\zeta}{kr_1}\right)} - \frac{M}{1 - \dfrac{t}{r}} - M\frac{P\zeta}{hr_1} - M\frac{T\zeta}{kr_1}\right\},$$

$$\epsilon T = \epsilon C' + \Theta P + \frac{1}{\zeta}\left\{\frac{M}{\left(1 - \dfrac{t}{r}\right)\left(1 - \dfrac{P\zeta}{hr_1}\right)\left(1 - \dfrac{T\zeta}{kr_1}\right)}\right.$$
$$\left. - \frac{M}{1 - \dfrac{t}{r}} - M\frac{P\zeta}{hr_1} - M\frac{T\zeta}{kr_1}\right\}.$$

Clearly

$$(\epsilon + \Theta) P = \epsilon (T - C'),$$

that is,

$$P = \frac{\epsilon}{\epsilon + \Theta} (T - C').$$

The value of P is a root of a cubic equation which, when $t = 0$ and $\zeta = 0$, has no term independent of P, and has a non-vanishing term involving the first power of P: so that it has one and only one root vanishing with t and ζ, and this root is a regular function of those variables. To obtain its expression without actually solving the cubic, we take

$$P = \Sigma\Sigma K_{mn} \zeta^m t^n,$$

where $K_{00} = 0$: we expand the right-hand side of the dominant equations as a regular function of t, ζ, P, T, and compare coefficients. The analysis, that leads to the values of κ_{mn}, ι_{mn}, can be used to obtain the value of K_{mn}, by making appropriate changes similar to those in the earlier corresponding case. These changes are now, as was the case before, such as to make

$$|\kappa_{mn}| < K_{mn}, \quad |\iota_{mn}| < K_{mn};$$

and therefore, as the series

$$\Sigma\Sigma K_{mn} \zeta^m t^n$$

converges, the series

$$\Sigma\Sigma k_{mn} \zeta^m t^n, \quad C + \Sigma\Sigma j_{mn} \zeta^m t^n,$$

also converge. The existence of the integrals, connected with the first stage, is therefore established.

182. Now writing

$$t_1 = \zeta\rho + \eta\Theta_1, \quad t_2 = \zeta\tau + \eta\Theta_2,$$

where ρ and τ are the regular functions of t and ζ as just determined, the equations for Θ_1 and Θ_2 are

$$t \frac{d\Theta_1}{dt} = f_1 (\Theta_1, \Theta_2, \zeta, \eta, t)$$

$$t \frac{d\Theta_2}{dt} = \theta\Theta_1 + f_2 (\Theta_1, \Theta_2, \zeta, \eta, t)$$

where f_1 and f_2 are regular functions of their arguments, which vanish when $\Theta_1 = 0$ and $\Theta_2 = 0$; the coefficients of the first powers of Θ_1 and Θ_2 vanish when $t = 0$; and any term, involving Θ_1 and Θ_2 in the form $\Theta_1{}^\lambda \Theta_2{}^\mu$, contains $\eta^{\lambda+\mu-1}$ as a factor.

The method of proof and its general course are the same as before (§ 176). The regular functions of ζ, η, t, which are the formal solution of the equations, are proved to converge, by being compared with the functions, which satisfy the dominant equations

$$
\epsilon \Phi_1 = \frac{1}{\eta} \left[\frac{M}{\left(1 - \dfrac{t}{r}\right)\left(1 - \dfrac{\zeta}{\rho}\right)\left(1 - \dfrac{\eta \Phi_1}{\sigma \alpha_1}\right)\left(1 - \dfrac{\eta \Phi_2}{\sigma \alpha_2}\right)} \right.
$$
$$
\left. - \frac{M}{\left(1 - \dfrac{t}{r}\right)\left(1 - \dfrac{\zeta}{\rho}\right)} - M \frac{\eta \Phi_1}{\sigma \alpha_1} - M \frac{\eta \Phi_2}{\sigma \alpha_2} \right],
$$

$$
\epsilon \Phi_2 = \epsilon \, |\, C \,| + |\, \theta \,| \, \Phi_1 + \frac{1}{\eta} \left[\frac{M}{\left(1 - \dfrac{t}{r}\right)\left(1 - \dfrac{\zeta}{\rho}\right)\left(1 - \dfrac{\eta \Phi_1}{\sigma \alpha_1}\right)\left(1 - \dfrac{\eta \Phi_2}{\sigma \alpha_2}\right)} \right.
$$
$$
\left. - \frac{M}{\left(1 - \dfrac{t}{r}\right)\left(1 - \dfrac{\zeta}{\rho}\right)} - M \frac{\eta \Phi_1}{\sigma \alpha_1} - M \frac{\eta \Phi_2}{\sigma \alpha_2} \right],
$$

and are such that, when $t = 0$, $\zeta = 0$, $\eta = 0$, then Φ_1 is zero and $\Phi_2 = |\, C \,|$. There exists a single quantity Φ_1, satisfying these equations and vanishing with t, which is expansible as a regular function of t, ζ, η in a non-infinitesimal region round t; the power-series, which is its expression, is consequently a converging series within that region. And therefore Φ_2, being given by

$$
\Phi_2 = |\, C \,| + \left(1 + \frac{|\, \theta \,|}{\epsilon}\right) \Phi_1,
$$

is also expressible as a regular function of t, ζ, η which, when $t = 0$, acquires the value $|\, C \,|$.

A comparison of the coefficients of $\zeta^l \eta^m t^n$ in Θ_1 and Θ_2, with those of the same combination of the variables in Φ_1 and Φ_2, is easily seen to lead to the inference, that the moduli of the former are less than the modulus of the latter; consequently the former series converge and therefore integrals of the equations, defined by the specified conditions, are proved to exist. Their explicit expressions, as power-series, are obtained as in § 176.

Ex. Consider a special (degenerate) form of the equations in Ex. 1, § 176, given by

$$z\frac{dx}{dz} = x + y - x^2 \left.\begin{array}{l} \\ \\ \end{array}\right\}.$$
$$z\frac{dy}{dz} = -\tfrac{1}{9}x + \tfrac{1}{3}y - xy + \kappa z$$

The critical quadratic is

$$(\Omega - 1)(\Omega - \tfrac{1}{3}) + \tfrac{1}{9} = 0,$$

that is,

$$(\Omega - \tfrac{2}{3})^2 = 0,$$

so that the roots are equal, the common value not being an integer.

To find explicit expression for the integrals, we can take

$$x = \frac{z}{u}\frac{du}{dz}, \quad y = \frac{z^2}{u}\frac{d^2u}{dz^2};$$

and both equations are satisfied by these values, provided u is determined as a solution of

$$z^3\frac{d^3u}{dz^3} + \tfrac{5}{2}z^2\frac{d^2u}{dz^2} + \tfrac{1}{9}z\frac{du}{dz} - \kappa z u = 0.$$

What is required is the complete solution of this equation, expressed in the vicinity of $z = 0$.

The fundamental determining equation is

$$m(m-1)(m-2) + \tfrac{5}{2}m(m-1) + \tfrac{1}{9}m = 0,$$

that is,

$$m(m - \tfrac{2}{3})^2 = 0.$$

Hence, by the usual theory of linear differential equations, three linearly independent solutions are given by

$$u_1 = R_1 = 1 + \frac{\kappa z}{\left(\frac{1}{3}\right)^2} + \frac{1}{2!}\frac{(\kappa z)^2}{\left(\frac{1.4}{3.3}\right)^2} + \frac{1}{3!}\frac{(\kappa z)^3}{\left(\frac{1.4.7}{3.3.3}\right)^2} + \dots;$$

$$u_2 = z^{\frac{2}{3}}R_2,$$

Here

$$u_3 = z^{\frac{2}{3}}(R_3 + R_2\log z).$$

$$R_2 = 1 + \frac{\kappa z}{\frac{5}{3}.1^2} + \frac{\kappa^2 z^2}{\frac{5.8}{3.3}(2!)^2} + \frac{\kappa^3 z^3}{\frac{5.8.11}{3.3.3}(3!)^2} + \dots;$$

and R_3 is a regular function of z that vanishes with z, its expression being

$$R_3 = B_1 z + B_2 z^2 + B_3 z^3 + \dots,$$

where

$$m^2(m + \tfrac{2}{3})B_m - \kappa B_{m-1} = (3m^2 + \tfrac{4}{3}m)\frac{\kappa^m}{\frac{5.8.11\dots 3m+2}{3^m}(m!)^2},$$

and $B_0 = 0$.

Accordingly, the complete solution of the equation in u is

$$u = AR_1 + Bz^{\frac{2}{3}}R_2 + Cz^{\frac{2}{3}}(R_3 + R_2\log z);$$

and therefore the integrals of the original equations, which satisfy the requirement of vanishing with z, are given by

$$x = \frac{z}{u}\frac{du}{dz}, \quad y = \frac{z^2}{u}\frac{d^2u}{dz^2}.$$

A brief discussion of these expressions shews that the theorem in the text is verified.

CASE II (b): *the critical quadratic has a repeated root, which is a positive integer.*

183. Denoting the repeated root by m, the equations are

$$t\frac{du}{dt} = mu + \alpha t + \theta(u, v, t)$$
$$t\frac{dv}{dt} = \kappa u + mv + \beta t + \phi(u, v, t)$$

where the functions θ, ϕ are regular; they vanish with u, v, t, and contain no terms of dimensions lower than two in their arguments.

The equations can be transformed as before (§ 171) by the appropriate substitutions; and this transformation can be effected $m - 1$ times, leading to new equations of the form

$$t\frac{dt_1}{dt} = t_1 + at + \theta_1(t_1, t_2, t)$$
$$t\frac{dt_2}{dt} = \kappa t_1 + t_2 + bt + \theta_2(t_1, t_2, t)$$

where t_1 and t_2 are to vanish with t; and θ_1, θ_2 are of the same type and properties as θ, ϕ in the first form.

There are two sub-cases according as κ is zero, or κ is not zero.

184. *First sub-case:* $\kappa = 0$. The equations can be taken in the form

$$t\frac{dx}{dt} = x + at + \theta_1(x, y, t)$$
$$t\frac{dy}{dt} = y + bt + \theta_2(x, y, t)$$

the integrals are to vanish with t; and the functions θ_1, θ_2 are regular functions of their arguments, which vanish when $x = 0$, $y = 0$, $t = 0$, and contain no terms of order lower than two in x, y, t combined.

The integrals, which vanish with t, are defined by the theorem:

The equations possess, in general, a double infinitude of non-regular integrals vanishing with t, which are regular functions of t and $t \log t$; and it is known that there are no regular integrals vanishing with t. If, however, both $a = 0$ and $b = 0$, the equations do not possess non-regular integrals vanishing with t; the only integrals vanishing with t are the double infinitude of regular integrals which the equations are known to possess.

This theorem can be established, as in other cases, by the construction of dominant equations and comparison with their integrals which are actually obtained in explicit expression.

For this purpose, consider the equations

$$t \frac{dX}{dt} - \sigma X + At = \Sigma\Sigma\Sigma A_{ijp} X^i Y^j t^p \left.\right\}$$
$$t \frac{dY}{dt} - \rho Y + Bt = \Sigma\Sigma\Sigma B_{ijp} X^i Y^j t^p \left.\right\},$$

where $i + j + p \geqslant 2$ in the two triple summations. The quantities σ and ρ are real, positive, and less than unity: ultimately, they will be made equal to unity. It follows, from the theorem of § 173, that there is a double infinite of integrals, which vanish with t, these integrals being regular functions of t, t^σ, t^ρ.

Let two new variables θ and ϕ be introduced such that

$$t^\sigma = t - (\sigma - 1)\,\theta + (\sigma - 1)^2\,\phi,$$
$$t^\rho = t - (\rho - 1)\,\theta + (\rho - 1)^2\,\phi;$$

we easily find

$$t \frac{d\theta}{dt} + t - \theta = (1 - \rho)(1 - \sigma)\,\phi = \beta\phi \left.\right\}$$
$$t \frac{d\phi}{dt} + \theta = (\sigma + \rho - 1)\,\phi = \alpha\phi \left.\right\},$$

where α and β are constants which, when $\rho = 1$, $\sigma = 1$, are equal to 1 and 0 respectively.

The regular functions of t, t^σ, t^ρ are expressible in the form of converging power-series; when t^σ and t^ρ are replaced by their values in terms of θ and ϕ, the new functions are regular functions of t, θ, ϕ. To obtain their expressions in this last form directly from the differential equations, we substitute

$$\left.\begin{array}{l} X = \Sigma\Sigma\Sigma\, h_{lmn}\, t^l\, \theta^m\, \phi^n \\ Y = \Sigma\Sigma\Sigma\, k_{lmn}\, t^l\, \theta^m\, \phi^n \end{array}\right\}$$

in the equations which are to be satisfied identically.   Now

$$t\frac{dX}{dt} = \left(t\frac{\partial}{\partial t} + t\frac{d\theta}{dt}\frac{\partial}{\partial \theta} + t\frac{d\phi}{dt}\frac{\partial}{\partial \phi}\right) X$$

$$= \Sigma\Sigma\Sigma\, \{(l + m + \alpha n)\, h_{lmn}\, t^l\, \theta^m\, \phi^n$$

$$- m h_{lmn}\, t^{l+1}\theta^{m-1}\phi^n - n h_{lmn}\, t^l\theta^{m+1}\phi^{n-1} + \beta m h_{lmn}\, t^l\theta^{m-1}\phi^{n+1}\};$$

hence, comparing coefficients of $t^l\,\theta^m\,\phi^n$ on the two sides, we have

$$(l + m + \alpha n - \sigma)\, h_{lmn} - (m + 1)\, h_{l-1, m+1, n} - (n + 1)\, h_{l, m-1, n+1}$$

$$+ (m + 1)\, \beta h_{l, m+1, n-1} = \alpha'_{lmn}.$$

Similarly

$$(l + m + \alpha n - \rho)\, k_{lmn} - (m + 1)\, k_{l-1, m+1, n} - (n + 1)\, k_{l, m-1, n+1}$$

$$+ (m + 1)\, \beta k_{l, m+1, n-1} = \beta'_{lmn}.$$

Here α'_{lmn}, β'_{lmn} are expressions which are linear in the coefficients A_{ijp}, B_{ijp} respectively, being an aggregate of terms of the form

$$A_{ijp}\, h_{l_1 m_1 n_1} \cdots h_{l_i m_i n_i}\, k_{l_1' m_1' n_1'} \cdots k_{l_j' m_j' n_j'},$$

$$B_{ijp}\, h_{l_1 m_1 n_1} \cdots h_{l_i m_i n_i}\, k_{l_1' m_1' n_1'} \cdots k_{l_j' m_j' n_j'},$$

respectively; the subscripts are subject to the relations

$$\left.\begin{array}{l} m_1 + \ldots + m_i + m_1' + \ldots + m_j' = m \\ n_1 + \ldots + n_i + n_1' + \ldots + n_j' = n \\ p + l_1 + \ldots + l_i + l_1' + \ldots + l_j' = l \end{array}\right\}.$$

In particular, we have

$$h_{000} = 0, \quad k_{000} = 0.$$

When $l + m + n = 1$, the equations for the coefficients in X are

$$(1 - \sigma)\, h_{100} - h_{010} = - A,$$

$$(1 - \sigma)\, h_{010} - h_{001} = 0,$$

$$(\alpha - \sigma)\, h_{001} + \beta h_{010} = 0,$$

which are satisfied by

$$h_{010} = (1 - \sigma)\, h_{100} + A \Big\rbrace,$$
$$h_{001} = (1 - \sigma)\, h_{010}$$

and h_{100} is arbitrary. Similarly,

$$k_{010} = (1 - \rho)\, k_{100} + B \Big\rbrace,$$
$$k_{001} = (1 - \rho)\, k_{010}$$

and k_{100} is arbitrary.

When $l + m + n = 2$, the equations for the coefficients in X are

$$(2 - \sigma)\, h_{020} - h_{011} = \alpha'_{020} \Big\rbrace$$
$$(1 + \alpha - \sigma)\, h_{011} + 2\beta h_{020} - 2h_{002} = \alpha'_{011} \Big\rbrace,$$
$$(2\alpha - \sigma)\, h_{002} + \beta h_{011} = \alpha'_{002} \Big\rbrace$$

$$(2 - \sigma)\, h_{110} - 2h_{020} - h_{101} = \alpha'_{110} \Big\rbrace$$
$$(1 + \alpha - \sigma)\, h_{101} - h_{011} + \beta h_{110} = \alpha'_{101} \Big\rbrace,$$
$$(2 - \sigma)\, h_{200} - h_{110} = \alpha'_{200}.$$

The first three equations, when solved, determine h_{020}, h_{011}, h_{002}; when the values of h_{020} and h_{011} are substituted in the next two equations, they determine h_{110}, h_{101}; the last equation then determines the form of h_{200}.

Similarly for the coefficients in Y.

For values of $l + m + n \geqslant 2$, the equations can be solved in a similar way. They are solved in groups for the successively increasing values of $l + m + n$. In each group, say that for which $l + m + n = N$ (so that the coefficients $h_{l'm'n'}$, $k_{l'm'n'}$, such that

$$l' + m' + n' \leqslant N - 1,$$

are supposed known), the convenient method is to arrange the equations in sets, determined by the values of l, and in sequence according to increasing values of l beginning with 0: in each set, the equations are arranged in sequence according to increasing values of n beginning with 0. In each set, we use the equations in succession, to express h_{lmn} in terms of $h_{l,N-l,0}$ and previously known coefficients and constants; when the first $N - l$ equations in the set have thus been used, the remaining equation, on substitution of the values of $h_{l,0,N-l}$, $h_{l,1,N-l-1}$, then determines $h_{l,N-l,0}$ and so also the values of all the coefficients $h_{l,m,n}$, such that $m + n = N - l$. Likewise for the coefficients k_{lmn}.

And then, as the solutions are known to be regular functions of t, θ, ϕ, the series

$$\Sigma\Sigma\Sigma\, h_{lmn}\, t^l \theta^m \phi^n, \quad \Sigma\Sigma\Sigma\, k_{lmn}\, t^l \theta^m \phi^n,$$

with the values of h_{lmn}, k_{lmn} which have just been obtained, converge.

As regards the forms of the coefficients h_{lmn}, k_{lmn}, they are the aggregates of positive terms T. The numerator of each term T is the sum of a number of positive quantities: it is a polynomial function of the coefficients A_{ijp}, B_{ijp}: it is also a polynomial function of those quantities h_{l+m+n}, k_{l+m+n}, for which $l+m+n=1$. The denominator of the term T is of the form

$$P + Q\beta,$$

where P is the product of factors of the types

$$l + m + \alpha n - \sigma, \quad l + m + \alpha n - \rho,$$

and where Q is an aggregate of quantities, each positive and similar to P but containing two factors fewer than P.

As regards the number of factors in P, being a part of a denominator in a term T in h_{lmn} or k_{lmn}, it can be proved, by an amplification of Jordan's argument used in § 71, that this number

$$\leqslant 3l + 2m + n.$$

It is known that, so long as σ and ρ are different from unity, the convergence of the power-series is absolute: hence this will be the case when

$$\sigma = 1 - \epsilon, \quad \rho = 1 - \epsilon,$$

where ϵ is a real positive quantity that can be taken as small as we please. Proceed therefore to the limit in which σ and ρ acquire the value unity, so that ϵ passes from small values to zero. The effect is to give to θ and ϕ the values

$$\theta = -t \log t, \quad \phi = \tfrac{1}{2} t (\log t)^2;$$

to change the differential equations to the forms

$$\left.\begin{array}{l} t\dfrac{dX}{dt} - X + At = \Sigma\Sigma\Sigma A_{ijp} X^i Y^j t^p \\[2mm] t\dfrac{dY}{dt} - Y + Bt = \Sigma\Sigma\Sigma B_{ijp} X^i Y^j t^p \end{array}\right\};$$

and to change the integrals to the forms

$$\left.\begin{array}{l} X = \Sigma\Sigma\Sigma h'_{lmn} t^l (-t \log t)^m \{\tfrac{1}{2} t (\log t)^2\}^n \\[2mm] Y = \Sigma\Sigma\Sigma k'_{lmn} t^l (-t \log t)^m \{\tfrac{1}{2} t (\log t)^2\}^n \end{array}\right\},$$

where h'_{lmn} and k'_{lmn} are the respective values of h_{lmn} and k_{lmn} when $\sigma = 1$, $\rho = 1$.

It is necessary to compare the coefficients h'_{lmn} and h_{lmn}: and likewise the coefficients k'_{lmn} and k_{lmn}. Let T be one of the terms in h_{lmn}, as explained above: and let T' be its value when $\sigma = 1$, $\rho = 1$. The effect of the change on the numerator is to replace $(1 - \sigma) h_{100} + A$ by A, h_{001} by 0, $(1 - \rho) k_{100} + B$ by B, k_{001} by 0, in every case a decrease: and therefore, as the numerator is a sum of positive terms, the whole effect on the numerator is to decrease it, that is,

$$\text{numerator of } T' < \text{numerator of } T.$$

As regards the denominator of T, in the form

$$P + Q\beta,$$

the quantity β is of the second order of small quantities; Q is an aggregate of a limited number of products, each containing a limited number of factors; hence $Q\beta$ is of the second order of small quantities. Let P' be the changed form of P, obtained from P by changing

$$l + m + \alpha n - \sigma \text{ into } l + m + n - 1,$$
and $\quad\quad l + m + \alpha n - \rho \text{ into } l + m + n - 1.$

Now $\quad l + m + \alpha n - \sigma - (l + m + n - 1) = -(2n - 1)\epsilon,$

a negative small quantity of the first order unless $n = 0$; so that, unless $n = 0$,

$$\frac{l + m + \alpha n - \sigma}{l + m + n - 1} = 1 - \gamma,$$

where γ is a positive small quantity of the first order. When $n = 0$,

$$l + m - \sigma - (l + m - 1)(2 - \sigma) = \epsilon(2 - l - m),$$

so that, as $l + m \geqslant 2$, we have

$$\frac{l + m - \sigma}{l + m - 1} = 2 - \sigma - \gamma',$$

where γ' is a positive small quantity of the first order, unless $l + m = 2$, and then $\gamma' = 0$. Hence

$$\frac{P'}{P} = \Pi \frac{1}{1 - \gamma} \Pi \frac{1}{2 - \sigma - \gamma'}$$

$$> \Pi \frac{1}{2 - \sigma}$$

$$> \frac{1}{(2 - \sigma)^{3l + 2m + n}},$$

the difference between the two sides being a small quantity of the first order. Also

$$\frac{Q\beta}{P'}$$

is a small quantity of the second order, that is, a quantity of an order less than the foregoing difference; consequently

$$\frac{P'}{P + Q\beta} > \frac{1}{(2 - \sigma)^{3l+2m+n}} \, .$$

The changes depreciated the numerator of T into that of T': hence

$$\frac{T'}{T} < \frac{P + Q\beta}{P'}$$

$$< (2 - \sigma)^{3l+2m+n}$$

$$< (2 - \sigma)^{3l+3m+3n}.$$

This result holds for every term in h_{lmn}; hence

$$\left| \frac{h'_{lmn}}{h_{lmn}} \right| < (2 - \sigma)^{3l+3m+3n}.$$

Similarly,

$$\left| \frac{k'_{lmn}}{k_{lmn}} \right| < (2 - \sigma)^{3l+3m+3n}.$$

Let the region of convergence of the power-series

$$\Sigma\Sigma\Sigma h_{lmn} t^l \theta^m \phi^n, \quad \Sigma\Sigma\Sigma k_{lmn} t^l \theta^m \phi^n,$$

be defined by the ranges

$$|t| \leqslant r, \quad |\theta| \leqslant r_1, \quad |\phi| \leqslant r_2;$$

and let M_1, M_2 be the maximum values of the moduli of the series respectively within this region; then

$$h_{lmn} < \frac{M_1}{r^l r_1^m r_2^n},$$

$$k_{lmn} < \frac{M_2}{r^l r_1^m r_2^n};$$

consequently

$$h'_{lmn} < \frac{M_1}{\left\{ \dfrac{r}{(2-\sigma)^3} \right\}^l \left\{ \dfrac{r_1}{(2-\sigma)^3} \right\}^m \left\{ \dfrac{r_2}{(2-\sigma)^3} \right\}^n},$$

$$k'_{lmn} < \frac{M_2}{\left\{ \dfrac{r}{(2-\sigma)^3} \right\}^l \left\{ \dfrac{r_1}{(2-\sigma)^3} \right\}^m \left\{ \dfrac{r_2}{(2-\sigma)^3} \right\}^n}.$$

Hence the series

$$\Sigma\Sigma\Sigma h'_{lmn}t^l\theta^m\phi^n, \quad \Sigma\Sigma\Sigma k'_{lmn}t^l\theta^m\phi^n,$$

converge for values of t such that $|t| < r$.

The existence of integrals of the equations

$$t\frac{dx}{dt} = x + at + \Sigma\Sigma\Sigma a_{ijp}x^iy^jt^p \left.\begin{array}{c} \\ \\ \\ \end{array}\right\}$$

$$t\frac{dy}{dt} = y + bt + \Sigma\Sigma\Sigma b_{ijp}x^iy^jt^p$$

can be deduced from the preceding result, by choosing

$$|a| = A, \quad |b| = B, \quad |a_{ijp}| = A_{ijp}, \quad |b_{ijp}| = B_{ijp},$$

as the quantities A, B, A_{ijp}, B_{ijp} which occur in those dominant equations. The expression for the integrals is

$$\left.\begin{array}{l} x = \Sigma\Sigma\Sigma H_{lmn}\,t^l\theta^m\phi^n \\ y = \Sigma\Sigma\Sigma K_{lmn}\,t^l\theta^m\phi^n \end{array}\right\},$$

where H_{lmn} is derived from h'_{lmn}, and K_{lmn} from k'_{lmn}, by changing A into $-a$, B into $-b$, A_{ijp} into a_{ijp}, and B_{ijp} into b_{ijp}. The effect of these changes is to give

$$|H_{lmn}| < h'_{lmn},$$

$$|K_{lmn}| < k'_{lmn};$$

and therefore the series for x and y converge.

The actual values are

$$\left.\begin{array}{l} x = at\log t + C_1t + \Sigma\Sigma\Sigma\,H_{lmn}t^l\theta^m\phi^n \\ y = bt\log t + C_2t + \Sigma\Sigma\Sigma\,K_{lmn}t^l\theta^m\phi^n \end{array}\right\},$$

where $\theta = -t\log t$, $\phi = \frac{1}{2}t(\log t)^2$, the summation is for values of l, m, n such that $l + m + n \geqslant 2$, and the coefficients C_1, C_2 are arbitrary constants.

But the formal expression is more general than the actual value. The equations determining the coefficients are

$$\left.\begin{array}{l} (l+m+n-1)H_{lmn} - (m+1)H_{l-1,\,m+1,\,n} - (n+1)H_{l,\,m-1,\,n+1} = E_{lmn} \\ (l+m+n-1)K_{lmn} - (m+1)K_{l-1,\,m+1,\,n} - (n+1)K_{l,\,m-1,\,n+1} = F_{lmn} \end{array}\right\};$$

also

$$H_{100} = C_1, \quad H_{010} = -a, \quad H_{001} = 0,$$

$$K_{100} = C_2, \quad K_{010} = -b, \quad K_{001} = 0.$$

It is clear that, when $l + m + n = 2$,

$$E_{lmn} = 0, \quad F_{lmn} = 0, \quad \text{if} \quad n = 1, \ 2 \ ;$$

hence H_{lmn}, K_{lmn} both vanish for $l + m + n = 2$, if $n = 1, \ 2$.

Thus for $l + m + n = 3$,

$$E_{lmn} = 0, \quad F_{lmn} = 0, \quad \text{if} \quad n = 1, \ 2, \ 3 \ ;$$

hence also H_{lmn}, K_{lmn} both vanish for $l + m + n = 3$, if $n = 1, 2, 3$. And so on: all the coefficients H_{lmn}, K_{lmn} vanish, if

$$n > 0 \ ;$$

that is, the quantity ϕ does not actually occur in the expressions for x and y, which accordingly are regular functions of t and $t \log t$.

The theorem is therefore established.

Note 1.   Any term in x and y is of the form

$$K t^m (t \log t)^n,$$

that is, $K t^{m+n} (\log t)^n$; and therefore the index of $\log t$ is never greater than the index of t.

If, however, the equations were

$$\left. \begin{aligned}
t \, \frac{dx}{dt} &= x + at + ct \log t + \Sigma\Sigma\Sigma\Sigma a_{ijpq} \, x^i y^j t^p (t \log t)^q \\
t \, \frac{dy}{dt} &= y + bt + c't \log t + \Sigma\Sigma\Sigma\Sigma b_{ijpq} \, x^i y^j t^p (t \log t)^q
\end{aligned} \right\} ,$$

where $i + j + p + q \geqslant 2$ for the summations, then the values of x and y satisfying the equations are

$$\left. \begin{aligned}
x &= -\tfrac{1}{2}ct (\log t)^2 + at \log t + C_1 t + \Sigma\Sigma\Sigma H_{lmn} t^l \theta^m \phi^n \\
y &= -\tfrac{1}{2}c't (\log t)^2 + bt \log t + C_2 t + \Sigma\Sigma\Sigma K_{lmn} t^l \theta^m \phi^n
\end{aligned} \right\} ,$$

where t, θ, ϕ have the same values as above, and the summations are for values of l, m, n, such that $l + m + n \geqslant 2$: the coefficients H_{lmn}, K_{lmn} are determinable as before.   Any term in x is

$$H t^{l+m+n} (\log t)^{m+2n},$$

that is, the index of $\log t$ is not greater than twice the index of t.

Note 2.   If a vanishes but not b, the solutions are still non-regular functions of t; likewise if b vanishes but not a.   In these cases, it is known that no regular integrals, which vanish with t, are possessed by the equation.

If $a = 0$, $b = 0$, then $H_{lm} = 0$, $K_{lm} = 0$, if $m \geqslant 1$: that is, $t \log t$ disappears from the expressions for x and y, which then become regular functions; they are the double infinitude of regular integrals that vanish with t. In this case, the regular integrals are the only integrals vanishing with t, which are possessed by the equation.

185. *Second sub-case: κ not zero.*

The theorem is:

The equations possess in general a double infinitude of non-regular integrals vanishing with t, which are regular functions of t, $t \log t$, $\frac{1}{2} t (\log t)^2$; and it is known that there are no regular integrals which vanish with t. If however $a = 0$, then the integrals can be arranged in two sets; one is a simple infinitude of non-regular integrals vanishing with t, which are regular functions of t and $t \log t$; the other is the simple infinitude of regular integrals vanishing with t, which the equation is known to possess. (It is necessary that the constant κ be different from zero: otherwise some of the coefficients in the second set are infinite unless b also is zero, in which form we revert to the first sub-case already considered.)

The method of establishment is similar to those which precede: it need therefore not be repeated after the many instances of it which have already been given.

The initial terms in the integrals of the equations, as taken in § 180, are

$$t_1 = a\theta + At + \ldots,$$

$$t_2 = \kappa a\phi + (\kappa A + b)\,\theta + Bt + \ldots,$$

the unexpressed terms being of higher order in t, θ, ϕ: here A and B are arbitrary, $\theta = t \log t$, and $\phi = \frac{1}{2} t (\log t)^2$. Any term in the expansion of t_1 or t_2, which involves ϕ, contains κ in its coefficient; the disappearance of the terms in ϕ from the integrals in the first sub-case is thus explained, for κ then is zero.

Ex. Consider the equations

$$t \frac{dx}{dt} = x + pt + (a, b, c \mathbb{X} x, y)^2 \left.\right\}$$
$$t \frac{dy}{dt} = y + p't + (a', b', c' \mathbb{X} x, y)^2 \left.\right\}, $$

where p and p' do not vanish.

Let θ denote $t \log t$. The equations have no regular integrals, which vanish with t. The non-regular integrals, which vanish with t, are regular functions of t and θ, in the form

$$x = at + p\theta + u_2 + u_3 + \ldots = u_1 + u_2 + u_3 + \ldots,$$
$$y = \beta t + p'\theta + v_2 + v_3 + \ldots = v_1 + v_2 + v_3 + \ldots,$$

where u_n and v_n denote the aggregate of terms in x and y, which are of dimensions n in t and θ; and a, β are arbitrary constants. Now

$$t \frac{d}{dt} = t \frac{\partial}{\partial t} + \theta \frac{\partial}{\partial \theta} + t \frac{\partial}{\partial \theta},$$

so that

$$pt + (a, b, c \langle u_1 + u_2 + \ldots, v_1 + v_2 + \ldots)^2$$

$$= -x + t \frac{\partial u_1}{\partial \theta} + t \frac{\partial u_2}{\partial \theta} + t \frac{\partial u_3}{\partial \theta} + \ldots$$

$$+ u_1 + 2u_2 + 3u_3 + \ldots$$

$$= t \frac{\partial u_1}{\partial \theta} + u_2 + t \frac{\partial u_2}{\partial \theta} + 2u_3 + t \frac{\partial u_3}{\partial \theta} + \ldots$$

Hence

$$(n-1) u_n + t \frac{\partial u_n}{\partial \theta} = \text{terms of order } n \text{ in } (a, b, c \langle u_1 + u_2 + \ldots, v_1 + v_2 + \ldots)^2;$$

and similarly

$$(n-1) v_n + t \frac{\partial v_n}{\partial \theta} = \ldots\ldots\ldots\ldots\ldots\ldots (a', b', c' \langle u_1 + u_2 + \ldots, v_1 + v_2 + \ldots)^2.$$

Therefore

$$u_2 + t \frac{\partial u_2}{\partial \theta} = (a, b, c \langle u_1, v_1)^2$$

$$= (a, b, c \langle p\theta + at, p'\theta + \beta t)^2,$$

so that, if

we have

$$u_2 = A_0 \theta^2 + 2A_1 \theta t + A_2 t^2,$$

$$A_0 = (a, b, c \langle p, p')^2,$$

$$A_1 + A_0 = (a, b, c \langle p, p' \rangle a, \beta),$$

$$A_2 + 2A_1 = (a, b, c \langle a, \beta)^2.$$

Similarly, if

we have

$$v_2 = A_0' \theta^2 + 2A_1' \theta t + A_2' t^2,$$

$$A_0' = (a', b', c' \langle p, p')^2,$$

$$A_1' + A_0' = (a', b', c' \langle p, p' \rangle a, \beta),$$

$$A_2' + 2A_1' = (a', b', c' \langle a, \beta)^2.$$

For the terms of order 3, we have

$$\left.\begin{aligned}
2u_3 + t \frac{\partial u_3}{\partial \theta} &= (a, b, c \langle u_1, v_1 \rangle u_2, v_2) \\
2v_3 + t \frac{\partial v_3}{\partial \theta} &= (a', b', c' \langle u_1, v_1 \rangle u_2, v_2)
\end{aligned}\right\};$$

and so on.

Concluding Note.

186.   Some sub-cases still remain over from Case I (a), when the roots ξ_1 and ξ_2 of the critical quadratic do not satisfy the conditions that (§ 173) prevent some one (or more) of the quantities

$$(\lambda - 1)\,\xi_1 + \mu\xi_2 + \nu, \qquad \lambda\xi_1 + (\mu - 1)\,\xi_2 + \nu,$$

from vanishing, for integer values of λ, μ, ν such that $\lambda + \mu + \nu \geqslant 2$. The real parts of ξ_1, ξ_2 are supposed to be positive.

The instances, that can occur, are obviously for $\lambda = 0$ in the first set, and $\mu = 0$ in the second set; both are included in the form

$$\xi = \mu\eta + \nu,$$

where ξ and η are the roots of the quadratic, and $\mu + \nu \geqslant 2$. The cases $\mu = 0$, $\mu = 1$, have already been discussed. For the remaining cases, we have the theorem: *The double infinitude of non-regular integrals, which vanish with t, are regular functions of t, t^η, $t^{\mu\eta+\nu} \log t$, where μ and ν are integers.* It can be established in the same manner as the similar theorems in the preceding sections.

Corresponding Results for a System of any number of Equations.

187.   It is clear that a number of theorems, merely the generalisation of those in the preceding investigations, can be stated concerning a system of n equations, which involve n dependent variables in the form

$$z\frac{dx_r}{dz} = c_r z + \sum_{m=1}^{n} a_{rm}x_m + \phi_r(z, x_1, \ldots, x_n), \qquad (r = 1, \ldots, n),$$

where the functions ϕ_r are regular functions of their arguments and contain no terms of dimensions lower than two. The general character of the integral system depends upon the roots of the critical equation

$$\begin{vmatrix} a_{11} - \Omega, & a_{12}, & a_{13}, & \ldots, & a_{1n} \\ a_{21}, & a_{22} - \Omega, & a_{23}, & \ldots, & a_{2n} \\ a_{31}, & a_{32}, & a_{33} - \Omega, & \ldots, & a_{3n} \\ \vdots & \vdots & \vdots & \ddots & \vdots \\ a_{n1}, & a_{n2}, & a_{n3}, & \ldots, & a_{nn} - \Omega \end{vmatrix} = 0,$$

of degree n, which will be supposed to have no zero roots.

I. If no one of the n roots be a positive integer, then the equations possess integrals, which are regular functions of z and vanish with z; such integrals are uniquely determinate.

II. If m (where $0 < m \leqslant n$) of the roots be positive integers, then in general the equations possess no integrals, which are regular functions of z and vanish with z. It may however happen that certain m relations among the coefficients of the differential equations are satisfied; if so, the equations then possess integrals, which are regular functions of z, which vanish with z, and which contain m arbitrary constants.

III. If the n roots of the equation are distinct from one another, say Ω_1, ..., Ω_n, and if Ω_1, ..., Ω_m have their real parts positive though they are not themselves positive integers, while the real parts of Ω_{m+1}, ..., Ω_n are negative or zero (provided that, in the latter case, there are imaginary parts of the roots), then in addition to the integrals which are regular, the equations possess a system of integrals, which vanish with z, which are regular functions of

$$z, \quad z^{\Omega_1}, \quad ..., \quad z^{\Omega_m},$$

and which contain m arbitrary constants.

The integer m may have any value from 0 to n.

IV. If the n roots of the equation are distinct from one another, say Ω_1, ..., Ω_n; if Ω_1, ..., Ω_m be positive integers, if Ω_{m+1}, ..., Ω_p have their real parts positive though they are not themselves positive integers, and if Ω_{p+1}, ..., Ω_n have their real parts negative or zero (in the last case, the imaginary parts must not vanish), then in general the equations (known by II. to possess no regular integrals vanishing with z) possess integrals, which vanish with z, which are regular functions of

$$z, \quad z \log z, \quad z^{\Omega_{m+1}}, \quad ..., \quad z^{\Omega_p},$$

and which contain p arbitrary constants.

If however the m relations among the coefficients, indicated in II., be satisfied, then the non-regular integrals, which vanish with z, are regular functions of

$$z, \quad z^{\Omega_{m+1}}, \quad ..., \quad z^{\Omega_p},$$

and contain $p - m$ arbitrary constants.

The integer m may have any value from 0 to n; the integer p (if distinct from m) may have any value from $m + 1$ to n.

V. If a root Ω have a multiplicity κ; if $\Omega_{\kappa+1}$, ..., Ω_m be positive integers, if Ω_{m+1}, ..., Ω_p have their real parts positive though they are not themselves positive integers, and if Ω_{p+1}, ..., Ω_n have their real parts negative or zero (in the last case, the imaginary parts must not vanish), then in general the equations possess integrals, which vanish with z. These integrals are regular functions of

$$z, \quad z^\Omega, \quad z^\Omega \log z, \quad z \log z, \quad z^{\Omega_{m+1}}, \quad ..., \quad z^{\Omega_p},$$

if the real part of Ω be positive though Ω is not itself a positive integer; these integrals contain p arbitrary constants. But if certain $m - \kappa$ relations among the coefficients of the differential equations be satisfied, so that there then exist integrals, which are regular functions of z and vanish with z, the non-regular integrals are regular functions of

$$z, \quad z^\Omega, \quad z^\Omega \log z, \quad z^{\Omega_{m+1}}, \quad ..., \quad z^{\Omega_p},$$

if the real part of Ω be positive though Ω itself be not a positive integer; these integrals contain $p - (m - \kappa)$ arbitrary constants.

If the real part of Ω be negative, then z^Ω and $z^\Omega \log z$ must be removed from the arguments of the non-regular integrals, when they are expressed as regular functions of non-regular arguments.

If Ω be a positive integer, then z^Ω and $z^\Omega \log z$ must be replaced by $z \log z$ among the arguments of the non-regular integrals, when they are expressed as regular functions of non-regular arguments.

Remaining typical forms of Chapter XI.

188. It is not proposed that any discussion of the alternative typical forms in Chapter XI shall here be made: a few remarks only will be devoted to one of them, as being associated with a possibility rejected in the preceding discussions.

The assumption has been made throughout, that no root of the critical quadratic $Q(s) = 0$ is zero.

Now suppose that one root is zero and that the other is κ; then the equations have the form (or can be changed to the form)

$$z \frac{dx}{dz} = az + b_1 y + \phi_1 (x, y, z) \Big\}$$
$$z \frac{dy}{dz} = a'z + \kappa y + \phi_2 (x, y, z) \Big\} ,$$

or the form

$$z \frac{dx}{dz} = az + \phi_1 (x, y, z) \Big\}$$
$$z \frac{dy}{dz} = a'z + a_2 x + \kappa y + \phi_2 (x, y, z) \Big\} ,$$

where ϕ_1 and ϕ_2 are regular functions of their arguments, and contain no terms of dimensions lower than two. By taking

$$X = \kappa x - b_1 y,$$

the former can be changed to

$$z \frac{dX}{dz} = a''z + \Phi_1 (X, y, z) \Big\}$$
$$z \frac{dy}{dz} = a'z + \kappa y + \Phi_2 (X, y, z) \Big\} ;$$

by taking

$$Y = a_2 x + \kappa y,$$

the latter can be changed to

$$z \frac{dx}{dz} = az + \Phi_1 (x, Y, z) \Big\}$$
$$z \frac{dY}{dz} = a''z + \kappa Y + \Phi_2 (x, Y, z) \Big\} ;$$

effectively the same as the transformation of the former.

Next, suppose that both roots of the critical quadratic are zero: the equations have the form

$$z \frac{dx}{dz} = az + cy + \phi_1 (x, y, z) \Big\}$$
$$z \frac{dy}{dz} = bz + \phi_2 (x, y, z) \Big\} ,$$

where ϕ_1 and ϕ_2 are regular functions of their arguments and contain no terms of dimensions lower than two.

In each of these cases, we have one equation of the type in § 159, as leading to distinct typical reduced forms: if, in the last case, $c = 0$, then both equations are of this type.

Ex. 1. Prove that the equations

$$z\frac{dx}{dz} = az + \phi_1(x, y, z) \left.\right\}$$
$$z\frac{dy}{dz} = bz + cy + \phi_2(x, y, z) \left.\right\}$$

possess integrals, which are regular functions of z and vanish with z, provided c is not a positive integer; but that, if c is a positive integer other than zero, then the equations possess no regular integrals, which vanish with z.

Prove also that, if the real part of c be negative, the regular integrals are the only integrals of the equations which vanish with z; while, if $b=0$, $c=1$, the equations possess an infinitude of regular integrals that vanish with z.

Also that, if the real part of c be positive, then the equations possess an infinitude of integrals, which vanish with z; they are regular functions of z and z^c, when c is not an integer, and are regular functions of z and $z \log z$, when c is an integer.

Ex. 2. Obtain the characteristic properties of the regular integrals and the non-regular integrals (if any), which vanish with z, of the equations

$$z\frac{dx}{dz} = az + \phi_1(x, y, z) \left.\right\}$$
$$z\frac{dy}{dz} = bz + cx + \phi_2(x, y, z) \left.\right\} .$$

Ex. 3. Discuss the integrals, which vanish with z, of the equations

$$z^\kappa \frac{dx}{dz} = pz + ax^2 + 2bxy + cy^2 \left.\right\}$$
$$z^\kappa \frac{dy}{dz} = qz + ax^2 + 2\beta xy + \gamma y^2 \left.\right\} ,$$

first, when $\kappa = 1$, secondly, when $\kappa = 2$.

Ex. 4. Discuss the integrals (if any), which vanish with z, of the equations

$$z\frac{dx}{dz} = \frac{(a, b, c, f, g, h \llbracket x, y, z)^2}{\lambda x + \mu y + \nu z} \left.\right\}$$
$$z\frac{dy}{dz} = \frac{(a', b', c', f', g', h' \llbracket x, y, z)^2}{\lambda' x + \mu' y + \nu' z} \left.\right\} .$$

CHAPTER XIII.

Systems of Equations with Multiform Values of the Derivatives; Singular Solutions*.

Integrals of Systems of Equations in the Vicinity of Branch-points.

189. The equations in the preceding chapters were such as to enable us to express the values of the derivatives explicitly in terms of the variables. The non-ordinary points that arose for consideration either were accidental singularities of the first kind or the second kind, or were essential singularities, of the values of the derivatives; in every instance, the derivatives were expressed as uniform functions.

We next proceed to the consideration of equations such that, in the explicit expressions for the derivatives, branch-points may occur among the non-ordinary points and, in consequence, those explicit expressions are multiform functions. It will be assumed, as in the corresponding instance for a single equation, that the number of forms for the derivatives is limited in number; the

---

* The theorems investigated in this chapter are established chiefly in connection with a system of two equations of the first order and any degree; it is manifest that the method is immediately applicable to systems of any number of equations, and corresponding theorems for such systems are stated in one or two instances. The main object is the discussion of the solutions as defined by initially assigned conditions. The development of the argument leads to the consideration of singular integrals; but such integrals (when they exist) are considered almost entirely from this point of view, and only slightly as derivable from a complete integral.

Reference may be made to papers by Mayer, *Math. Ann.*, t. xxii (1883), pp. 368—392; Goursat, *Amer. Journ. Math.*, t. xi (1889), pp. 329—372; Fine, *Amer. Journ. of Math.*, t. xii (1890), pp. 295—322; and Dixon, *Phil. Trans.* (1895, A), pp. 523—566, who gives numerous illustrations.

most general case, on this supposition, is that in which the dependent variables, say y and z, the independent variable x, and the derivatives p and q, where

$$p = \frac{dy}{dx}, \quad q = \frac{dz}{dx},$$

satisfy two equations

$$\left. \begin{array}{l} F(x, y, z, p, q) = 0 \\ G(x, y, z, p, q) = 0 \end{array} \right\}.$$

The equations $F = 0$, $G = 0$, will therefore be regarded as rational in both p and q; they usually (though not necessarily) will be regarded as rational also in y and z; they may be merely analytical in x. Further, it will be assumed that the system is irreducible: that is, on the one hand, that no equation of the form

$$E(x, y, z) = 0,$$

involving only the variables and not their derivatives, can be obtained by algebraical elimination between $F = 0$ and $G = 0$: and on the other hand, that the system is not resoluble into a number of systems of equations of lower degrees in p and q.

The variables are taken to be complex. But when the variables are restricted to have only real values, it is possible to associate geometrical interpretations of configurations in ordinary space with the differential equations; this sometimes will be done, but chiefly for the subsidiary purpose of illustration.

190. From the general existence-theorem it follows that, if the equations

$$F(a, b, c, \beta, \gamma) = 0, \quad G(a, b, c, \beta, \gamma) = 0,$$

determine values of β and γ, whether they be simple or multiple roots, there exist solutions of the equations

$$F(x, y, z, p, q) = 0, \quad G(x, y, z, p, q) = 0,$$

determined by the condition that y and z acquire assigned values b and c respectively, when $x = a$. In order to understand the character of the point $x = a$ in relation to the integrals, we consider their expression in the immediate vicinity. For this purpose, take

$$x = a + \xi, \quad y = b + \eta, \quad z = c + \zeta,$$

so that η and ζ are functions of ξ, which vanish when $\xi = 0$. Because $\dfrac{d\eta}{d\xi}$ and $\dfrac{d\zeta}{d\xi}$ are equal to β and γ respectively, when $\xi = 0$, we have

$$\eta = \beta\xi + H, \quad \zeta = \gamma\xi + Z,$$

where H and Z are of order greater than unity in ξ, for values of $|\xi|$ that are sufficiently small. Also

$$p = \beta + \frac{dH}{dx} = \beta + H',$$

$$q = \gamma + \frac{dZ}{dx} = \gamma + Z',$$

where H' and Z' vanish when $\xi = 0$, and are of small modulus for sufficiently small values of $|\xi|$: also H and Z are of order higher by unity than H' and Z' respectively in powers of ξ. Substituting these in the original equations and noting the equations that determine β and γ, we have

$$0 = \xi \frac{\partial F}{\partial a} + \eta \frac{\partial F}{\partial b} + \zeta \frac{\partial F}{\partial c} + H' \frac{\partial F}{\partial \beta} + Z' \frac{\partial F}{\partial \gamma} + \ldots,$$

$$0 = \xi \frac{\partial G}{\partial a} + \eta \frac{\partial G}{\partial b} + \zeta \frac{\partial G}{\partial c} + H' \frac{\partial G}{\partial \beta} + Z' \frac{\partial G}{\partial \gamma} + \ldots,$$

where $\dfrac{\partial F}{\partial a}$ is the value of $\dfrac{\partial}{\partial x} F(x, y, z, p, q)$ when, after differentiation, the initial values a, b, c, β, γ are substituted for x, y, z, p, q: and similarly for the other coefficients. Write

$$\begin{aligned} f_1 &= \frac{\partial F}{\partial a} + \beta \frac{\partial F}{\partial b} + \gamma \frac{\partial F}{\partial c} \\ g_1 &= \frac{\partial G}{\partial a} + \beta \frac{\partial G}{\partial b} + \gamma \frac{\partial G}{\partial c} \end{aligned} \Bigg\};$$

then, if f_1 and g_1 are distinct from zero, the equations take the form

$$0 = f_1 \xi + H' \frac{\partial F}{\partial \beta} + Z' \frac{\partial F}{\partial \gamma} + \ldots,$$

$$0 = g_1 \xi + H' \frac{\partial G}{\partial \beta} + Z' \frac{\partial G}{\partial \gamma} + \ldots$$

It may happen that f_1 and g_1 are zero. In any case, let

$$\begin{aligned} f_n &= \frac{1}{n!}\left(\frac{\partial}{\partial a} + \beta \frac{\partial}{\partial b} + \gamma \frac{\partial}{\partial c}\right)^n F \\ g_n &= \frac{1}{n!}\left(\frac{\partial}{\partial a} + \beta \frac{\partial}{\partial b} + \gamma \frac{\partial}{\partial c}\right)^n G \end{aligned} \Bigg\},$$

for all values of n: so that the equations take the form

$$0 = \sum_{n=1} f_n \xi^n + \text{H}' \frac{\partial F}{\partial \beta} + \text{Z}' \frac{\partial F}{\partial \gamma} + \dots \left.\begin{array}{c} \\ \\ \end{array}\right\}$$

$$0 = \sum_{n=1} g_n \xi^n + \text{H}' \frac{\partial G}{\partial \beta} + \text{Z}' \frac{\partial G}{\partial \gamma} + \dots \left.\begin{array}{c} \\ \\ \end{array}\right\}$$,

the unexpressed terms involving H', Z', H, Z, and being of order higher than H', Z' in the small quantities.

It will be assumed that not all the coefficients f and g vanish. Such evanescent coefficients would arise, for parametric values of a, b, c, β, γ, if the original differential equations were of the forms

$$F(y - xp, \ z - xq, \ p, \ q) = 0,$$
$$G(y - xp, \ z - xq, \ p, \ q) = 0,$$

a generalised form of Clairaut's equation; and they might arise for special values of a, b, c, β, γ, or of some of them, e.g. if β and γ were zero, and F and G were explicitly independent of x. The latter cases will be discussed separately.

There is a fundamental distinction between the forms of the integrals according as J, where

$$J = \frac{\partial F}{\partial \beta} \frac{\partial G}{\partial \gamma} - \frac{\partial F}{\partial \gamma} \frac{\partial G}{\partial \beta},$$

is not zero, or is zero.

191.   When J is not zero, the equations can be solved linearly for H' and Z', so that

$$J\text{H}' + \sum_{n=1} \left(f_n \frac{\partial G}{\partial \gamma} - g_n \frac{\partial F}{\partial \gamma} \right) \xi^n + \dots = 0,$$

$$J\text{Z}' + \sum_{n=1} \left(- f_n \frac{\partial G}{\partial \beta} + g_n \frac{\partial F}{\partial \beta} \right) \xi^n + \dots = 0.$$

Because not all the coefficients f and g can vanish, and because J is not zero, not all the coefficients of powers of ξ alone in the last forms of the equations can vanish. Suppose that ξ^m is the lowest term in the first equation which survives, and ξ^n is the lowest term in the second; then our equations become

$$\begin{aligned} \text{H}' &= A\xi^m + \dots \\ \text{Z}' &= B\xi^n + \dots \end{aligned} \left.\begin{array}{c} \\ \\ \end{array}\right\},$$

where on the right-hand side higher powers of H' and Z' may occur, and also the first powers of H' and Z', if multiplied by coefficients that vanish with ξ, H, Z. Solving, therefore, so as to obtain H$'$ and Z$'$ explicitly, we have

$$H' = A\xi^m + f(\xi, H, Z) \Big\}$$
$$Z' = B\xi^n + g(\xi, H, Z) \Big\} \, ,$$

as the general type of equations for determining H and Z: the functions f and g are regular functions of their arguments, and they contain only terms of higher order, in powers of ξ and small quantities, than are H$'$, Z$'$, ξ^m, ξ^n.

The general existence-theorem can be applied to them, the determining conditions being that both H and Z must vanish when $\xi = 0$; hence

$$H = \xi^{m+1} P(\xi), \quad Z = \xi^{n+1} Q(\xi),$$

where $P(\xi)$ and $Q(\xi)$ are regular functions of ξ, which do not vanish with ξ. Accordingly

$$y - b = \beta(x - a) + (x - a)^{m+1} P(x - a) \Big\}$$
$$z - c = \gamma(x - a) + (x - a)^{n+1} Q(x - a) \Big\} \, ,$$

so that y and z are regular functions: the assumptions made are that J does not vanish, and (tacitly) that β and γ are finite.

Manifestly there are as many regular integrals of this type, as there are distinct sets of values β and γ, which keep J different from zero. For each of them, the combination of values a, b, c of the variables x, y, z is ordinary for the equations.

Secondly, suppose that one of the two quantities β and γ is infinite and the other finite, say $\beta = \infty$; then at $\xi = 0$, we have

$$\frac{d\eta}{d\xi} = \infty, \quad \frac{d\zeta}{d\xi} = \text{finite},$$

and therefore, at $\xi = 0$,

$$\frac{d\xi}{d\eta} = 0, \quad \frac{d\zeta}{d\eta} = 0.$$

Accordingly, we take η as the independent variable; using the result in the preceding case and assuming that the corresponding J is not zero, we have

$$x - a = (y - b)^{m+1} P_1(y - b) \Big\}$$
$$z - c = (y - b)^{n+1} Q_1(y - b) \Big\} \, ,$$

and therefore

$$y - b = (x - a)^{\frac{1}{m+1}} P_2 \{(x - a)^{\frac{1}{m+1}}\},$$

$$z - c = (x - a)^{\frac{n+1}{m+1}} Q_2 \{(x - a)^{\frac{1}{m+1}}\},$$

where P_2 and Q_2 are regular functions of their argument, and do not vanish with it. Manifestly $x = a$ is an algebraical branch-point for both the integral functions, because $m \geqslant 1$.

Thirdly, suppose that both the quantities β and γ are infinite, so that, at $\xi = 0$, we have

$$\frac{d\eta}{d\xi} = \infty, \quad \frac{d\zeta}{d\xi} = \infty,$$

and therefore

$$\frac{d\xi}{d\eta} = 0, \quad \frac{d\zeta}{d\eta} = \kappa,$$

where κ is determined from the equations on making η the independent variable. If κ should be infinite, then we make ζ the independent variable: and we have, at $\xi = 0$,

$$\frac{d\xi}{d\zeta} = 0, \quad \frac{d\eta}{d\zeta} = 0.$$

It therefore is sufficient, in the former case, to take $\kappa = 0$, or $\kappa = a$ finite quantity; to secure this, we choose ζ or η as the new independent variable, according as $\frac{d\zeta}{d\xi}$ or $\frac{d\eta}{d\xi}$ is found to be the greater infinity. Again using the result of the first case, and assuming that the corresponding J is not zero, we have

$$x - a = (y - b)^{m+1} P_1 (y - b),$$

$$z - c = \kappa (y - b) + (y - b)^{n+1} Q_1 (y - b),$$

and therefore

$$y - b = (x - a)^{\frac{1}{m+1}} P_2 \{(x - a)^{\frac{1}{m+1}}\},$$

$$z - c = (x - a)^{\frac{1}{m+1}} Q_2 \{(x - a)^{\frac{1}{m+1}}\},$$

where P_2 is a regular function of its argument, which does not vanish when its argument is zero, and Q_2 is also a regular function of its argument, which may or may not vanish when its argument is zero. Manifestly, $x = a$ is an algebraical branch-point for both the integral functions, because $m \geqslant 1$.

192. Next, suppose that $J = 0$ for the initial values considered: there are several sub-cases.

(i) Let $J = 0$, while none of the first derivatives of F and G with regard to β and γ vanish. Then we may take

$$\frac{\partial F}{\partial \beta} = \theta \frac{\partial F}{\partial \gamma}, \quad \frac{\partial G}{\partial \beta} = \theta \frac{\partial G}{\partial \gamma},$$

where θ is a finite quantity; and the equations for H' and Z' thus become

$$
\left.
\begin{aligned}
0 &= \sum_{n=1} f_n \xi^n + (\theta H' + Z') \frac{\partial F}{\partial \gamma} + \text{terms of higher orders in } H', Z' \\
&\qquad\qquad\qquad\qquad\qquad\qquad + \text{other higher terms} \\
0 &= \sum_{n=1} g_n \xi^n + (\theta H' + Z') \frac{\partial G}{\partial \gamma} + \dots\dots\dots\dots\dots\dots\dots\dots
\end{aligned}
\right\} .
$$

Hence

$$0 = \sum_{n=1} \left(f_n \frac{\partial G}{\partial \gamma} - g_n \frac{\partial F}{\partial \gamma} \right) \xi^n + \text{terms in } H', Z' \text{ of order} > 1$$
$$+ \text{ other higher terms.}$$

In this last equation, let the term of lowest order in H', which survives when Z' is replaced by $-\theta H'$, be of degree l; and of the coefficients $f_n \dfrac{\partial G}{\partial \gamma} - g_n \dfrac{\partial F}{\partial \gamma}$, let that which is given by $n = l'$ be the first which does not vanish for successive values of n. (For simplicity, it will be assumed that $f_n = 0$, $g_n = 0$, for $n = 1, \dots, l' - 1$: in point of fact, all that is necessary is that

$$f_n \frac{\partial G}{\partial \gamma} - g_n \frac{\partial F}{\partial \gamma} = 0,$$

for $n = 1, \dots, l' - 1$. The difference between the two assumptions lies in the more complicated analysis, which leads to the expressions in the latter case.) We then have

$$
\left.
\begin{aligned}
Z' + \theta H' &= A \xi^{l'} + \text{higher powers} + \text{terms in } H, Z \\
H' &= B \xi^{l} + \dots\dots\dots\dots + \text{terms in } H, Z
\end{aligned}
\right\},
$$

so that we infer

$$
\left.
\begin{aligned}
H' &= \xi^{\frac{l'}{l}} P(\xi^{\frac{1}{l}}) + f(\xi^{\frac{1}{l}}, H, Z) \\
Z' &= \xi^{\frac{l'}{l}} Q(\xi^{\frac{1}{l}}) + g(\xi^{\frac{1}{l}}, H, Z)
\end{aligned}
\right\},
$$

where P and Q are regular functions of their argument, which do not vanish with it, and which are such that

$$Q(0) = - \theta P(0);$$

and f and g are regular functions of their arguments. Now let

$$\xi = t^l;$$

then

$$\frac{dH}{dt} = l t^{l'+l-1} P(t) + l t^{l-1} f(t, H, Z)\Bigg|$$

$$\frac{dZ}{dt} = l t^{l'+l-1} Q(t) + l t^{l-1} g(t, H, Z)\Bigg| \; ;$$

also, H and Z must vanish when $x = a$, that is, when $t = 0$. Hence

$$H = t^{l'+l} P_1(t)$$

$$= (x - a)^{\frac{l'}{l}+1} P_1 \{(x-a)^{\frac{1}{l}}\},$$

$$Z = t^{l'+l} Q_1(t)$$

$$= (x - a)^{\frac{l'}{l}+1} Q_1 \{(x-a)^{\frac{1}{l}}\},$$

where P_1 and Q_1 are regular functions of their argument, which do not vanish with it, and which are such that

$$Q_1(0) = - \theta P_1(0).$$

Consequently

$$y - b = \beta (x - a) + (x-a)^{\frac{l'}{l}+1} P_1 \{(x-a)^{\frac{1}{l}}\},$$

$$z - c = \gamma (x - a) + (x-a)^{\frac{l'}{l}+1} Q_1 \{(x-a)^{\frac{1}{l}}\}.$$

Because $l \geqslant 2$, the point $x = a$ is manifestly an algebraical branch-point for the integral functions.

The simplest case arises when $l = 2$; and then

$$y - b = \beta (x - a) + \kappa (x - a)^{\frac{3}{2}} + \lambda_1 (x - a)^2 + \ldots,$$

$$z - c = \gamma (x - a) - \theta \kappa (x - a)^{\frac{3}{2}} + \lambda_2 (x - a)^2 + \ldots.$$

As the integrals now are known, by the existence-theorem, to occur in this form, the coefficients can be determined by actual substitution of these expressions in the equations.

(ii) Next, supposing still that J vanishes, let one of the first derivatives of F or G vanish: say

$$\frac{\partial F}{\partial \beta} = 0;$$

then, since $J = 0$, we have either $\dfrac{\partial G}{\partial \beta} = 0$ or $\dfrac{\partial F}{\partial \gamma} = 0$: so that there are really two alternatives, viz.

$\dfrac{\partial F}{\partial \beta} = 0,\ \dfrac{\partial G}{\partial \beta} = 0$: the results of this case can be applied, *mutatis mutandis*, to the case when

$$\dfrac{\partial F}{\partial \gamma} = 0,\quad \dfrac{\partial G}{\partial \gamma} = 0;$$

$\dfrac{\partial F}{\partial \beta} = 0,\ \dfrac{\partial F}{\partial \gamma} = 0$: the results of this case can be applied, *mutatis mutandis*, to the case when

$$\dfrac{\partial G}{\partial \beta} = 0,\quad \dfrac{\partial G}{\partial \gamma} = 0.$$

In the first of the alternatives, the equations determining H' and Z' are

$$\left.\begin{aligned}
0 &= \sum_{n=1} f_n \xi^n + \frac{\partial F}{\partial \gamma} Z' + f\,(\xi,\, H,\, Z,\, H',\, Z') \\
0 &= \sum_{n=1} g_n \xi^n + \frac{\partial G}{\partial \gamma} Z' + g\,(\xi,\, H,\, Z,\, H',\, Z')
\end{aligned}\right\},$$

where the terms in the regular functions f and g are of order higher than Z' in powers of ξ. There are many varieties of forms that arise from relations among literal coefficients, or from vanishing coefficients : they all can be treated in a manner sufficiently indicated for the simplest of them, which is as follows. Suppose that no one of the quantities $f_1,\ g_1,\ f_1\dfrac{\partial G}{\partial \gamma} - g_1\dfrac{\partial F}{\partial \gamma}$ vanishes, and suppose that, in the regular function

$$f\frac{\partial G}{\partial \gamma} - g\frac{\partial F}{\partial \gamma},$$

H'^2 is the lowest power of H', which occurs free from the other variables; then solving the equations for H' and Z' under these conditions, we find

$$\left.\begin{aligned}
H' &= \mu \xi^{\frac12} + \text{higher powers of } \xi^{\frac12}, \text{ and powers of } H \text{ and of } Z \\
Z' &= \rho \xi\ +\ \dots\dots\dots\dots\ \xi^{\frac12},\ \dots\dots\dots\dots H \dots\dots\dots Z
\end{aligned}\right\}.$$

Taking $\xi = t^2$, these give

$$\frac{dH}{dt} = 2\mu t^2 + \text{higher powers of } t, \text{ and powers of } H \text{ and of } Z,$$

$$\frac{dZ}{dt} = 2\rho t^3 + \dots\dots\dots\dots\dots\dots\dots\dots\dots\dots\dots\dots\dots,$$

the right-hand sides being regular functions of t, H, Z. The variables H and Z are to vanish when $x = a$, that is, when $t = 0$; hence solutions of these equations exist, and they have the form

$$H = \tfrac{2}{3}\mu t^3 P(t)$$
$$= \tfrac{2}{3}\mu (x - a)^{\frac{3}{2}} P\{(x - a)^{\frac{1}{2}}\},$$
$$Z = \tfrac{1}{2} t^4 Q(t)$$
$$= \tfrac{1}{2}(x - a)^2 Q\{(x - a)^{\frac{1}{2}}\},$$

where P is a regular function of its argument such that $P(0) = 1$, and Q is a regular function of its argument such that $Q(0)$ is not zero. The solution of the original equations is then given by

$$y - b = \beta(x - a) + \tfrac{2}{3}\mu (x - a)^{\frac{3}{2}} P\{(x - a)^{\frac{1}{2}}\},$$
$$z - c = \gamma(x - a) + \tfrac{1}{2}(x - a)^2 Q\{(x - a)^{\frac{1}{2}}\}.$$

The point $x = a$ is manifestly an algebraical branch-point for the functions y and z, defined by the differential equations.

As already stated, many of the cases included in the general form adopted can be treated almost exactly as by the foregoing method. Others, however, arise which can be dealt with only by a subsequent method (§ 195): a remark which applies also to what follows immediately.

In the second of the alternatives, the equations determining H′ and Z′ are

$$0 = \sum_{n=1} f_n \xi^n \qquad\qquad\qquad + f(\xi, H, Z, H', Z')$$
$$0 = \sum_{n=1} g_n \xi^n + H' \frac{\partial G}{\partial \beta} + Z' \frac{\partial G}{\partial \gamma} + g(\xi, H, Z, H', Z')$$

In this instance, as in the last, we shall not deal with all the possible forms that may arise. We shall discuss only the simplest of them, viz., that in which f_1 and g_1 are not zero, and the terms in f and g, of lowest dimensions in H′ and Z′ alone, are of the second order, say

$$f(\xi, H, Z, H', Z') = \alpha H'^2 + 2\beta H'Z' + \gamma Z'^2$$
$$\qquad\qquad + \text{terms in } \xi, H, Z, H', Z' \text{ of higher order,}$$
$$g(\xi, H, Z, H', Z') = \alpha' H'^2 + 2\beta' H'Z' + \gamma' Z'^2$$
$$\qquad\qquad + \dots\dots\dots\dots\dots\dots\dots\dots\dots\dots\dots$$

Then we have
$$H' = \lambda \xi^{\frac{1}{2}} + \mu \xi + \dots,$$
$$Z' = l \xi^{\frac{1}{2}} + m \xi + \dots,$$
where
$$\frac{\lambda}{q} = \frac{l}{-p} = \left\{ \frac{-f_1}{\alpha q^2 - 2\beta pq + \gamma p^2} \right\}^{\frac{1}{2}},$$

$$\frac{\mu}{\gamma p - \beta q} = \frac{m}{-\beta p + \alpha q}$$
$$= \frac{g_1 (\alpha q^2 - 2\beta pq + \gamma p^2) - f_1 (\alpha' q^2 - 2\beta' pq + \gamma' p^2)}{(\alpha q^2 - 2\beta pq + \gamma p^2)^2},$$

and $p = \dfrac{\partial G}{\partial \beta}$, $q = \dfrac{\partial G}{\partial \gamma}$; thus H' and Z' are regular functions of $\xi^{\frac{1}{2}}$, H, and Z.   Take $\xi = t^2$, so that

$$\left. \begin{aligned} \frac{dH}{dt} &= 2\lambda t^2 + \text{regular function of } t, \text{ H, Z} \\[2mm] \frac{dZ}{dt} &= 2l t^2 + \dots\dots\dots\dots\dots\dots\dots \end{aligned} \right\}$$

Now H and Z are to vanish when $t = 0$; hence
$$H = \tfrac{2}{3}\lambda t^3 P(t) = \tfrac{2}{3}\lambda (x - a)^{\frac{3}{2}} P \{(x - a)^{\frac{1}{2}}\},$$
$$Z = \tfrac{2}{3} l t^3 Q(t) = \tfrac{2}{3} l (x - a)^{\frac{3}{2}} Q \{(x - a)^{\frac{1}{2}}\},$$

where P and Q are regular functions of their argument, and each of them becomes equal to unity when their argument vanishes. The solution of the original equations is then given by

$$y - b = \beta (x - a) + \tfrac{2}{3}\lambda (x - a)^{\frac{3}{2}} P \{(x - a)^{\frac{1}{2}}\},$$
$$z - c = \gamma (x - a) + \tfrac{2}{3} l (x - a)^{\frac{3}{2}} Q \{(x - a)^{\frac{1}{2}}\}.$$

The point $x = a$ is manifestly an algebraical branch-point for the functions y and z, defined by the differential equations.

(iii)   Lastly, suppose that J vanishes because both the first derivatives of F and of G vanish. The equations for H' and Z' then take the form

$$0 = \sum_{n=1} f_n \xi^n + f(\xi, \text{H, Z, H', Z'}),$$
$$0 = \sum_{n=1} g_n \xi^n + g(\xi, \text{H, Z, H', Z'}),$$

where f and g are regular functions as before, now containing no linear terms in H' and Z' alone.   Again, we consider only the simplest case, viz. that in which f_1 and g_1 are not zero, and f and g

contain terms of the second order in H′ and Z′ alone; then we have

$$H' = \rho \xi^{\frac{1}{2}} + \dots,$$
$$Z' = \sigma \xi^{\frac{1}{2}} + \dots,$$

where the expressions for H′ and Z′ are regular functions of $\xi^{\frac{1}{2}}$, H, and Z. The condition, that H and Z vanish when $x = a$, determines the solution of these equations in the form

$$H = \tfrac{2}{3}\rho\,(x-a)^{\frac{3}{2}} P\,\{(x-a)^{\frac{1}{2}}\},$$
$$Z = \tfrac{2}{3}\sigma\,(x-a)^{\frac{3}{2}} Q\,\{(x-a)^{\frac{1}{2}}\}.$$

The solution of the original equations is then given by

$$y - b = \beta\,(x-a) + \tfrac{2}{3}\rho\,(x-a)^{\frac{3}{2}} P\,\{(x-a)^{\frac{1}{2}}\},$$
$$z - c = \gamma\,(x-a) + \tfrac{2}{3}\sigma\,(x-a)^{\frac{3}{2}} Q\,\{(x-a)^{\frac{1}{2}}\},$$

where P and Q are regular functions of their argument, and each of them becomes equal to unity when $x = a$. The point $x = a$ is an algebraical branch-point of the functions, defined by the original differential equations.

It has been tacitly assumed that the values β and γ, which make F, G, J, simultaneously vanish, are finite. For values which are infinite, we change the independent variable as in § 191, say to y; and write

$$\frac{dx}{dy} = P, \quad \frac{dz}{dy} = Q,$$

so that the values of P and Q, which make F, G, J vanish, are finite: (one of them is zero). Then

$$p = \frac{1}{P}, \quad q = \frac{Q}{P};$$

so that, if

$$P^m F(x, y, z, p, q) = \phi(x, y, z, P, Q),$$
$$P^n G(x, y, z, p, q) = \gamma(x, y, z, P, Q),$$

we have

$$J\left(\frac{\phi, \gamma}{P, Q}\right) = -P^{m+n-3} J\left(\frac{F, G}{p, q}\right).$$

Thus when one of the original equations is of the second degree at least, so that, in order to make ϕ and γ integral in P and Q, $m + n$ must be equal to 3 at least, the new Jacobian is zero because $J = 0$. The same investigations shew, as before, that in the respective cases the point is an algebraical branch-point of the functions, defined by the differential equations.

193. These results may be summarised as follows :

Solutions of the equations

$$F(x, y, z, p, q) = 0, \quad G(x, y, z, p, q) = 0,$$

exist, determined by the condition of assuming values $y = b$, $z = c$, when $x = a$.

If $\xi = \beta$, $\eta = \gamma$, be finite simultaneous roots of

$$F(a, b, c, \xi, \eta) = 0, \quad G(a, b, c, \xi, \eta) = 0,$$

then $x = a$ is an ordinary point of the functions y and z in those solutions, provided $\xi = \beta$, $\eta = \gamma$ does not make J vanish, where

$$J = \frac{\partial F}{\partial \xi} \frac{\partial G}{\partial \eta} - \frac{\partial F}{\partial \eta} \frac{\partial G}{\partial \xi};$$

but if $\xi = \beta$, $\eta = \gamma$, does make J vanish, then $x = a$ is an algebraical branch-point of those functions.

If the equations

$$F(a, b, c, \xi, \eta) = 0, \quad G(a, b, c, \xi, \eta) = 0,$$

possess roots ξ and η, one at least of which is infinite, then $x = a$ is a branch-point of the functions y and z in those solutions; but taking that variable as independent which is associable with the greater magnitude of the roots ξ and η under consideration, say y, then (unless the Jacobian J vanishes) $y = b$ is an ordinary point of the functions x and z, determined by the equations as functions of y. If however the Jacobian J does vanish, then the set of initial values (whichever be taken as the independent variable) constitute an algebraical branch-point of the functions.

In every case where the point is an algebraical branch-point for the functions, the first derivatives of the various branches, which circulate round the point in a cycle, are equal to one another at the point.

Geometrical interpretation when the variables are real.

194. When the variables x, y, z are restricted so as to have only real values, they can be taken as the coordinates of a point in ordinary space ; the differential equations then are equations of twisted curves in three dimensions.

Between the three equations

$$F(x, y, z, p, q) = 0, \quad G(x, y, z, p, q) = 0,$$

$$J = \frac{\partial (F, G)}{\partial (p, q)} = 0,$$

let p and q be eliminated. This can be done by obtaining the simultaneous solutions of $F = 0$, $G = 0$, say N in number, regarded as algebraical equations in p and q: these are to be substituted in J, giving values

$$J_1, J_2, \dots, J_N,$$

say: then the eliminant is

$$\Theta = J_1 J_2 \dots J_N,$$

so that, after the symmetric functions on the right-hand side have been evaluated, Θ is a function of x, y, z *. Thus $\Theta = 0$ is the equation of a surface, which may be called the *focal surface*. If a, b, c be a point on it, so that $\Theta (a, b, c) = 0$, then one at least of the quantities $J_1, \dots, J_N$ vanishes; if a, b, c be a point not on the surface, then no one of the quantities $J_1, \dots, J_N$ vanishes.

First, let a, b, c be a point not lying upon the surface $\Theta = 0$; no simultaneous roots β and γ of the equations

$$F(a, b, c, \xi, \eta) = 0, \quad G(a, b, c, \xi, \eta) = 0,$$

make

$$J, = \frac{\partial (F, G)}{\partial (\xi, \eta)},$$

vanish: the point represents an ordinary combination of values for the differential equations. Through such a point, there pass N curves in space; their directions at the point are independent of one another, being determined by the equations

$$\frac{dy}{dx} = \beta, \quad \frac{dz}{dx} = \gamma.$$

As the point a, b, c moves about in space, this cluster of N curves also moves about; and so long as the point remains off the focal surface, the directions of the N curves remain distinct from one another, while they change as the point moves.

Secondly, let a, b, c be a point lying upon the surface $\Theta = 0$; then at least one of the quantities J vanishes, and more than one

* See also, for another method. Cayley, *Collected Math. Papers*, vol. I, pp. 370—374.

may vanish. Let $J_1, \ldots, J_\kappa$ vanish; and let $J_{\kappa+1}, \ldots, J_{N'}$, not vanish, where, if m_i represent the multiplicity of the roots making J_i vanish, we have

$$N' - \kappa + \sum_{i=1}^{\kappa} m_i = N.$$

Let the value J_t arise through the values $\xi = \beta_t, \eta = \gamma_t$; so that the sets of roots, which make J vanish, are $\beta_1, \gamma_1 : \beta_2, \gamma_2 : \ldots : \beta_\kappa, \gamma_\kappa$. Through the point there still passes the cluster of N curves; but for the m_i curves determined by β_i and γ_i, the m_i branches touch one another, the common tangent at the point of contact being distinct in direction from all other tangents to curves through the point. We thus have

(i) m_i branches of curves through the point, with a common direction at the point given by

$$\frac{dy}{dx} = \beta_i, \quad \frac{dz}{dx} = \gamma_i :$$

this holds for $i = 1, \ldots, \kappa$, the respective common tangents having different directions:

(ii) $N - \sum_{i=1}^{\kappa} m_i$ other branches through the point, each with its own direction, which is different from directions of all the other branches, and from the common direction of the sets of branches above.

Further, the normal to the surface $\Theta = 0$ at the point a, b, c has its direction-cosines proportional to

$$\frac{\partial \Theta}{\partial a}, \quad \frac{\partial \Theta}{\partial b}, \quad \frac{\partial \Theta}{\partial c} :$$

and any curve of the cluster through the point has its direction-cosines proportional to

$$1, \quad \beta, \quad \gamma.$$

Hence, unless

$$\frac{\partial \Theta}{\partial a} + \beta \frac{\partial \Theta}{\partial b} + \gamma \frac{\partial \Theta}{\partial c} = 0,$$

the direction of that curve of the cluster does not lie in the tangent-plane to $\Theta = 0$ at the point. In general, this is not the case; and therefore, in general, the directions of the curves of the cluster are not tangent-lines to the focal surface.

But denoting

$$\frac{\partial \Theta}{\partial x} + p\,\frac{\partial \Theta}{\partial y} + q\,\frac{\partial \Theta}{\partial z}$$

by K, and the values of K for the N roots of $F = 0$, $G = 0$, by $K_1, \ldots, K_n$, let

$$\Phi = K_1 K_2 \ldots K_n,$$

so that, when the symmetric functions are evaluated, Φ is a function of x, y, z. If then a, b, c be a point on the surface $\Phi = 0$, some one at least of the quantities K vanishes; that is, some one at least of the quantities

$$\frac{\partial \Theta}{\partial a} + \beta\,\frac{\partial \Theta}{\partial b} + \gamma\,\frac{\partial \Theta}{\partial c}$$

vanishes. Hence at any point on the curve of intersection of the focal surface by the surface $\Phi = 0$, some one (or more than one) of the directions of curves of the cluster through the point must be a tangent-line to the focal surface.

When the moving point a, b, c passes off the focal surface, the directions of all the curves of the cluster again become different from one another.

Note 1. It was seen that, if either β or γ or both (as simultaneous roots of the algebraical equations) be infinite, the value $x = a$ is an algebraical branch-point for the functions y and z: but that y or z, as the case may be, is an ordinary point for x and the other of the two, when J does not vanish. The geometrical significance is, that the corresponding curve of the cluster has its direction at a, b, c perpendicular to one (or to two) of the coordinate axes; and if the point lie off the focal surface, that curve is not touched at the point by any other curve of the cluster.

Note 2. The simplest case of all, when the algebraical roots of $F = 0$, $G = 0$, satisfy also $J = 0$, is that in which the multiplicity of a root-pair is merely double. For each such root-pair, two branches of the cluster of curves touch: for the two branches, being then part of one and the same curve, the point is a cusp. There then are as many cusps as there are root-pairs of double multiplicity; the cuspidal tangents are different from one another in direction, and they are different from the tangents to all the curves of the cluster arising through simple root-pairs. The focal surface then is a locus of cusps of the integral twisted curves.

Note 3. In connection with the integral of the two differential equations considered, there are a couple of arbitrary quantities, viz. the values arbitrarily assigned to y and z for a specific value of x. If the integral can be obtained by quadratures, it will have the form

$$f(x, y, z, \alpha, \beta) = 0, \quad g(x, y, z, \alpha, \beta) = 0,$$

where α and β are the arbitrary constants in the integration, determinable by the conditions

$$f(a, b, c, \alpha, \beta) = 0, \quad g(a, b, c, \alpha, \beta) = 0.$$

The equations

$$f(x, y, z, \alpha, \beta) = 0, \quad g(x, y, z, \alpha, \beta) = 0,$$

define a congruence of curves ; the further equation

$$J = \frac{\partial (f, g)}{\partial (\alpha, \beta)} = 0$$

defines points on each curve of the congruence, which are called *focal points*. The locus of the focal points for all the curves of the congruence is called the focal surface, and its equation is obtained[*] by eliminating α and β between

$$f = 0, \quad g = 0, \quad J = 0.$$

The parallelism of the analytical processes, which lead to the surface $\Theta = 0$, is the ground on which that surface in the geometrical interpretation is called the focal surface of the differential equations : but the parallelism is restricted to the analytical processes, and does not, in general, extend to the relations between the curves and the focal surface.

Note 4. The whole discussion of the integral of the equation has been made to depend upon the assignment of initial values and their relation to $\Theta = 0$. Nothing has yet been indicated as to singular solutions, if they exist : they will be considered immediately after some examples of the preceding general theory have been given.

---

[*] Darboux, *Theorie générale des surfaces*, t. ii, pp. 1—6; Goursat, *American Journ. Math.*, t. xi, pp. 343—344.

Method of determining a solution when the initial values make the equations evanescent.

195. In all the preceding forms, a tacit assumption has been made that the values of β and γ can be actually determined in association with assigned initial values a, b, c. This assumption would be justified by the event when a, b, c are parametric values, which have no special relation to the form of the differential equations; but when they have particular values, it may happen that many instances can arise—allusion has been made to some of them—in which the determination is difficult or even impossible by the methods suggested. Such an instance would occur when the values assigned, as initial values for x, y, z, make the equation evanescent without regard to the values of β and γ.

The method of space-diagrams (§ 157) can be applied to this case, in a manner similar to that by which (plane) Puiseux-diagrams were applied to the corresponding situation in the case of a single equation. For simplicity, we assume that 0, 0, 0 are the initial values assigned to the variables: in effect, this is merely replacing them by $x - a$, $y - b$, $z - c$ respectively. In order to secure the most general form of equations, we take them to be

$$\left. \begin{array}{l} \Sigma\Sigma\Sigma\Sigma\Sigma\, A_{r m_1 m_2 n_1 n_2} x^r y^{m_1} p^{m_2} z^{n_1} q^{n_2} = 0 \\ \Sigma\Sigma\Sigma\Sigma\Sigma\, B_{r' m_1' m_2' n_1' n_2'} x^{r'} y^{m_1'} p^{m_2'} z^{n_1'} q^{n_2'} = 0 \end{array} \right\} ;$$

solutions y and z of these equations are required, which vanish when $x = 0$. We shall assume that there is no term in either equation which is free from all the variables; for when there are constant terms in the equations, the discussion at the beginning of this chapter is adequate for the full determination of the integrals.

Since y and z, if they exist, are to vanish with x, it may be that, for sufficiently small values of $|x|$, they can be asymptotically represented by their most important terms in the forms

$$y = \rho x^\lambda, \quad z = \sigma x^\mu,$$

where λ and μ are quantities having their real parts positive. When the method is effective, it determines λ and μ as real positive quantities; should such determination be impossible, the inference is that integrals of the form indicated do not exist for sufficiently small values of $|x|$. The values of p and q are

$$p = \lambda \rho x^{\lambda - 1}, \quad q = \mu \sigma x^{\mu - 1};$$

but these are not necessarily finite or zero when $x = 0$. When these values are substituted, the order of the typical term in the summation in the first equation is

$$r - m_2 - n_2 + \lambda (m_1 + m_2) + \mu (n_1 + n_2),$$

$= M$, say; we suppose that the order of no term in the equation after substitution is lower than M, though several terms may be of this order. Gathering together the terms which involve the same power of x, we have

$$x^M \sum\sum\sum\sum A_{r m_1 m_2 n_1 n_2} \lambda^{m_2} \rho^{m_1 + m_2} \mu^{n_2} \sigma^{n_1 + n_2} + \text{higher powers of } x = 0,$$

where the summation in the coefficient of x^M extends over all those terms in the original equation for which the substitutions give the lowest index M.

Similarly, for the second equation, let

$$r' - m_2' - n_2' + \lambda (m_1' + m_2') + \mu (n_1' + n_2'),$$

$= N$, say, be the lowest index for the substitutions, so that the equation takes the form

$$x^N \sum\sum\sum\sum B_{r' m_1' m_2' n_1' n_2'} \lambda^{m_2'} \rho^{m_1' + m_2'} \mu^{n_2'} \sigma^{n_1' + n_2'} + \text{higher powers of } x = 0.$$

To obtain the proper values of λ and μ, we take three perpendicular coordinate axes; we mark all the points

$$X = r - m_2 - n_2, \quad Y = m_1 + m_2, \quad Z = n_1 + n_2,$$

for the first equation, and all the points

$$X = r' - m_2' - n_2', \quad Y = m_1' + m_2', \quad Z = n_1' + n_2',$$

for the second equation. By means of these points, we construct a configuration, as in § 157, consisting of a broken-plane figure everywhere convex to the origin. Each plane portion determines positive quantities λ and μ by its direction-coordinates, and contains all the points of the tableau, which correspond to the terms giving rise to the power of x that is lowest for the substitutions

$$y \propto x^\lambda, \quad z \propto x^\mu.$$

Having thus obtained λ and μ, the values of ρ and σ are given by

$$\left. \begin{array}{l} \sum\sum\sum\sum A_{r m_1 m_2 n_1 n_2} \lambda^{m_2} \rho^{m_1 + m_2} \mu^{n_2} \sigma^{n_1 + n_2} = 0 \\ \sum\sum\sum\sum B_{r' m_1' m_2' n_1' n_2'} \lambda^{m_2'} \rho^{m_1' + m_2'} \mu^{n_2'} \sigma^{n_1' + n_2'} = 0 \end{array} \right\},$$

where the summation extends over the terms that give rise to the index M. Denoting a pair of non-zero roots by ρ_1, σ_1, we write

$$y = x^\lambda (\rho_1 + u), \quad z = x^\mu (\sigma_1 + v),$$

substitute in the original equations, and proceed in the customary manner to determine the critical characteristics of u and v.

If the resulting equations for u and v indicate that functions and v exist, which vanish with x, branches of the integrals will thus arise, in connection with each root-pair ρ_1 and σ_1, and with the associated values of λ and μ. In similar circumstances, groups of branches will arise for each determination of λ and μ.

The tableau of noted points may be such that definite determinations of λ and μ are not possible. It then frequently occurs that the equations, which determine ρ and σ in general, cease to determine those quantities, but may give values of λ and μ; then the quantities ρ_1 and σ_1 are arbitrary.

Examples.

196. *Ex.* 1. Consider the equations

$$x + py + qz = 0,$$
$$\kappa^2 \{(qx - z)^2 + (px - y)^2\} = (qy - pz)^2.$$

Taking

$$F = \kappa^2 \{(qx - z)^2 + (px - y)^2\} - (qy - pz)^2,$$
$$G = x + py + qz,$$

we have

$$\tfrac{1}{2} J = \tfrac{1}{2} \frac{\partial (F,\, G)}{\partial (p,\, q)}$$
$$= (qy - pz)(y^2 + z^2 - \kappa^2 x^2),$$

after slight reduction. Solving $F = 0$, $G = 0$ for p and q, we find

$$qy - pz = \kappa \frac{x^2 + y^2 + z^2}{(y^2 + z^2 - \kappa^2 x^2)^{\frac{1}{2}}},$$

so that

$$\tfrac{1}{2} J = \kappa (x^2 + y^2 + z^2)(y^2 + z^2 - \kappa^2 x^2)^{\frac{1}{2}},$$

and therefore

$$\Theta = (x^2 + y^2 + z^2)^2 (y^2 + z^2 - \kappa^2 x^2).$$

There are two sets of values of p and q; there are therefore two branches of the integral function. Let arbitrary initial values b and c be assigned to y and z, when $x = a$.

If a, b, c be such that neither of the equations

$$a^2 + b^2 + c^2 = 0,$$
$$b^2 + c^2 - \kappa^2 a^2 = 0,$$

is satisfied, then the two branches of the integral are distinct from one another; they are given by the equations

$$\left.\begin{aligned}x^2+y^2+z^2 &=a^2+b^2+c^2\\(cy-bz)^2 &=\kappa^2\{(az-cx)^2+(bx-ay)^2\}\end{aligned}\right\}.$$

If a, b, c be such that the equation

$$a^2+b^2+c^2=0$$

is satisfied, the two branches of the integral touch at the point a, b, c: in fact, they become one and the same, given by

$$\left.\begin{aligned}x^2+y^2+z^2 &=0\\ax+by+cz &=0\end{aligned}\right\}.$$

If a, b, c be such that the equation

$$b^2+c^2-\kappa^2a^2=0$$

is satisfied, the two branches of the integral touch at the point a, b, c: in fact, they become one and the same, given by

$$\left.\begin{aligned}x^2+y^2+z^2 &=a^2+b^2+c^2\\by+cz-\kappa^2ax &=0\end{aligned}\right\}.$$

The geometrical interpretation, that arises when x, y, z, a, b, c all are real, is quite simple. Noting that the equation

$$(cy-bz)^2=\kappa^2\{(az-cx)^2+(bx-ay)^2\}$$

is equivalent to the two linear equations

$$\frac{\kappa^2}{b^2}\left(\frac{x}{a}-\frac{z}{c}\right)-\frac{1}{a^2}\left(\frac{y}{b}-\frac{z}{c}\right)=\frac{\pm\kappa}{abc}\left(\frac{x}{a}-\frac{y}{b}\right)(b^2+c^2-\kappa^2a^2)^{\frac{1}{2}},$$

we see that, in the most general case, the two integral curves through the point a, b, c are the sections of the sphere

$$x^2+y^2+z^2=a^2+b^2+c^2$$

by these two planes through its centre. In connection with this sphere, construct the cone

$$y^2+z^2-\kappa^2x^2=0:$$

the intersection of the sphere and the cone is composed of two small circles. When the point a, b, c on the surface of the sphere lies in the belt between the small circles, the integral curves are the two great circles drawn through a, b, c touching the small circles: this is the first case. When the point a, b, c lies outside the belt, the integral curves become imaginary; they are conjugate imaginaries; but the impossibility of drawing them, in spite of their functional existence, is one drawback in limiting the variables to have only real values. When the point a, b, c lies on the cone, and therefore on one or the other of the two small circles, the integral curve becomes the single great circle touching the small circle at the point (and touching also the other small circle at the diametrically opposite point): the tangent to the great circle is a tangent-line of the cone. This is the third case. When the point a, b, c is the origin, the integral curve becomes a single point-circle at the origin: this is the second case.

Ex. 2. Consider the equations

$$\left. \begin{aligned} y &= px + \lambda pq \\ z &= qx + \mu pq \end{aligned} \right\},$$

where λ and μ are constants.

The values of p and q are

$$\left. \begin{aligned} 2\lambda qx &= -x^2 - \mu y + \lambda z + \Theta^{\frac{1}{2}} \\ 2\mu px &= -x^2 + \mu y - \lambda z + \Theta^{\frac{1}{2}} \end{aligned} \right\},$$

where

$$\Theta = (x^2 + \mu y + \lambda z)^2 - 4\lambda\mu yz.$$

Also taking

$$F = -y + px + \lambda pq, \quad G = -z + qx + \mu pq,$$

we have

$$J = \frac{\partial(F, G)}{\partial(p, q)}$$
$$= x(x + q\lambda + p\mu)$$
$$= \Theta^{\frac{1}{2}}.$$

The values of p and q which make $J=0$, there being two sets of them, must be determined by initial values satisfying

$$\Theta = 0,$$

which accordingly is the focal equation.

If a, b, c be initial values such that

$$(a^2 + \mu b + \lambda c)^2 - 4\lambda\mu bc$$

is not zero, then two branches of the integral of the equation are given by

$$\left. \begin{aligned} y &= Ax + \lambda AB \\ z &= Bx + \mu AB \end{aligned} \right\},$$

where A and B are determined by the equations

$$\left. \begin{aligned} b &= Aa + \lambda AB \\ c &= Ba + \mu AB \end{aligned} \right\}.$$

But if a, b, c satisfy the equation

$$(a^2 + \mu b + \lambda c)^2 - 4\lambda\mu bc = 0,$$

the two branches are one and the same. The values of A and B are given in the former case by

$$2\lambda Ba = -a^2 - \mu b + \lambda c + \{(a^2 + \mu b + \lambda c)^2 - 4\lambda\mu bc\}^{\frac{1}{2}},$$
$$2\mu Aa = -a^2 + \mu b - \lambda c + \{(a^2 + \mu b + \lambda c)^2 - 4\lambda\mu bc\}^{\frac{1}{2}},$$

the same irrational value being taken for A and B in each determination, and the two irrational values leading to the two determinations. The single set of values of A and B in the latter case is given by

$$2\lambda Ba = -a^2 - \mu b + \lambda c,$$
$$2\mu Aa = -a^2 + \mu b - \lambda c.$$

The geometrical interpretation can be obtained as before, when x, y, z have a, b, c, as assigned initial values.　We have

$$\Theta = (x^2 + \mu y + \lambda z)^2 - 4\lambda\mu yz = 0,$$

which is a quartic surface.　The equations

$$\left.\begin{array}{l} y = Ax + \lambda AB \\ z = Bx + \mu AB \end{array}\right\}$$

represent a straight line; this straight line touches the surface $\Theta = 0$ in the point

$$x = 0, \quad y = \lambda AB, \quad z = \mu AB,$$

and in the point

$$x = -\mu A - \lambda B, \quad y = -\mu A^2, \quad z = -\lambda B^2:$$

that is, it is a bitangent of the quartic surface.

When the point a, b, c, through which the integral (line) curve passes, lies off the surface, there are two sets of values of A and B; that is, two different bitangents can be drawn to the surface through a point, which does not lie on it; and their equations satisfy the differential equations.

When the point a, b, c, lies on the surface, there is only one set of values of A and B; only one bitangent can be drawn to the surface through the point and satisfying the equations; it touches the surface in the initial point a, b, c, and in the point

$$x = 0, \quad y = \frac{1}{2\mu}(a^2 + \mu b + \lambda c), \quad z = \frac{1}{2\lambda}(a^2 + \mu b + \lambda c).$$

Ex. 3.　Consider the equations

$$\left.\begin{array}{l} F = y - xp = 0 \\ G = x^2 q^2 - (\lambda x^2 + \mu y^2)^2 (\lambda x^2 + \mu y^2 + \rho) = 0 \end{array}\right\},$$

where λ, μ, ρ are constants.

The values of p and q are immediately obtainable: there are, in general, two sets of them.　Also

$$J = \frac{\partial(F, G)}{\partial(p, q)} = -2x^3 q,$$

so that the values of p and q, which make $J = 0$ as well as $F = 0$, $G = 0$, are given by

$$q = 0, \quad p = \frac{y}{x},$$

and, at the same time,

$$\Theta = (\lambda x^2 + \mu y^2)^2 (\lambda x^2 + \mu y^2 + \rho) = 0.$$

The integral equations are

$$\left.\begin{array}{l} y = Ax \\ (3z - B)^2 = (\lambda x^2 + \mu y^2 + \rho)^3 \end{array}\right\},$$

where the arbitrary constants A and B are determined by means of the values b and c, which are assigned to y and z, when $x = a$, and satisfy the equations

$$\left.\begin{array}{l} b = Aa \\ (3c - B)^2 = (\lambda a^2 + \mu b^2 + \rho)^3 \end{array}\right\}.$$

If the initial values a, b, c be such that neither of the equations

$$\lambda a^2 + \mu b^2 \quad\ = 0,$$

$$\lambda a^2 + \mu b^2 + \rho = 0,$$

is satisfied, there are two branches of the integral of the equation.

If the values a, b, c be such as to satisfy

$$\lambda a^2 + \mu b^2 = 0,$$

the two branches of the integral of the equation have the same values of p and q for x, y, $z = a$, b, c: and these values satisfy the equation in p and q derived from

$$\lambda x^2 + \mu y^2 = 0,$$

for the initial values of the variables.

If the values a, b, c be such as to satisfy

$$\lambda a^2 + \mu b^2 + \rho = 0,$$

the two branches of the integral of the equation have the same values of p and q for x, y, $z = a$, b, c: but these values do not satisfy the equation in p and q derived from

$$\lambda x^2 + \mu y^2 + \rho = 0.$$

The geometrical interpretation is obvious. The integral curves are plane sextics. Through any point in space, not lying on the cylinder

$$\lambda x^2 + \mu y^2 + \rho = 0,$$

nor on either of the two planes

$$\lambda x^2 + \mu y^2 = 0,$$

two sextics can be drawn not touching one another and satisfying the equation.

When a point is taken on either of the planes

$$\lambda x^2 + \mu y^2 = 0,$$

two sextics can be drawn through it touching one another at the point: their common tangent lies in the plane.

When a point is taken on the cylinder

$$\lambda x^2 + \mu y^2 + \rho = 0,$$

the two sextics become branches of the same curve having the point for a cusp: the cuspidal tangent does not lie in the tangent-plane of the cylinder at the point.

The cylinder

$$\lambda x^2 + \mu y^2 + \rho = 0$$

is, in fact, a locus of cusps of the sextics, which satisfy the differential equations.

Ex. 4. Discuss similarly the equations

$$\left.\begin{array}{l} y - xp = 0 \\ x^2 q^2 = x^2 + y^2 - 1 \end{array}\right\}.$$

(Goursat.)

Ex. 5. As an instance of equations, which are rendered evanescent by the assignment of initial conditions, so that a merely algebraical solution for associable values of p and q ceases to be possible, consider

$$\left.\begin{array}{r} hx^3 + gx^2p + fy^2zpq + x^2yp^2q = 0 \\ ay^3 + bxyq + cz^2p + y^2zpq^2 = 0 \end{array}\right\},$$

with the initial conditions that $y = 0$ and $z = 0$ when $x = 0$. The points to be marked in the space-diagram are

for the first equation $3, 0, 0$; $1, 1, 0$; $-2, 3, 2$; $-1, 3, 1$:

say A ; B ; C ; D :

and for the second equation $0, 3, 0$; $0, 1, 1$; $-1, 1, 2$; $-3, 3, 3$:

say E , F , G , H .

On constructing the tableau in perspective, it is easy to see that the points B, F, G lie in one straight line: that the points E, D, C, H lie in one straight line: and that these two straight lines are parallel. There are accordingly only two plane portions to take account of, viz. the plane through the lines $HCDE, EB, BFG$, its equation being

$$\xi + \tfrac{1}{2}\eta + \zeta = \tfrac{3}{2},$$

and the plane through the lines GFB, BA, its equation being

$$\xi + 2\eta + \zeta = 3.$$

Hence there are two sets of values of λ and μ, viz.

$$\left.\begin{array}{l}\lambda = \tfrac{1}{2} \\ \mu = 1\end{array}\right\}, \quad \left.\begin{array}{l}\lambda = 2 \\ \mu = 1\end{array}\right\};$$

and the terms giving rise to the lowest terms for the respective substitutions are those, which correspond to points in the tableau lying on the planes that determine the substitutions.

First, consider the orders $\lambda = 2$, $\mu = 1$; so that the most important terms in y and z are

$$y = \rho x^2, \quad z = \sigma x,$$

ρ and σ being constants different from zero. Substituting in the two equations, retaining only the lowest powers of x in each (they are the third powers for each), and equating their coefficients to zero, we have

$$h + 2\rho g = 0, \quad b + 2\sigma c = 0,$$

so that

$$\rho = -\frac{h}{2g}, \quad \sigma = -\frac{b}{2c}.$$

Now take

$$y = x^2(\rho + Y), \quad z = x(\sigma + Z),$$

where ρ and σ have the values just obtained, and Y, Z must vanish when $x = 0$: so that

$$p = x\left(2\rho + 2Y + x\frac{dY}{dx}\right),$$

$$q = \sigma + Z + x\frac{dZ}{dx},$$

and $x\dfrac{dY}{dx}$, $x\dfrac{dZ}{dx}$ are of the same orders in x as Y and Z respectively. Substituting in the first of the equations, dividing by x^3, and reducing, we have

$$x\frac{dY}{dx} + 2Y + \frac{1}{16}\frac{bh^3}{c^2g^4}(4c - bf)x^3 + \text{higher terms} = 0.$$

Substituting in the second equation and proceeding in a similar way, we find

$$x\frac{dZ}{dx} - Z - \frac{1}{2}\frac{bg}{ch}x\frac{dY}{dx} + \frac{1}{4}\frac{h^2}{bg^2}\left(a - \frac{b^3}{4c^3}\right)x^3 + \text{higher terms} = 0.$$

The other terms in each of these forms are regular functions of x, $x\dfrac{dZ}{dx}$, $x\dfrac{dY}{dx}$, Y, Z; the orders of these terms as small quantities, when $|x|$ is sufficiently small, are manifestly higher than the orders of the terms retained. Solving the modified equations, we find

$$\left.\begin{aligned}
x\frac{dY}{dx} &= \frac{1}{16}\frac{bh^3}{c^2g^4}(bf-4c)x^3 - 2Y + \theta(x,\ Y,\ Z)\\
x\frac{dZ}{dx} &= \left\{\frac{1}{4}\frac{h^2}{bg^2}\left(\frac{b^3}{4c^3} - a\right) + \frac{1}{32}\frac{b^2h^2}{c^3g^3}(bf-4c)\right\}x^3 - \frac{bg}{ch}Y + Z + \phi(x,\ Y,\ Z)
\end{aligned}\right\},$$

where θ and ϕ are regular functions of their arguments, which contain no term in x alone of order less than four, and no term in Y and Z free from x of order less than two.

These equations are of the class considered in the last chapter. The critical quadratic is

$$(\Omega + 2)(\Omega - 1) = 0,$$

so that one root is unity and the other is -2; and the initial conditions are that $Y = 0$, $Z = 0$. It therefore follows

(i) that the equations possess integrals, which are regular functions of x and vanish with x; their expressions are

$$Y = \frac{1}{80}\frac{bh^3}{c^2g^4}(bf - 4c)x^3 P(x),$$

$$Z = \left\{\frac{1}{8}\frac{h^2}{bg^2}\left(\frac{b^3}{4c^3} - a\right) + \frac{3}{320}\frac{b^2h^2}{c^3g^3}(bf - 4c)\right\}x^3 Q(x),$$

where $P(x)$ and $Q(x)$ are regular functions of x, which become equal to 1 when x vanishes :

(ii) that the equations possess an infinitude of integrals, which vanish with x, and are regular functions of x and $x\log x$.

The corresponding integrals of the original equations, which vanish with x, are

$$y = x^2\left(-\frac{h}{2g} + Y\right), \qquad z = x\left(-\frac{b}{2c} + Z\right).$$

Secondly, consider the orders $\lambda = \tfrac{1}{2}$, $\mu = 1$. Take

$$x = t^2,$$

so that, when y and z are expressed in terms of t, their most important terms for sufficiently small values of $|t|$ are

$$y = \rho t, \qquad z = \sigma t^2,$$

where ρ and σ are constants : and then

$$p = \tfrac{1}{2}\frac{\rho}{t}, \qquad q = \sigma.$$

Substituting in the equations, retaining only the lowest power of t—it is t^3 in each case—and making the coefficients vanish, we have

$$\tfrac{1}{2}g\rho + \tfrac{1}{2}f\rho^3\sigma^2 + \tfrac{1}{4}\rho^3\sigma = 0,$$

from the first equation, and

$$a\rho^3 + b\rho\sigma + \tfrac{1}{2}c\rho\sigma^2 + \tfrac{1}{2}\rho^3\sigma^3 = 0,$$

from the second. Rejecting zero values of the coefficients ρ and σ, we find

$$\frac{g}{f\sigma^2 + \tfrac{1}{2}\sigma} = -\rho^2 = \frac{b\sigma + \tfrac{1}{2}c\sigma^2}{a + \tfrac{1}{2}\sigma^3} \; ;$$

so that there are four values of σ, being the roots of

$$\sigma^2(c\sigma + 2b)(2f\sigma + 1) - 2g(\sigma^3 + 2a) = 0,$$

and there are two values of ρ for each value of σ.

Denoting by ρ and σ any one of these sets of simultaneous values, let

$$y = t(\rho + Y), \qquad z = t^2(\sigma + Z),$$

so that

$$p = \frac{1}{2t}\left(\rho + Y + t\frac{dY}{dt}\right),$$

$$q = \sigma + Z + \tfrac{1}{2}t\frac{dZ}{dt}.$$

After substitution and reduction, the first equation gives

$$4ht^3 - 4gY + \rho^2\sigma t\frac{dY}{dt} - g\frac{\rho}{\sigma}t\frac{dZ}{dt} + \rho^3 Z(4f\sigma + 1) + \text{higher terms} = 0 \; ;$$

and the second equation gives

$$(c + \rho^2\sigma)\sigma^2 t\frac{dY}{dt} + (b + \rho^2\sigma^2)\rho t\frac{dZ}{dt}$$

$$- \sigma Y(4b + 2c\sigma) + \rho Z(2b + 2c\sigma + 3\rho^2\sigma^2) + \text{higher terms} = 0,$$

there being no term involving t only in the latter. Solving these for $t\dfrac{dY}{dt}$ and $t\dfrac{dZ}{dt}$, we find

$$\left.\begin{aligned}
t\frac{dY}{dt} &= aY + \beta Z + \kappa t^3 + \theta_1(Y, Z) \\
t\frac{dZ}{dt} &= \gamma Y + \delta Z + \lambda t^3 + \theta_2(Y, Z)
\end{aligned}\right\},$$

where θ_1 and θ_2 are regular functions of their arguments, which contain no terms of dimensions less than 2 ; and a, β, γ, δ, κ, λ are constants, which depend upon the coefficients of the original equations.

10—2

Hence, in general, solutions of these equations exist, which are regular functions of t and vanish with t: a result which holds for each of the (eight) sets of values of ρ and σ. Consequently there are in general eight sets of integrals of the original equations, which can be arranged in four pairs of sets: they are regular functions of $x^{\frac{1}{2}}$, and they vanish with x. The forms of a, β, γ, δ determine the possibility of the existence of non-regular solutions of the equations, which vanish with x.

Ex. 6.   Obtain integrals of the equations

$$\begin{aligned}
ay^2z^3 + (y^2q^2 + z^2p^2)\,a^2 &= x^4 \\
\beta y^2 z^3 + \tfrac{5}{7}\,(yq + zp)\,a &= x^2
\end{aligned} \Bigg\},$$

such as to vanish when $x = 0$; likewise of the equations

$$\begin{aligned}
x^2(p^2 + q^2 + 1) &= ayz + bzx + cxy \\
ayq + \beta zp &= x
\end{aligned} \Bigg\},$$

and of the equations

$$\begin{aligned}
az^2p^2 + 2bzypq + cy^2q^2 &= x^2 \\
ay^3p^3 + 3\beta y^2zp^2q + 3\gamma yz^2pq^2 + \delta z^3q^3 &= x^4
\end{aligned} \Bigg\},$$

in each case with the same initial conditions.

Singular Solutions of Systems of Equations.

197.   It has been seen that, if values of p and q given by

$$F = 0, \quad G = 0,$$

are such as to satisfy

$$J = \frac{\partial(F,\,G)}{\partial(p,\,q)} = 0$$

at a particular point, then two or more integral curves touch at the point; and that, if $\Theta(x,\,y,\,z) = 0$, be the result of eliminating p and q between $F = 0$, $G = 0$, $J = 0$, then the common tangent of those integral curves does not, in general, lie in the tangent-plane to $\Theta = 0$ at the point.

It may, however, happen that the common tangent does lie in the tangent-plane to $\Theta = 0$ at the point; then, at the point, $\Theta = 0$ provides an integral of the original equations, although it does not provide a full solution, because two equations are necessary for that purpose.   It might happen that, at every point on $\Theta = 0$, the integral curves touch, having their common tangent in the tangent-plane of $\Theta = 0$ there; in that case $\Theta = 0$ is an envelope of curves, and it provides an integral of the original equations, though (for the same reason as before) it does not provide the full solution. It might even happen that $\Theta = 0$ is a surface on which integral curves lie.

It thus is clear that, in some cases or under some conditions, integrals of the equations exist, which are not included in the set hitherto considered. Some of the integrals indicated may be of the class of particular integrals; others will be of a class which are called *singular*. We proceed to their consideration.

Instead of taking Θ, which (§ 194) is a product of values of J for the sets of values of p and q satisfying $F = 0$ and $G = 0$, and which therefore vanishes when J vanishes, we take

$$F = 0, \quad G = 0, \quad J = 0,$$

as the three equations. When the values of p and q, given by $F = 0$ and $G = 0$, are substituted in these two equations, they then are satisfied identically; but $J = 0$ is not necessarily satisfied identically, and usually it is satisfied only because the point at which the values are obtained lies on $\Theta = 0$, say, in consequence of $\Theta = 0$. Writing

$$\left.\begin{aligned}
\frac{dF}{dx} &= \frac{\partial F}{\partial x} + \frac{dy}{dx}\frac{\partial F}{\partial y} + \frac{dz}{dx}\frac{\partial F}{\partial z} \\
\frac{dG}{dx} &= \frac{\partial G}{\partial x} + \frac{dy}{dx}\frac{\partial G}{\partial y} + \frac{dz}{dx}\frac{\partial G}{\partial z} \\
F_p &= \frac{\partial F}{\partial p}, \quad F_q = \frac{\partial F}{\partial q} \\
G_p &= \frac{\partial G}{\partial p}, \quad G_q = \frac{\partial G}{\partial q}
\end{aligned}\right\},$$

we have

$$\left.\begin{aligned}
U = 0 &= \frac{dF}{dx} + F_p\frac{dp}{dx} + F_q\frac{dq}{dx} \\
V = 0 &= \frac{dG}{dx} + G_p\frac{dp}{dx} + G_q\frac{dq}{dx}
\end{aligned}\right\},$$

and therefore

$$\left.\begin{aligned}
G_qU - F_qV &= \frac{dF}{dx}G_q - \frac{dG}{dx}F_q + J\frac{dp}{dx} \\
-G_pU + F_pV &= -\frac{dF}{dx}G_p + \frac{dG}{dx}F_p + J\frac{dq}{dx}
\end{aligned}\right\}.$$

Now let the values p and q satisfy $J = 0$: and assuming the most general case, let

$$F_q = \Omega F_p, \quad G_q = \Omega G_p,$$

where Ω is not zero, so that

$$\frac{dF}{dx} G_q - \frac{dG}{dx} F_q = \Omega \left(\frac{dF}{dx} G_p - \frac{dG}{dx} F_p \right).$$

Then the two equations, from which the terms $J\frac{dp}{dx}$, $J\frac{dq}{dx}$ respectively have disappeared, become one only, say

$$\frac{dF}{dx} G_p - \frac{dG}{dx} F_p = 0.$$

Let

$$\left. \begin{aligned} F_x &= \frac{\partial F}{\partial x} + p\frac{\partial F}{\partial y} + q\frac{\partial F}{\partial z} \\ G_x &= \frac{\partial G}{\partial x} + p\frac{\partial G}{\partial y} + q\frac{\partial G}{\partial z} \end{aligned} \right\} ;$$

and let

$$K = F_x G_p - G_x F_p.$$

Hence if values of y and z are such that the values of their derivatives, given by $F = 0$ and $G = 0$, satisfy also $J = 0$, then they satisfy also the equation

$$K = 0.$$

198. The converse of this result is not necessarily valid: it cannot be claimed that quantities p and q, which algebraically satisfy the equations

$$F = 0, \quad G = 0, \quad J = 0, \quad K = 0,$$

are necessarily such as to be derivatives of the variables y and z with regard to x. In order to determine their relations to the solution of the equation, we suppose them substituted in the equations $F = 0$ and $G = 0$, which are satisfied identically: and in $J = 0$, which is satisfied, but in such a way as to admit of derivation with regard to x, so as to give a relation among the derivatives. We thus have

$$\frac{dF}{dx} + F_p \frac{dp}{dx} + F_q \frac{dq}{dx} = 0,$$

$$\frac{dG}{dx} + G_p \frac{dp}{dx} + G_q \frac{dq}{dx} = 0,$$

$$\frac{dJ}{dx} + J_p \frac{dp}{dx} + J_q \frac{dq}{dx} = 0.$$

The first two equations lead to the single equation, free from $\dfrac{dp}{dx}$ and $\dfrac{dq}{dx}$, as already obtained in the form

$$\frac{dF}{dx} G_p - \frac{dG}{dx} F_p = 0 ;$$

the third equation, combined with either of the first two equations, then gives the values of $\dfrac{dp}{dx}$ and $\dfrac{dq}{dx}$. In other words, in order that the three equations may be consistent in determining $\dfrac{dp}{dx}$ and $\dfrac{dq}{dx}$, we must have

$$\begin{vmatrix} \dfrac{dF}{dx}, & F_p, & F_q \\[2mm] \dfrac{dG}{dx}, & G_p, & G_q \\[2mm] \dfrac{dJ}{dx}, & J_p, & J_q \end{vmatrix} = 0,$$

so that

$$\frac{dF}{dx}(G_p J_q - G_q J_p) + \frac{dG}{dx}(J_p F_q - J_q F_p) + J\frac{dJ}{dx} = 0.$$

Now

$$G_p J_q - G_q J_p = G_p(J_q - \Omega J_p),$$
$$J_p F_q - J_q F_p = F_p(-J_q + \Omega J_p),$$

and $J = 0$: so that the equation is

$$\left(G_p \frac{dF}{dx} - F_p \frac{dG}{dx}\right)(J_q - \Omega J_p) = 0,$$

which is satisfied in virtue of

$$G_p \frac{dF}{dx} - F_p \frac{dG}{dx} = 0.$$

Substituting the full expressions for $\dfrac{dF}{dx}$ and $\dfrac{dG}{dx}$, the equation $\dfrac{dF}{dx} G_p - \dfrac{dG}{dx} F_p = 0$ becomes

$$0 = \frac{\partial(F, G)}{\partial(x, p)} + \frac{dy}{dx}\frac{\partial(F, G)}{\partial(y, p)} + \frac{dz}{dx}\frac{\partial(F, G)}{\partial(z, p)}.$$

Also

$$0 = K = F_x G_p - G_x F_p$$
$$= \frac{\partial(F, G)}{\partial(x, p)} + p\frac{\partial(F, G)}{\partial(y, p)} + q\frac{\partial(F, G)}{\partial(z, p)},$$

and therefore

$$\left(\frac{dy}{dx} - p\right) \frac{\partial (F, G)}{\partial (y, p)} + \left(\frac{dz}{dx} - q\right) \frac{\partial (F, G)}{\partial (z, p)} = 0.$$

This is the only equation of the kind that can be obtained: it does not prove that $\frac{dy}{dx} = p$ and $\frac{dz}{dx} = q$: but it proves that, if one of these equalities holds, then the other necessarily holds. It is therefore necessary to establish one of the equalities.

Now from $F = 0$, $G = 0$, $J = 0$, it is possible, in general, to eliminate any two of the five variables. Instead of eliminating p and q, so as to obtain $\Theta = 0$, let q and z be eliminated: and let the result be

$$H(x, y, p) = 0,$$

so that a value of p (derived from $H = 0$), with the associated value of q, satisfies

$$G_p \frac{dF}{dx} - F_p \frac{dG}{dx} = 0,$$

and therefore also

$$\left(\frac{dy}{dx} - p\right) \frac{\partial (F, G)}{\partial (y, p)} + \left(\frac{dz}{dx} - q\right) \frac{\partial (F, G)}{\partial (z, p)} = 0.$$

The equation $H = 0$ thus far is only an algebraical consequence of the three equations

$$F = 0, \quad G = 0, \quad J = 0,$$

which involve x, y, z, p, q: and it is satisfied without any regard to (possible) functional relations between p, q, y, z. In order that $H = 0$ may have an added significance, we associate with it the equation

$$\frac{dy}{dx} - p = 0,$$

that is, we postulate the differential equation

$$H\left(x, y, \frac{dy}{dx}\right) = 0,$$

and we infer from this equation, and from $H(x, y, p) = 0$, the relation

$$\frac{dy}{dx} - p = 0.$$

Consequently, also $\dfrac{dz}{dx} = q$: so that the quantities p and q then are the derivatives of y and z with regard to x. Hence an integral of

$$H\left(x, y, \frac{dy}{dx}\right) = 0,$$

and the consequent value of z given by

$$\Theta\,(x, y, z) = 0,$$

constitute a solution of the original equations. It is a singular solution: when the most general integral of $H = 0$ is used, this singular solution involves one arbitrary constant. Accordingly, we have the theorem:

If values of p and q satisfy the equations

$$F\,(x, y, z, p, q) = 0, \quad G\,(x, y, z, p, q) = 0,$$

$$0 = \frac{\partial\,(F,\,G)}{\partial\,(x,\,p)} + p\,\frac{\partial\,(F,\,G)}{\partial\,(y,\,p)} + q\,\frac{\partial\,(F,\,G)}{\partial\,(z,\,p)},$$

$$0 = \frac{\partial\,(F,\,G)}{\partial\,(x,\,q)} + p\,\frac{\partial\,(F,\,G)}{\partial\,(y,\,q)} + q\,\frac{\partial\,(F,\,G)}{\partial\,(z,\,q)},$$

$$0 = J = \frac{\partial\,(F,\,G)}{\partial\,(p,\,q)},$$

the third and fourth of which are equivalent to one another in virtue of $J = 0$, then the equations $F = 0$, $G = 0$ possess a singular solution involving an arbitrary constant; and the singular solution is constituted by the combination of the general integral of

$$H\left(x, y, \frac{dy}{dx}\right) = 0,$$

which is the eliminant (in z and q) of $F = 0$, $G = 0$, $J = 0$, with

$$\Theta\,(x, y, z) = 0,$$

which is the eliminant (in p and q) of the same three equations.

199. The analysis, leading to the proposition that has just been enunciated, shews that any solution of $H = 0$ (whether a complete integral or not), combined with $\Theta = 0$, satisfies the original equations. As $H = 0$ is an equation of the first order, it may possess a singular solution of its own, which of course is not included in its complete integral; and therefore solutions of the original equations, distinct from the singular solutions already

indicated, are constituted by the equation $\Theta = 0$ and the singular solution (if any) of the equation $H = 0$.

The conditions, which can be associated with the original equations, and are alike necessary and sufficient to secure the existence of this further class of singular solutions, are obtainable in simple form. That $H = 0$ should possess a singular integral, it is necessary and sufficient that the equations

$$H = 0, \quad \frac{\partial H}{\partial p} = 0, \quad \frac{\partial H}{\partial x} + \frac{\partial H}{\partial y} p = 0,$$

should be satisfied by values of y and p, the third equation securing that the values of y and p, as determined by the first two equations, satisfy the relation $\frac{dy}{dx} = p$. Now by the theory of elimination*, since H is an eliminant of F, G, J, we have

$$H = AF + BG + CJ,$$

where A, B, C, are functions of x, y, z, p, q; and this relation is an identity. Hence, as $H = 0$, we have

$$0 = AF + BG + CJ;$$

because $\dfrac{\partial H}{\partial p} = 0$, we have

$$0 = AF_p + BG_p + CJ_p,$$

the terms $F\dfrac{\partial A}{\partial p} + G\dfrac{\partial B}{\partial p} + J\dfrac{\partial C}{\partial p}$ vanishing, because the values of p and q satisfy $F = 0$, $G = 0$, $J = 0$; and because $H_x = \dfrac{\partial H}{\partial x} + p\dfrac{\partial H}{\partial y} = 0$, we have

$$0 = AF_x + BG_x + CJ_x,$$

the terms $FA_x + GB_x + JC_x$ vanishing, because F, G, J vanish for the values considered. From the last two, we find

$$0 = A\left(F_pG_x - G_pF_x\right) + C\left(J_pG_x - G_pJ_x\right),$$

or, because $F_pG_x - G_pF_x = 0$, we have

$$L = J_pG_x - G_pJ_x = 0,$$

a new equation.

---

* Cayley, *Coll. Math. Papers*, vol. I, pp. 370—374; Forsyth, *Phil. Trans.* (1883, I), pp. 324—329.

Further, we have

$$\frac{\partial H}{\partial p} = AF_p + BG_p + CJ_p :$$

and

$$0 = AF_q + BG_q + CJ_q,$$

because H is explicitly independent of q; hence

$$G_q \frac{\partial H}{\partial p} = AJ + C(J_p G_q - J_q G_p) = C(J_p G_q - J_q G_p),$$

$$F_q \frac{\partial H}{\partial p} = -BJ + C(J_p F_q - J_q F_p) = C(J_p F_q - J_q F_p).$$

Accordingly $\dfrac{\partial H}{\partial p} = 0$ is satisfied, if either of the equations

$$\frac{\partial(J, G)}{\partial(p, q)} = 0, \quad \frac{\partial(J, F)}{\partial(p, q)} = 0,$$

(equivalent to one another in virtue of $J = 0$) is satisfied: or as may easily be proved, if the equation

$$F_q(G_q{}^2 F_{pp} - 2G_q G_p F_{pq} + G_p{}^2 F_{qq}) = G_q(F_q{}^2 G_{pp} - 2F_p F_q G_{pq} + F_p{}^2 G_{qq})$$

is satisfied. This last equation, taken with $J = 0$, also leads to the (alternative) forms which are the simplest, viz.

$$\frac{F_q}{F_p} = \frac{G_q}{G_p} = \frac{J_q}{J_p} (= \Omega).$$

Thus the aggregate of equations is

$$F = 0, \quad G = 0, \quad J = 0, \quad K = 0,$$

$$L = 0, \quad \frac{J_q}{J_p} = \frac{G_q}{G_p} \left(= \frac{F_q}{F_p} \right):$$

and this aggregate must be satisfied, if a singular solution of the kind specified is possessed by the original equations.

200. We can prove that this condition is also sufficient. The values of p and q satisfy $F = 0$, $G = 0$ identically; they satisfy $J = 0$, in virtue of $H = 0$; hence

$$\frac{dF}{dx} + F_p \frac{dp}{dx} + F_q \frac{dq}{dx} = 0,$$

$$\frac{dG}{dx} + G_p \frac{dp}{dx} + G_q \frac{dq}{dx} = 0,$$

$$\frac{dJ}{dx} + J_p \frac{dp}{dx} + J_q \frac{dq}{dx} = 0.$$

From the first two, we have

$$\frac{dF}{dx}G_p - \frac{dG}{dx}F_p = J\frac{dq}{dx} = 0;$$

and from the second and third, we have

$$\frac{dG}{dx}J_p - \frac{dJ}{dx}G_p = (G_pJ_q - J_pG_q)\frac{dq}{dx} = 0.$$

Combining the former of these with $K = 0$, we have (after subtraction)

$$\frac{\partial(F, G)}{\partial(p, y)}\left(\frac{dy}{dx} - p\right) + \frac{\partial(F, G)}{\partial(p, z)}\left(\frac{dz}{dx} - q\right) = 0;$$

combining the latter of them with $L = 0$, we have (also after subtraction)

$$\frac{\partial(G, J)}{\partial(p, y)}\left(\frac{dy}{dx} - p\right) + \frac{\partial(G, J)}{\partial(p, z)}\left(\frac{dz}{dx} - q\right) = 0.$$

The determinant of the coefficients of $\frac{dy}{dx} - p$ and $\frac{dz}{dx} - q$ in the last two equations is

$$= \frac{\partial(F, G)}{\partial(p, y)}\frac{\partial(G, J)}{\partial(p, z)} - \frac{\partial(F, G)}{\partial(p, z)}\frac{\partial(G, J)}{\partial(p, y)},$$

$$= \frac{\partial G}{\partial p}\frac{\partial(F, G, J)}{\partial(p, y, z)},$$

which in general is distinct from zero*; hence

$$\frac{dy}{dx} - p = 0, \quad \frac{dz}{dx} - q = 0:$$

that is, the values of y, z, p, q provided by the equations constitute a solution of the original equations. We therefore have a theorem :—

If values of p, q, y, z are found to satisfy the equations

$$F(x, y, z, p, q) = 0, \quad G(x, y, z, p, q) = 0$$

$$J = \frac{\partial(F, G)}{\partial(p, q)} = 0$$

$$\frac{F_x}{F_p} = \frac{G_x}{G_p} = \frac{J_x}{J_p}, \quad \left(\frac{F_q}{F_p} = \right)\frac{G_q}{G_p} = \frac{J_q}{J_p}$$

* Even if it is zero, another combination of equations can be constructed such that the corresponding determinant is $\frac{\partial F}{\partial p}\frac{\partial(F, G, J)}{\partial(p, y, z)}$: this could be distinct from zero. But not all the cases are dealt with in the text.

then the values of y and z, as functions of x, constitute a singular solution of the equations $F = 0$, $G = 0$, distinct in character from the singular solutions, which are determined by the equations

$$F = 0, \quad G = 0, \quad J = 0, \quad F_x G_p - F_p G_x = 0,$$

and involve an arbitrary constant. They are called singular solutions of the *second class.*

It thus appears that, for a system of two equations in two dependent variables, there may be two classes of singular solutions: one of them involves an arbitrary constant, the other of them does not. A solution which arises from the most general solution, by giving particular values to one or to both of the arbitrary constants, must be regarded as a particular solution; a solution which arises from the first class of singular solutions, by giving a particular value to the arbitrary constant, must be regarded as a particular case of the first class of singular solutions, and not as a singular solution of the second class.

Further, it appears from the analysis that the singular solutions of the first class are derivable, partly from the original equations, but partly also through the solution of a deduced differential equation; and that the singular solution of the second class is derivable either (i), entirely from the original equations without the construction of the intermediate differential equation required for the first class: or (ii), partly from the original equations and partly from the intermediate equation indicated.

Note 1. The preceding method of discussing the singular solutions of a system of equations and, incidentally, the various classes of integrals which may be possessed by a system, is based upon a direct study of the differential equations, beginning with the relation of the non-ordinary points of the equations to the integrals that may exist.

In connection with the deduction of the singular integrals from the differential equations themselves, memoirs by Mayer[*], Goursat[†], and Dixon[‡] may be consulted with advantage. The last of these memoirs discusses also, with ample illustrations, the

---

[*] *Math. Ann.*, t. xxii (1883), pp. 368—392.

[†] *Amer. Journ. Math.*, t. xi (1889), pp. 329—372.

[‡] *Phil. Trans.*, (1895, Part i, A), pp. 523—565.

derivation of the various classes of integrals from a complete primitive, supposed given: this aspect of the subject is naturally to be compared with the process of § 207, which leads to the classification of integrals. All the memoirs just quoted deal with the singular solutions (if any) of a single equation in one dependent variable of an order higher than the first.

Note 2. Throughout the whole discussion of the singular integrals (if any) of two equations $F = 0$, $G = 0$, it has been assumed that $J = 0$ is an independent equation, so that it is not satisfied identically, in virtue of the values of p and q derived from $F = 0$, $G = 0$: and so also therefore that the elimination of p and q, between $F = 0$, $G = 0$, $J = 0$, leads to a definite relation between x, y, z of the form $\Theta = 0$. If the equations possess singular integrals, then

$$\Theta_x = 0,$$

in virtue of $G_x F_p - F_x G_p = 0$, which is satisfied.

It may, however, be the fact that $J = 0$ is satisfied identically in virtue of $F = 0$, $G = 0$. Even when this is not the fact and when the system possesses singular integrals, it is possible to construct an associated system for which this possibility actually occurs. For instance, let $\Theta = 0$ be the result of eliminating p and q between $F = 0$, $G = 0$, $J = 0$. Let $G = 0$, $J = 0$ be solved so as to give p and q in terms of x, y, z; and let the values be substituted in F, the result being, of course, either Θ or a (possibly irrational) factor of Θ: suppose it the former, and now consider the associated system

$$F - \Theta = 0, \quad G = 0.$$

In order that this system may possess singular solutions, we must have

$$\text{Jacobian} \left(\frac{F - \Theta,\ G}{p,\ q} \right) = 0,$$

that is, because Θ is a function of x, y, z, which does not explicitly involve p and q,

$$J = 0;$$

and also

$$G_p (F_x - \Theta_x) - (F_p - \Theta_p) G_x = 0.$$

But $\Theta_x = 0$, because $G_p F_x - F_p G_x = 0$; and $\Theta_p = 0$. Hence this equation certainly is satisfied. Also the three equations

$$F - \Theta = 0, \quad G = 0, \quad J = 0$$

coexist; they merely determine p and q, and they do not lead to any relation between x, y, z: that is, the original equations, of themselves and without any limitations by way of added conditions, satisfy the requirements for the possession of the first class of singular solutions.

The examination and the significance of this result are left as an exercise.

201. *Ex.* 1. Consider the equations

$$F = -y + xp + p^2 + q = 0 \atop G = -z + xq + pq \quad = 0 \Big\} ,$$

which were first discussed by Serret[*], and are again discussed by Mayer[†], and Goursat[‡]. We have

$$J = (x + 2p)(x + p) - q.$$

Hence, if the equations possess singular solutions, values of p and q must exist which make $J = 0$. But also

$$F_x = \frac{\partial F}{\partial x} + p \frac{\partial F}{\partial y} + q \frac{\partial F}{\partial z} = 0,$$

$$G_x = \frac{\partial G}{\partial x} + p \frac{\partial G}{\partial y} + q \frac{\partial G}{\partial z} = 0,$$

so that the equation

$$G_p F_x - F_p G_x = 0$$

is satisfied. Accordingly, the equation $H = 0$ is to be formed by the elimination of z and q between $F = 0$, $G = 0$, $J = 0$: it is

$$H(x, y, p) = -y + 4px + x^2 + 3p^2 = 0.$$

The complete integral of $H = 0$ is easily found to be

$$y = a^2 - \tfrac{1}{4}(x + a)^2.$$

The associable value of z is obtainable by algebraic resolution of $F = 0$, $G = 0$, $J = 0$: we have, from $G = 0$,

$$\frac{z}{x + p} = q$$

$$= (x + p)(x + 2p),$$

from $J = 0$; that is,

$$z = (x + p)^2 (x + 2p)$$

$$= -\tfrac{1}{4} a (x - a)^2,$$

because $p = -\tfrac{1}{2}(x + a)$. Hence

$$z = -\tfrac{1}{4} a (x - a)^2 \atop y = -\tfrac{1}{4}(x + a)^2 + a^2 \Big\}$$

* *Liouville*, 1ʳᵉ Sér., t. xviii (1853), p. 29.
† *Math. Ann.*, t. xxii, p. 382.
‡ *Amer. Journ. Math.*, vol. xi, p. 360.

constitute a singular solution of the original equations, and the solution is of the first class.

The further equations to determine a singular solution of the second class, if it exists, are (as connected with the original equations)

$$J_x G_p - J_p G_x = 0,$$

that is,

$$(2x + 3p) q = 0 ;$$

and

$$\frac{J_q}{J_p} = \frac{F_q}{F_p},$$

that is,

$$\frac{-1}{4p + 3x} = \frac{1}{2p + x},$$

that is,

$$6p + 4x = 0 ;$$

hence both the additional equations are satisfied by

$$3p + 2x = 0.$$

This gives

$$p = -\tfrac{2}{3}x ;$$

hence, from $J = 0$,

$$q = -\tfrac{1}{3}x^2,$$

from $F = 0$,

$$y = -\tfrac{1}{3}x^2,$$

and from $G = 0$,

$$z = -\tfrac{1}{27}x^3.$$

Accordingly

$$y = -\tfrac{1}{3}x^2, \quad z = -\tfrac{1}{27}x^3,$$

constitute a singular solution of the second class.

The solution $q = 0$, of $J_x G_p - J_p G_x = 0$, does not satisfy $J_q F_p - J_p F_q = 0$ and the simultaneous equations; hence it cannot lead to a singular solution of the second class. (Taken with the other equations, it will be found to lead to a solution

$$z = 0, \quad y = -\tfrac{1}{4}x^2,$$

which is a particular instance of the singular solution of the first class.)

The complete integral of the original equations is

$$\left. \begin{array}{l} y = ax + a^2 + \beta \\ z = \beta x + a\beta \end{array} \right\},$$

where a and β are arbitrary constants.

The equation $\Theta = 0$, obtained by the elimination of p and q between $F = 0$, $G = 0$, $J = 0$, is

$$\{27 (z - xy) - 2x (x^2 - 9y)\}^2 = 4 (x^2 + 3y)^3.$$

Ex. 2.   Discuss the equations

$$\left. \begin{array}{l} q + p + \tfrac{1}{2}p^2 = 0 \\ z + y (1 + p) = \tfrac{1}{2}xp^2 + \tfrac{1}{3}p^3 \end{array} \right\}.$$

(Mayer.)

Ex. 3.   Shew that the equations

$$y=px+p^2+pq+q \\ z=qx+pq+q^2 \Bigg\}$$

possess singular solutions of both classes ; and obtain them.

(Dixon.)

Ex. 4.   Discuss the equations

$$(x^2-1)\,p^2=y^2-1 \\ (x^2-1)\,q^2=z^2-1 \Bigg\}\,,$$

in reference to the various classes of solutions that can be deduced.

(Dixon.)

Ex. 5.   Obtain the singular solutions of the equations

$$y=px+\tfrac{1}{3}q^3 \\ z=qx+\tfrac{1}{2}p^2 \Bigg\}\,.$$

(Dixon.)

Ex. 6.   Shew that, in general, singular solutions of the first class exist for equations of the form

$$\Phi\,(y-px,\ z-qx,\ p,\ q)=0 \\ \Psi\,(y-px,\ z-qx,\ p,\ q)=0 \Bigg\}\,,$$

where Φ and Ψ are any functions of their arguments.

(Goursat.)

Obtain the equations which must be satisfied, in order that the singular solutions of the second class may exist.

Ex. 7.   The complete integral of the equations

$$x+py+qz=0 \\ (p^2+q^2-c^2)\,(y^2+z^2)=x^2\,(1+c^2) \Bigg\}\,,$$

is given by

$$x^2+y^2+z^2=A^2 \\ B+\tan^{-1}\frac{y}{z}=\tfrac{1}{2}c\log\frac{A+x}{A-x} \Bigg\}\,.$$

Discuss the relation, to this complete integral, of the solution

$$y=ax,\quad z=bx,$$

where $a^2+b^2+1=0$.

(The differential equations can be interpreted as the equations of rhumb-lines on a family of concentric spheres.)

Ex. 8.   In the general investigation, it has been assumed (for analytical simplicity) that $J=0$, while no one of the quantities $\dfrac{\partial F}{\partial p}$, $\dfrac{\partial F}{\partial q}$, $\dfrac{\partial G}{\partial p}$, $\dfrac{\partial G}{\partial q}$ vanishes.   It would be useful to discuss some of the more important cases, when at least one of these quantities does vanish.

SYSTEM OF ANY NUMBER OF EQUATIONS.

202. The preceding results, which belong to a system of two equations in two dependent variables, suggest the results, which belong to a similar system of n equations in n independent variables, in the form

$$F_r(x, y_1, ..., y_n, p_1, ..., p_n) = 0, \qquad (r = 1, 2, ..., n),$$

where $p_i = \dfrac{dy_i}{dx}$.

Such equations are known, by the general existence theorem, to possess complete integrals, that is, a set of n equations among the variables $x, y_1, ..., y_n$, involving n arbitrary constants.

It can be proved, as in §§ 197, 198 for the system of two equations, that, if values $p_1, p_2, ..., p_n$ satisfy the equations

$$F_1 = 0, \quad F_2 = 0, \quad ..., \quad F_n = 0,$$

$$J = \frac{\partial(F_1, F_2, ..., F_n)}{\partial(p_1, p_2, ..., p_n)} = 0,$$

$$\begin{vmatrix} F_{1x}, & F_{1p_1}, & F_{1p_2}, & ..., & F_{1p_n} \\ F_{2x}, & F_{2p_1}, & F_{2p_2}, & ..., & F_{2p_n} \\ \hdotsfor{5} \\ F_{nx}, & F_{np_1}, & F_{np_2}, & ..., & F_{np_n} \end{vmatrix} = 0$$

where

$$F_{ix} = \frac{\partial F_i}{\partial x} + \sum_{m=1}^{n} p_m \frac{\partial F_i}{\partial y_m},$$

$$F_{ip_j} = \frac{\partial F_i}{\partial p_j},$$

and if among the $n+1$ equations $F_1 = 0, ..., F_n = 0, J = 0$, the variables p_n and y_n be algebraically eliminated, with the result

$$G_s(x, y_1, ..., y_{n-1}, p_1, ..., p_{n-1}) = 0, \quad (s = 1, 2, ..., n-1),$$

and if

$$\Theta(x, y_1, ..., y_n) = 0$$

denote the result of the algebraical elimination of $p_1, ..., p_n$ among the same $n+1$ equations, then any set of integrals of the equations $G_s = 0$, combined with $\Theta = 0$, constitutes a set of integrals of the original equations, not included among (and therefore distinct from) the complete integrals of the original equations.

We know that, in the instance of a single equation in a single dependent variable, there may be one class of solution not included in the complete solution, viz. the singular solution. We have seen that, in the instance of two equations in two dependent variables, there may be two classes of solutions not included in the complete solution, viz. one class, involving a single arbitrary constant, another class, involving no arbitrary constant and not included in the former. We thus infer inductively that, in addition to the complete integrals belonging to a set of n equations in n dependent variables, there may (but not necessarily must) exist n other classes of solutions not included in the complete integrals. The first of these classes contains $n-1$ arbitrary constants: the second of them contains $n-2$ arbitrary constants and is not included in the first; the sth contains $n-s$ arbitrary constants and is not included in any of the preceding $s-1$ classes; and the last class contains no arbitrary constant, and is not included in any of the earlier classes.

In order that the first class of these additional solutions, which will be called *singular* solutions, may be possessed by the equations, certain conditions must be satisfied. For the system of n equations, the conditions are that the values of $p_1, p_2, \ldots, p_n$, which satisfy

$$F_1 = 0, \quad F_2 = 0, \quad \ldots, \quad F_n = 0,$$

must also satisfy the equations

$$\| F_{ix}, \quad F_{ip_1}, \quad F_{ip_2}, \quad \ldots, \quad F_{ip_n} \| = 0.$$

The last equations are obtained by equating to zero the determinants of order n that can be framed from this array: it is not difficult to shew that they are equivalent to a couple of independent conditions.

In order that the second class of singular solutions may be possessed by the original equations, certain additional conditions must be satisfied. Let

$$G_1 = 0, \quad G_2 = 0, \quad \ldots, \quad G_{n-1} = 0,$$

denote the result of eliminating p_n and y_n between

$$F_1 = 0, \quad F_2 = 0, \quad \ldots, \quad F_n = 0,$$

$$J = \frac{\partial (F_1, \ldots, F_n)}{\partial (p_1, \ldots, p_n)} = 0 ;$$

the second class of singular solutions of the original equations is the first class of singular solutions of the equations $G = 0$, and in order that these may exist, two independent conditions attaching to the equations $G = 0$ must be satisfied : these are the additional conditions for the original equations.

And so on, for each class of singular solutions in succession : the existence of each additional class requires that two additional conditions shall be satisfied. If at any stage the two additional conditions are not satisfied, then not merely does the corresponding class of singular solutions not exist, but no one of the remaining classes of singular solutions in the succession exists.

Ex. 1.   In connection with the equations

$$F(x, y_1, y_2, y_3, p_1, p_2, p_3) = 0$$
$$G(x, y_1, y_2, y_3, p_1, p_2, p_3) = 0$$
$$H(x, y_1, y_2, y_3, p_1, p_2, p_3) = 0$$

denote $\dfrac{\partial F}{\partial x} + \sum\limits_{i=1}^{3} p_i \dfrac{\partial F}{\partial y_i}$ by F_x, $\dfrac{\partial F}{\partial p_i}$ by F_i; and so for G and for H : and let J denote

$$\frac{\partial (F, G, H)}{\partial (p_1, p_2, p_3)}.$$

Shew that, if the equations are to possess the first class of singular solutions (involving two arbitrary constants), it is necessary and sufficient that the conditions

$$\begin{vmatrix} F_x, & F_1, & F_2, & F_3 \\ G_x, & G_1, & G_2, & G_3 \\ H_x, & H_1, & H_2, & H_3 \end{vmatrix} = 0,$$

equivalent to two independent conditions, shall be satisfied by values of p_1, p_2, p_3, which satisfy $F = 0$, $G = 0$, $H = 0$.

Shew that, if the equations are to possess the second class of singular solutions (involving one arbitrary constant), it is necessary and sufficient that every first minor of the determinant

$$\begin{vmatrix} J_x, & F_x, & G_x, & H_x \\ J_1, & F_1, & G_1, & H_1 \\ J_2, & F_2, & G_2, & H_2 \\ J_3, & F_3, & G_3, & H_3 \end{vmatrix}$$

shall vanish, for values of p_1, p_2, p_3, which satisfy $F = 0$, $G = 0$, $H = 0$; and that the conditions are equivalent to four independent conditions.

Lastly, denoting any one of the three Jacobians

$$\frac{\partial (J, F, G)}{\partial (p_1, p_2, p_3)}, \qquad \frac{\partial (J, G, H)}{\partial (p_1, p_2, p_3)}, \qquad \frac{\partial (J, H, F)}{\partial (p_1, p_2, p_3)},$$

by I, shew that, if the equations are to possess the third class of singular solutions (involving no arbitrary constant), it is necessary and sufficient that every first minor in each of the determinants included in the array

$$\begin{Vmatrix} I_x, & J_x, & F_x, & G_x, & H_x \\ I_1, & J_1, & F_1, & G_1, & H_1 \\ I_2, & J_2, & F_2, & G_2, & H_2 \\ I_3, & J_3, & F_3, & G_3, & H_3 \end{Vmatrix}$$

shall vanish, for values of p_1, p_2, p_3 that satisfy $F=0$, $G=0$, $H=0$: and that the conditions are equivalent to six independent conditions.

Ex. 2. Discuss the singular solutions (if any) of the equations of motion, of a heavy body moving round a fixed point, when they are taken in the form

$$0 = \left(\frac{d\nu}{dt}\right)^2 - 4\kappa^2 H,$$

$$0 = \frac{d\rho}{dt} - R,$$

$$0 = \left\{(\nu\sigma - \rho^2)\frac{d\mu}{dt} - (\lambda\sigma - \mu\rho)R\right\}^2 - H\left\{\sigma(2\kappa\mu + \lambda) - \rho(ap + \beta q + \gamma r)\right\}^2,$$

where

$$\nu = A^2 p^2 + B^2 q^2 + C^2 r^2,$$
$$\rho = Aap + B\beta q + C\gamma r,$$
$$\kappa\mu = Ap^2 + Bq^2 + Cr^2 - h,$$
$$R = a(B-C)qr + \beta(C-A)rp + \gamma(A-B)pq,$$
$$H = \begin{vmatrix} 1, & \lambda, & \mu \\ \lambda, & \nu, & \rho \\ \mu, & \rho, & \sigma \end{vmatrix},$$

and A, B, C, a, β, γ, h, κ, λ, σ are constants.

(Mayer.)

Ex. 3. Shew that, in general, there are six distinct sets of integrals of the equations

$$\left.\begin{aligned} y_1 &= p_1 x + p_2 p_3 \\ y_2 &= p_2 x + p_3 p_1 \\ y_3 &= p_3 x + p_1 p_2 \end{aligned}\right\},$$

determined by the conditions that y_1, y_2, y_3 must assume values b_1, b_2, b_3 respectively, when $x = a$.

Shew further that the equations possess the first class of singular solutions : and associate them with the equations

$$Y = Px + \tfrac{1}{2}P^2 - \tfrac{1}{5}\frac{Q}{x} - \tfrac{3}{10}x^2,$$

$$2Z = \tfrac{8}{5}Qx - \tfrac{3}{5}x^4 + (Y - Px)^2 + \tfrac{2}{5}P(x^3 - Q),$$

where

$$Y = y_1 + y_2 + y_3, \quad 2Z = y_1^2 + y_2^2 + y_3^2, \quad P = \frac{dY}{dx}, \quad Q = \frac{dZ}{dx}.$$

Do the equations possess any of the other classes of singular solutions ?

Ex. 4. Discuss the classes of solutions possessed by the equations

$$y_1 = p_1 x + p_2{}^2 + p_3{}^2$$
$$y_2 = p_2 x + p_3{}^2 + p_1{}^2$$
$$y_3 = p_3 x + p_1{}^2 + p_2{}^2$$

Ex. 5. Shew that, in general, the equations

$$y_i = p_i x + f_i(p_1, p_2, p_3), \qquad\qquad (i = 1, 2, 3),$$

possess the first class of singular solutions, provided the functions f_i are not linear functions of their arguments.

ANALYTICAL RELATION BETWEEN SINGULAR INTEGRALS AND COMPLETE INTEGRAL.

203. The relation of the singular integrals, if they exist, to the complete integral can be exhibited in a different analytical form : and some tests can be obtained, in order to settle whether a given solution is a particular case of some class of integrals, or the general case of some more special class of integrals. It will be sufficient to deal with the case of a couple of dependent variables, determined by a couple of equations which may be taken in the form

$$F(x, y, z, p, q) = 0, \quad G(x, y, z, p, q) = 0.$$

In association with these, we take

$$J = \frac{\partial(F, G)}{\partial(p, q)} = 0;$$

and assuming that J does not vanish solely in consequence of $F = 0$, $G = 0$, denote by $H(x, y, p) = 0$, the result of eliminating z and q between $F = 0$, $G = 0$, $J = 0$. Let

$$y = \eta, \quad p = P = \frac{d\eta}{dx},$$

denote a solution of $H = 0$; then since $H = 0$ is a consequence of $F = 0$, $G = 0$, $J = 0$, it follows that, when a solution of $H = 0$ is actually used, the other three equations are equivalent only to two, say $F = 0$, $G = 0$. Accordingly, they then determine the two outstanding unknown quantities z and q, say in the forms

$$z = \zeta, \quad q = Q,$$

where Q is not necessarily equal to $\dfrac{d\zeta}{dx}$. We have to discriminate, as will be seen, between the case, when Q is not equal to $\dfrac{d\zeta}{dx}$, and the case, when Q is equal to $\dfrac{d\zeta}{dx}$.

Now take

$$y - \eta = Y, \quad z - \zeta = Z,$$
$$p - P = u, \quad q - Q = v;$$

then, in order that y and z may satisfy the original equations, we must have

$$F(x,\ \eta + Y,\ \zeta + Z,\ P + u,\ Q + v) = 0,$$
$$G(x,\ \eta + Y,\ \zeta + Z,\ P + u,\ Q + v) = 0.$$

But also

$$F(x,\ \eta,\ \zeta,\ P,\ Q) = 0, \quad G(x,\ \eta,\ \zeta,\ P,\ Q) = 0;$$

so that all the terms in the former free from Y, Z, u, v vanish. Thus the two equations determine u and v in terms of Y and Z, and also of x; among the values thus determined, there must be some, which vanish when $Y = 0$, $Z = 0$. But when $Y = 0$, so that zero is a possible value of u because

$$H(x,\ \eta,\ P) = 0,$$

and when $Z = 0$ so that $z = \zeta$, then, because $F = 0$, $G = 0$, $H = 0$, we have also $J = 0$: that is, the values, which are obtained for u and v as vanishing when $Y = 0$, $Z = 0$, must be of multiple occurrence.

Let

$$u = A(x,\ Y,\ Z)$$
$$v = B(x,\ Y,\ Z)$$

denote a root-pair which is of simple occurrence; the functions A and B are regular functions of Y and Z, which have uniform functions of x for the coefficients: let $x = b$ be any ordinary point of these coefficients, so that A and B are regular functions also of $x - b$: also A, B do not vanish when $Y = 0$, $Z = 0$.

Let

$$u = f(x,\ Y,\ Z)$$
$$v = g(x,\ Y,\ Z)$$

denote a root-pair which is of multiple occurrence; both f and g vanish when $Y = 0$, $Z = 0$: and they are multiform functions of Y and Z. Let $x = b$ denote an ordinary point of f and g, regarded solely for the moment as a function of x; then f and g can be expressed as regular functions of $x - b$, having for their coefficients multiform functions of Y and Z which vanish with Y and Z.

The point b will, in every case, be regarded as parametric. What is wanted is a pair of values of Y and Z, which vanish at $x = b$, so that y and η, z and ζ, have the same value there: the argument for this requirement being similar to that in § 104. Also

$$u = \frac{dY}{dx},$$

$$v = \frac{dZ}{dx} + \frac{d\zeta}{dx} - Q.$$

When $y = \eta$, $z = \zeta$, do not constitute a singular solution, so that Q is not equal to $\frac{d\zeta}{dx}$, the point b is an ordinary non-zero point of $\frac{d\zeta}{dx} - Q$.

204.   The general character of the functions determined by the equations

$$\left. \begin{aligned} \frac{dY}{dx} &= A\,(x,\ Y,\ Z) \\ \frac{dZ}{dx} &= B\,(x,\ Y,\ Z) + Q - \frac{d\zeta}{dx} \end{aligned} \right\},$$

whether η, ζ constitute a singular solution or not, is at once given by the application of the existence-theorem. Let $A\,(b,\ 0,\ 0) = \beta_1$, and let β_2 denote the value of

$$B\,(x,\ Y,\ Z) + Q - \frac{d\zeta}{dx},$$

when $x = b$, $Y = 0$, $Z = 0$; then solutions Y and Z exist as regular functions of $x - b$, in the form

$$\left. \begin{aligned} Y &= \beta_1\,(x - b)\,R\,(x - b) \\ Z &= \beta_2\,(x - b)\,S\,(x - b) \end{aligned} \right\},$$

where R and S are regular functions of their argument, which become unity when the argument vanishes. The corresponding integrals of the original equation are

$$\left. \begin{aligned} y - \eta &= \beta_1\,(x - b)\,R\,(x - b) \\ z - \zeta &= \beta_2\,(x - b)\,S\,(x - b) \end{aligned} \right\}.$$

There is no special interest attaching to the relation between the solutions η, ζ, and this branch of the general integral.

205. The general character of the functions determined by the equations

$$\frac{dY}{dx} = f(x, Y, Z)$$

$$\frac{dZ}{dx} = g(x, Y, Z) + Q - \frac{d\zeta}{dx}$$

is more difficult to obtain, because no definite algebraical solution of the simultaneous equations in u and v is always possible: though it may be effected in particular cases.

When Q is not equal to $\frac{d\zeta}{dx}$, so that η, ζ do not then constitute a singular solution, one inference from the above equations can be made. Since Y and Z are to vanish when $x = b$, and since $f(x, Y, Z)$, $g(x, Y, Z)$ vanish when $Y = 0$, $Z = 0$, it follows that

$$\frac{dY}{dx} = 0, \qquad \frac{dZ}{dx} = \left(Q - \frac{d\zeta}{dx}\right)_{x=b} = \gamma \text{ say,}$$

when $x = b$; hence we shall have

$$Y = \epsilon (x - b)^{1+\lambda} + \dots$$
$$Z = \gamma (x - b) + \dots$$

where λ is a positive quantity (greater than zero): and so

$$Z - \gamma (x - b) = \eta (x - b)^{1+\mu} + \dots,$$

where μ is a positive quantity (greater than zero).

When Q is equal to $\frac{d\zeta}{dx}$, not merely at $x = b$ but for all values of x, so that η, ζ constitute a singular solution, then the equations are

$$\frac{dY}{dx} = f(x, Y, Z), \qquad \frac{dZ}{dx} = g(x, Y, Z).$$

Now Y and Z are to vanish when $x = b$; also f and g vanish, when $Y = 0$, $Z = 0$; hence $\frac{dY}{dx}$, $\frac{dZ}{dx}$ vanish when $x = b$. Accordingly, we may take

$$Y = \alpha (x - b)^{1+\lambda}, \qquad Z = \beta (x - b)^{1+\mu},$$

for sufficiently small values of $|x - b|$, as the most important terms of Y and Z, the quantities λ and μ being real and greater than zero. But after this inference as regards the general type, it is unnecessary actually to solve the equations

$$F(x, \eta + Y, \zeta + Z, P + u, Q + v) - F(x, \eta, \zeta, P, Q) = 0,$$

$$G(x, \eta + Y, \zeta + Z, P + u, Q + v) - G(x, \eta, \zeta, P, Q) = 0,$$

so as to express u and v algebraically in terms of Y and Z. With the above values of Y and Z, we have

$$u = \frac{dY}{dx} = \alpha(1 + \lambda)(x - b)^\lambda,$$

$$v = \frac{dZ}{dx} = \beta(1 + \mu)(x - b)^\mu :$$

substituting, and securing that the lowest power of $x - b$ has a vanishing coefficient in each equation, we obtain two indicial relations (which determine λ and μ), and two coefficient relations (which determine α and β).

The equations for u and v are

$$u \frac{\partial F}{\partial P} + v \frac{\partial F}{\partial Q} + \dots = 0,$$

$$u \frac{\partial G}{\partial P} + v \frac{\partial G}{\partial Q} + \dots = 0 ;$$

also, we have

$$J = \frac{\partial(F, G)}{\partial(P, Q)} = 0,$$

and we suppose that no one of the quantities $\dfrac{\partial F}{\partial P}, \dfrac{\partial F}{\partial Q}, \dfrac{\partial G}{\partial P}, \dfrac{\partial G}{\partial Q}$ vanishes. Then, when we take

$$\frac{\partial F}{\partial P} = c \frac{\partial F}{\partial Q}, \quad \frac{\partial G}{\partial P} = c \frac{\partial G}{\partial Q},$$

the above equations give $v + cu$ as a quantity of higher order in powers of $x - b$ than either v or u is: that is, in this case $\lambda = \mu$, $\beta + c\alpha = 0$. This form is the simplest that occurs; other forms, in which even all the four first derivatives of F and G vanish, can occur, and then the method of proceeding is as suggested above.

The general theory might be developed somewhat on the lines of Hamburger's investigations, as sketched in §§ 103—108; this, however, will be left undiscussed, and we proceed to consider a couple of examples.

206. *Ex.* 1. We shall consider the equations

$$y = px + p^2 + q, \quad z = qx + pq,$$

which are known (§ 201) to possess singular solutions.

These singular solutions satisfy $F = 0$, $G = 0$, $J = 0$, $H = 0$; accordingly, with the notation of § 205, and using the results of the Ex. 1 in § 201, we take

$$\eta = -\tfrac{1}{4}(x+a)^2 + a^2, \quad P = -\tfrac{1}{2}(x+a),$$
$$\zeta = -\tfrac{1}{4}a(x-a)^2, \quad Q = -\tfrac{1}{2}a(x-a);$$

and we write

$$y - \eta = Y, \quad z - \zeta = Z,$$
$$p - P = u, \quad q - Q = v.$$

Then on substitution and reduction, we find

$$\left.\begin{matrix} Y = -au + u^2 + v \\ Z = \tfrac{1}{2}(x-a)(v-au) + uv \end{matrix}\right\};$$

and therefore solutions for u and v are derivable in the form

$$v = au - u^2 + Y,$$

u denoting any root of the equation

$$u^3 + \tfrac{1}{2}(x-3a)u^2 - uY + Z - \tfrac{1}{2}(x-a)Y = 0.$$

Clearly the root-pair $u = 0$, $v = 0$ (when $Y = 0$, $Z = 0$) is of multiple occurrence.

Let Δ denote the discriminant of the equation in u, viz.

$$\Delta = \tfrac{1}{54}(x-3a)^3 Z - \tfrac{1}{108}(x-a)(x-3a)^3 Y$$
$$+ Z^2 - \tfrac{2}{3}xYZ + \tfrac{1}{27}(2x^2 + 6ax - 9a^2)Y^2 - \tfrac{4}{27}Y^3;$$

and let

$$\Gamma = \tfrac{1}{216}(x-3a)^3 + \tfrac{1}{2}Z - \tfrac{1}{6}xY.$$

Then if

$$\xi = [-\Gamma + \tfrac{1}{2}\Delta^{\frac{1}{2}}]^{\frac{1}{3}},$$
$$\eta = [-\Gamma - \tfrac{1}{2}\Delta^{\frac{1}{2}}]^{\frac{1}{3}};$$

the three values of $u + \tfrac{1}{6}(x-3a)$ are $\xi + \eta$, $\omega\xi + \omega^2\eta$, $\omega^2\xi + \omega\eta$, where ω is a cube root of unity. We thus find for the three roots

$$u_1 = -\frac{1}{2}(x-3a)\left\{1 + 8\frac{Z - aY}{(x-3a)^3} + \dots\right\},$$

$$u_2 = \frac{i\,6\sqrt{3}\,\Delta^{\frac{1}{2}}}{(x-3a)^2}\{1 + A(x, Y, Z)\} + B(x, Y, Z),$$

$$u_3 = -\frac{i\,6\sqrt{3}\,\Delta^{\frac{1}{2}}}{(x-3a)^2}\{1 + A(x, Y, Z)\} + B(x, Y, Z),$$

where $A(x, Y, Z)$, $B(x, Y, Z)$, and the unexpressed parts of u_1, are regular functions of Y and Z, which vanish with Y and Z, and have uniform functions of x for their coefficients. The associable values of v are

$$v_r = au_r - u_r^2 + Y,$$

for $r = 1, 2, 3$.

Now

$$u = -P + \frac{d}{dx}(Y+\eta) = \frac{dY}{dx},$$

$$v = -Q + \frac{d}{dx}(Z+\zeta) = \frac{dZ}{dx},$$

in the present case; so that we have

$$\frac{dY_r}{dx} = u_r, \qquad \frac{dZ_r}{dx} = v_r,$$

for $r = 1, 2, 3$. The various sets must be examined in turn.

First, for $r = 1$, we have

$$\frac{dY_1}{dx} = u_1 = -\frac{1}{2}(x-3a) + E(x, Y_1, Z_1),$$

$$\frac{dZ_1}{dx} = v_1 = -\frac{1}{4}(x-a)(x-3a) + C(x, Y_1, Z_1),$$

where E and C are regular functions of their arguments in the vicinity of $x = b$, $Y_1 = 0$, $Z_1 = 0$. Of these equations, solutions Y_1 and Z_1 exist in the form of regular functions of $x - b$, vanishing when $x = b$: and

$$\left.\begin{aligned}
\left(\frac{dY_1}{dx}\right)_{x=b} &= -\frac{1}{2}(b-3a) \\
\left(\frac{dZ_1}{dx}\right)_{x=b} &= -\frac{1}{4}(b-a)(b-3a)
\end{aligned}\right\}.$$

The corresponding solutions of the original equations are

$$y - \eta = Y_1, \qquad z - \zeta = Z_1;$$

they are such that, for a general value b of x, they are equal to $\eta(b)$ and $\zeta(b)$, but the values of their derivatives at the point are not equal to the derivatives of η and ζ.

Secondly, for $r = 2$, we have

$$\frac{dY_2}{dx} = u_2, \qquad \frac{dZ_2}{dx} = v_2;$$

and values of Y_2 and Z_2 are required, which vanish when $x = b$. To find the forms of these values, we take

$$Y_2 = \lambda(x-b)^n, \qquad Z_2 = \mu(x-b)^n,$$

for sufficiently small values of $|x-b|$: in order to be effective, we must have $n > 0$. The terms in u_2 and v_2, which correspond to the parts $A(x, Y, Z)$ and $B(x, Y, Z)$, are of too high an order; and the most important term in Δ is

$$\frac{i}{108}\{2(b-3a)^3\mu - (b-a)(b-3a)^3\lambda\}(x-b)^n.$$

Substituting, and equating indices, we have

$$n - 1 = \tfrac{1}{2}n,$$

so that $n=2$. Equating coefficients, we find

$$2\lambda = i \left\{ \frac{2\mu - (b-a)\lambda}{b-3a} \right\}^{\frac{1}{2}},$$

and therefore

$$\mu = a\lambda,$$

that is,

$$2\lambda = i(-\lambda)^{\frac{1}{2}} ;$$

Hence

$$\lambda = +\tfrac{1}{4}, \quad \mu = +\tfrac{1}{4}a.$$

$$\left. \begin{aligned} Y_2 &= \tfrac{1}{4}(x-b)^2 + \ldots \\ Z_2 &= \tfrac{1}{4}a(x-b)^2 + \ldots \end{aligned} \right\}.$$

The corresponding solutions of the original equations are

$$y - \eta = Y_2, \quad z - \zeta = Z_2 ;$$

they are such that, for a general value b of x, they are equal to $\eta(b)$ and $\zeta(b)$, and also the values of their derivatives at the point are equal to the values of the derivatives of η and ζ.

Similarly, for $r=3$, we find

$$\left. \begin{aligned} Y_3 &= \tfrac{1}{4}(x-b)^2 + \ldots \\ Z_3 &= \tfrac{1}{4}a(x-b)^2 + \ldots \end{aligned} \right\},$$

and the corresponding solutions of the original equations are

$$y - \eta = Y_3, \quad z - \zeta = Z_3.$$

There is the same inference as in the last case. Moreover, it follows that the values of the derivatives of the last two sets of solutions are equal to one another at $x = b$.

To obtain this result by direct verification from the known forms, we proceed as follows. The complete integral is known to be

$$y = ax + a^2 + \beta, \quad z = \beta x + a\beta ;$$

the singular solution under consideration is

$$\eta = -\tfrac{1}{4}(x+a)^2 + a^2, \quad \zeta = -\tfrac{1}{4}a(x-a)^2.$$

To determine the complete integral which, at $x = b$, coincides in value (not in form) with the singular integral, we have

$$\left. \begin{aligned} ab + a^2 + \beta &= -\tfrac{1}{4}(a+b)^2 + a^2 \\ \beta b + a\beta &= -\tfrac{1}{4}a(b-a)^2 \end{aligned} \right\}.$$

These equations, when solved, give

$$a = a - b, \qquad \beta = -\tfrac{1}{4}(b-a)^2 ;$$

and

$$a = -\tfrac{1}{2}(a+b), \quad \beta = -\tfrac{1}{2}(b-a)a,$$

the latter pair being repeated. From the above values

$$\begin{aligned} y - \eta &= ax + a^2 + \beta + \tfrac{1}{4}(x+a)^2 - a^2 \\ &= ax + \tfrac{1}{4}(x+a)^2 - ab - \tfrac{1}{4}(a+b)^2 \\ &= \left\{ a + \tfrac{1}{2}(a+b) \right\}(x-b) + \tfrac{1}{4}(x-b)^2 \\ z - \zeta &= \left\{ \beta + \tfrac{1}{2}(b-a)a \right\}(x-b) + \tfrac{1}{4}a(x-b)^2 \end{aligned} \left. \right\}.$$

Taking the simple root-pair, we have

$$y-\eta=\tfrac{1}{2}(3a-b)(x-b)+\tfrac{1}{4}(x-b)^2=Y_1 \atop z-\zeta=\tfrac{1}{4}(b-a)(3a-b)(x-b)+\tfrac{1}{4}a(x-b)^2=Z_1 \Big\} ;$$

taking the multiple root-pair, we have

$$y-\eta=\tfrac{1}{4}(x-b)^2=Y_2,\ Y_3 \atop z-\zeta=\tfrac{1}{4}a(x-b)^2=Z_2,\ Z_3 \Big\}.$$

As a matter of fact, Y_2 and Y_3 are equal to one another for this equation; and likewise Z_2 and Z_3: not merely their own values and the values of their first derivatives at $x=b$.

In the same way it can be shewn that, taking the singular solution

$$\eta=-\tfrac{1}{3}x^2,\quad \zeta=-\tfrac{1}{27}x^3,$$

and choosing those values of the constants in the complete integrals

$$y=ax+a^2+\beta,\quad z=\beta x+a\beta,$$

which make them, at $x=b$, equal to the singular integral, we have only a single root-pair: and the relation is

$$y-\eta=\tfrac{1}{3}(x-b)^2 \atop z-\zeta=\tfrac{1}{3}b(x-b)^2+\tfrac{1}{27}(x-b)^3 \Big\}.$$

Ex. 2.   Discuss similarly the equations

$$y=px+\tfrac{1}{2}c(p^2+q^2) \atop z=qx+cpq \Big\}.$$

Ex. 3.   Consider the equations

$$y=2px+apq \atop z=2qx+bpq+cx \Big\},$$

where a, b, c are constants. These equations are easily proved to possess no singular solutions.

We have

$$J=4x^2+2x(aq+bp);$$

so that, taking $J=0$, we obtain

$$aq+bp+2x=0.$$

Combining this with the initial equations, so as to eliminate z and q from the set of three, we have

$$H(x,y,p)=y+bp^2=0.$$

Accordingly

$$\eta=-\frac{1}{4b}(x+a)^2,$$

$$P=-\frac{1}{2b}(x+a):$$

when these are substituted for y and p, the other quantities are determined in the form

$$\zeta=cx-\frac{1}{4a}(3x-a)^2,$$

$$Q=-\frac{1}{2a}(3x-a).$$

Now take

$$y - \eta = Y, \quad z - \zeta = Z,$$
$$p - P = u, \quad q - Q = v,$$

and substitute in the original equations : we find, after simple reductions

$$\left. \begin{array}{l} Y = \dfrac{1}{2b}(x+a)(bu-av)+auv \\[3mm] Z = -\dfrac{1}{2a}(3x-a)(bu-av)+buv \end{array} \right\}.$$

Let Δ denote the expression

$$(bY-aZ)^2 + 4bxY(3x-a) + 4axZ(x+a) \,;$$

then

$$u = \frac{1}{4bx}(bY-aZ+\Delta^{\frac{1}{2}}),$$

$$v = \frac{1}{4ax}(-bY+aZ+\Delta^{\frac{1}{2}}) \,;$$

also

$$u = \frac{dY}{dx},$$

$$v = \frac{dZ}{dx} + c - \frac{1}{a}(3x-a),$$

so that the equations for Y and Z are

$$\left. \begin{array}{l} \dfrac{dY}{dx} = \dfrac{1}{4bx}(bY-aZ+\Delta^{\frac{1}{2}}) \\[3mm] \dfrac{dZ}{dx} = \dfrac{1}{a}(3x-a)-c+\dfrac{1}{4ax}(-bY+aZ+\Delta^{\frac{1}{2}}) \end{array} \right\}.$$

Let f denote an ordinary point for the various coefficients ; that is, f must be different from zero : further, as in § 203, we choose f, so that $3f-a-ac$ is not zero ; and we determine solutions Y and Z such that $Y=0$, $Z=0$ when $x=f$. Then we have

$$Y = \mu(x-f)^{\frac{3}{2}}R\{(x-f)^{\frac{1}{2}}\},$$
$$Z = \rho(x-f) + \kappa(x-f)^{\frac{3}{2}}S\{(x-f)^{\frac{1}{2}}\},$$

where R and S are regular functions of their argument, which are equal to 1 when the argument vanishes, and the constants are given by the equations

$$\left. \begin{array}{l} a\rho = 3f-a-ac \\[2mm] 3af\kappa = \{a\rho(f+a)\}^{\frac{1}{2}} \\[2mm] 3bf\mu = \{a\rho(f+a)\}^{\frac{1}{2}} \end{array} \right\}.$$

The integrals of the original equations are given by

$$\left. \begin{array}{l} y = \eta + Y \\ z = \zeta + Z \end{array} \right\},$$

being the integrals which, at $x=f$, are equal to η and ζ. It is clear that, at $x=f$, the derivative of y is equal to that of η, but the derivative of z is not equal to that of ζ : in other words, the quantities η and ζ derived through $J=0$, $H=0$ do not satisfy the original equations, and so do not constitute a singular integral.

CLASSIFICATION OF INTEGRALS.

207. The investigations of §§ 197—202 have shewn that, in some cases when definite conditions are satisfied by the forms of the original equations, those equations may possess more than one kind of solution—the number of kinds, other than the complete integral, not exceeding the number of dependent variables (§ 202): and in each case, the different kinds, in their construction, were made to depend upon the differential equations themselves.

Now it is well known that, in the case of a single equation of the first order, the singular solution (if any exists) can be deduced from the complete primitive: though, in the course of deduction, other equations may be associated with it which are not solutions. The argument briefly (and incompletely) stated is as follows. Let

$$\phi(x, y, c) = 0,$$

where c is an arbitrary constant, be the complete primitive of the equation

$$\Phi(x, y, p) = 0;$$

so that $\Phi = 0$ is the result of eliminating c between $\phi = 0$ and

$$\frac{\partial \phi}{\partial x} + p \frac{\partial \phi}{\partial y} = 0.$$

As regards elimination, the constancy of c is not essential to the process: the forms of the equations from which the elimination is to be made must be the same, if hypotheses as to other forms of c be adopted. Suppose, then, that c is such a function of x, that no formal change is caused in the equation which expresses p: thus the equation

$$\frac{\partial \phi}{\partial x} + p \frac{\partial \phi}{\partial y} + \frac{\partial \phi}{\partial c} \frac{dc}{dx} = 0$$

must lead to the same form for p as before, and therefore

$$\frac{\partial \phi}{\partial c} \frac{dc}{dx} = 0,$$

so that, when c is not a constant, we have

$$\frac{\partial \phi}{\partial c} = 0.$$

This equation must, for the purpose in question, coexist with $\phi(x, y, c) = 0$: eliminating c, let the result be

$$\phi_1(x, y) = 0.$$

If ϕ_1 be resoluble, let

$$\psi(x, y) = 0$$

denote a member of the aggregate of equations, cumulatively equivalent to $\phi_1 = 0$.

Now the original equation does not necessarily possess a singular solution: so that the equation $\psi = 0$ does not necessarily provide a solution. Without entering on the general analytical investigation as to the conditions, which must be satisfied in order that $\psi = 0$ should provide a solution, it will suffice here to point out that the simplest plan is actually to substitute in the differential equation, so that the equation

$$\Phi\left(x, y, -\frac{\dfrac{\partial \psi}{\partial x}}{\dfrac{\partial \psi}{\partial y}}\right) = 0$$

must (if ψ provide a solution) be satisfied either identically or in virtue of $\psi = 0$. Moreover, the functional relations between $\phi_1 = 0$, the discriminant of the complete integral, and the discriminant of Φ in regard to p, will not here be considered[*].

Ex. 1.   The primitive of

$$p^2 x - py + a = 0$$

is

$$y = cx + \frac{a}{c},$$

where c is an arbitrary constant.   Hence

$$\phi = cx - y + \frac{a}{c},$$

and so

$$\frac{\partial \phi}{\partial c} = x - \frac{a}{c^2}.$$

The inferred equation $\phi_1(x, y) = 0$, the eliminant of $\phi = 0$ and $\frac{\partial \phi}{\partial c} = 0$, which possibly provides a new solution, is

$$\phi_1 = y^2 - 4ax = 0.$$

---

[*] There is a vast amount of literature upon the subject; see Hill, *Proc. Lond. Math. Soc.*, vol. xix (1889), pp. 561—589, and Hamburger, *Crelle*, t. cxii (1893), pp. 205—246, where many references are given, and the relations between the discriminants are discussed.

Thus, from $\phi_1 = 0$, the value of p is $\dfrac{2a}{y}$: substituting, the original equation is satisfied in virtue of $\phi_1 = 0$, and so there is a singular solution.

Ex. 2.   Denoting $x + y$ by θ, the primitive of

$$p^2 (2\theta - 3\theta^2) - p (3 - 8\theta + 6\theta^2) + 2\theta - 3\theta^2 = 0,$$

is

$$\phi (x, y, c) = 3 (x - c) (y + c) - (x + y)^3 = 0.$$

The eliminant of $\phi = 0$ and $\dfrac{\partial \phi}{\partial c} = 0$ is

$$\phi_1 (x, y) = (x + y)^2 (\tfrac{3}{4} - x - y).$$

Taking

$$\psi_1 (x, y) = x + y = 0,$$

we have $p = -1$; the differential equation is not satisfied.

Taking

$$\psi_2 (x, y) = x + y - \tfrac{3}{4} = 0,$$

we have $p = -1$; the differential equation is satisfied in virtue of $\psi_2 = 0$.

Hence one of the equations equivalent to $\phi_1 = 0$ provides a solution ; the other of those equations does not provide a solution.

208.   The inference, that the complete primitive of a single equation of the first order does not, in all cases, include in itself all possible solutions of the equation, raises the corresponding question as to the comprehensibility of the complete primitive of a system of equations.   Adopting the natural generalisation of the method used for the single equation of the first order, we consider two equations of the form

$$F(x, y, z, p, q) = 0, \quad G(x, y, z, p, q) = 0,$$

with the customary notation: and we assume that the complete primitive exists in some form

$$\left. \begin{array}{l} f(x, y, z, a, b) = 0 \\ g(x, y, z, a, b) = 0 \end{array} \right\},$$

where a and b are arbitrary constants which, if necessary, will be determinable by means of initial assigned values.   In order that they may constitute a solution, then the elimination of a and b between

$$\left. \begin{array}{l} \dfrac{\partial f}{\partial x} + p \dfrac{\partial f}{\partial y} + q \dfrac{\partial f}{\partial z} = 0 \\[2mm] \dfrac{\partial g}{\partial x} + p \dfrac{\partial g}{\partial y} + q \dfrac{\partial g}{\partial z} = 0 \\[2mm] f = 0, \quad g = 0 \end{array} \right\},$$

must lead to a couple of differential equations which, if not actually in the forms $F = 0$, $G = 0$, must be equivalent to these equations.

As regards the eliminant differential equations, there is no final difference, whether the quantities to be eliminated are parametric or variable in their nature; if therefore, a and b are made functions of x, such that the forms of the equations expressing p and q are the same as before, then the equations from which the elimination is made are unchanged in form, and the final eliminant is unaltered. In order that this may be the result, a and b must be functions of x such that

$$\left.\begin{array}{c} \dfrac{\partial f}{\partial a}\dfrac{da}{dx} + \dfrac{\partial f}{\partial b}\dfrac{db}{dx} = 0 \\[2ex] \dfrac{\partial g}{\partial a}\dfrac{da}{dx} + \dfrac{\partial g}{\partial b}\dfrac{db}{dx} = 0 \end{array}\right\}.$$

Since constant values of a and b, which certainly satisfy these equations, merely give the complete integral, we pass to other possibilities. Because a and b are to be functions of x, we may regard b as a function of a: so that the two new equations can be replaced by

$$J = \frac{\partial (f, g)}{\partial (a, b)} = 0,$$

and by either

$$\frac{\partial f}{\partial a}da + \frac{\partial f}{\partial b}db = 0, \text{ or } \frac{\partial g}{\partial a}da + \frac{\partial g}{\partial b}db = 0,$$

the last two being equivalent to one another in virtue of $J = 0$.

Now from $f = 0$, $g = 0$, $J = 0$, which, in general, constitute a set of three independent equations, it is possible to consider x, y, z as explicitly expressible in terms of a and b; so that, when their values are substituted in either

$$\frac{\partial f}{\partial a} + \frac{\partial f}{\partial b}\frac{db}{da} = 0, \text{ or } \frac{\partial g}{\partial a} + \frac{\partial g}{\partial b}\frac{db}{da} = 0,$$

the result is a single differential equation of the first order between a and b. Let

$$h (a, b) = 0$$

represent an integral of this new equation; if it be the complete primitive, it will contain an arbitrary constant; if it be a singular solution of that new equation, no such arbitrary constant will occur.

There are now four equations, viz.

$$f = 0, \quad g = 0, \quad J = 0, \quad h = 0,$$

involving a, b, x, y, z; when a and b are eliminated, we have two relations between x, y, z, which involve an arbitrary constant if $h = 0$ is a complete primitive, and involve no arbitrary constant if $h = 0$ is a singular solution. But it does not follow that the two relations thus obtained constitute a solution of the original equations: the easiest plan of settling the doubt is actually to substitute in the equations.

One simple kind of source for equations, which may be thus obtained and do not furnish solutions of the original equation, can be inferred by considering an associable geometrical interpretation. In the case of a single equation, which has

$$\phi(x, y, c) = 0$$

for its primitive, the equation, which results from the elimination of c between

$$\phi = 0 \quad \text{and} \quad \frac{\partial \phi}{\partial c} = 0,$$

includes the locus (if any) of the nodes of the system of curves; and though at any such point the values of x and y are the same for the node-locus as for the curve, the value of $\frac{dy}{dx}$ for the node-locus is not necessarily the same as for the curve, because the two need not touch. In the case of a couple of equations, the general primitive represents a congruence of curves in space; the equation $J = 0$ is satisfied at every node, and the eliminant of $f = 0$, $g = 0$, $J = 0$, will include the nodes of the system. Thus the integrals derived by the further operations may be connected with a curve-locus of nodes, the directions of which are not necessarily one of the directions of the curves at the node, and the equations of which therefore will not, in general, constitute an integral of the equations.

Ex. 1.   Consider the equations

$$\left. \begin{aligned} -y + ax + a^2 + b &= 0 \\ -z + bx + ab &= 0 \end{aligned} \right\},$$

where a and b are arbitrary (Ex. 1, § 201). We have

$$J = \begin{vmatrix} x + 2a, & 1 \\ b, & x + a \end{vmatrix} = x^2 + 3ax + 2a^2 - b = 0;$$

the two equations (equivalent to one another in virtue of $J=0$) are

$$(x+2a)\,da+db=0,$$
$$bda+(x+a)\,db=0.$$

Hence

$$-\frac{db}{da}=x+2a,\text{ by the first equation,}$$

$$=a-\frac{b}{\dfrac{db}{da}},\text{ by the second equation :}$$

thus

$$b=a\frac{db}{da}+\left(\frac{db}{da}\right)^2,$$

so that

$$b=a\kappa+\kappa^2.$$

Hence

$$x+2a=-\frac{db}{da}=-\kappa,$$

so that

$$a=-\tfrac{1}{2}(x+\kappa),$$
$$b=\tfrac{1}{2}\kappa(\kappa-x).$$

Substituting in the initial equations, we have

$$\left.\begin{array}{l}y=\kappa^2-\tfrac{1}{4}(x+\kappa)^2\\z=-\tfrac{1}{4}\kappa(x-\kappa)^2\end{array}\right\},$$

which are a couple of equations constructed by the method indicated.

The differential equations of the original system are

$$\left.\begin{array}{l}y=px+p^2+q\\z=qx+pq\end{array}\right\}:$$

on substituting the values just obtained, it appears that the differential equations are satisfied, and therefore that the new equations provide a solution not included (because b and a are functions of x for the purpose) in the original system.

The equation between b and a possesses also a solution, singular to itself, viz.

$$a^2+4b=0,$$

so that $\dfrac{db}{da}=-\tfrac{1}{2}a$, and therefore

$$x+2a=\tfrac{1}{2}a\;;$$

whence

$$a=-\tfrac{2}{3}x,$$
$$b=-\tfrac{1}{9}x^2.$$

Then

$$y=-\tfrac{1}{3}x^2,\quad z=-\tfrac{1}{27}x^3:$$

on substitution, these are found to satisfy the differential equations of the original system.

Ex. 2.   Consider the equations whose primitive is

$$\left.\begin{array}{l}f=x^2+y^2+z^2-a^2=0\\g=ax^2+by^2+2abxy-a^3=0\end{array}\right\},$$

where a and b are arbitrary. We have

so that
$$J = -2a\,(y^2 + 2axy),$$

$$y = 0, \quad x^2 = a^2, \quad z = 0,$$

are values of x, y, z, derived from $f = 0$, $g = 0$, $J = 0$. Now the two equations for variations of a and b are

$$0 = a\,da,$$

$$0 = (x^2 + 2bxy - 3a^2)\,da + (y^2 + 2axy)\,db,$$

which, when the values of x, y, z are substituted, are satisfied by

Thus
$$a = \text{constant}, \quad b = \text{any quantity.}$$

$$y = 0, \quad z = 0,$$

is a locus obtained by the method of § 208 : it is a locus of nodes of the curves defined by the primitive. That it does not provide a solution of the corresponding differential equations can be seen at once : for one of them is

$$x + yp + zq = 0,$$

which is not satisfied in connection with $y = 0$, $z = 0$.

Ex. 3. Discuss the alternative relation $y + 2ax = 0$, arising in the last example out of $J = 0$.

Ex. 4. Shew that, in connection with a complete primitive

$$y = ax + f(a, b), \quad z = bx + F(a, b),$$

the singular solution (if it exists) is given by the association of the equations

$$x = -\frac{\partial f}{\partial a} - \frac{\partial f}{\partial b}\frac{db}{da} = -\frac{\partial F}{\partial a}\frac{da}{db} - \frac{\partial F}{\partial b},$$

$$\frac{\partial f}{\partial b}\left(\frac{db}{da}\right)^2 + \left(\frac{\partial f}{\partial a} - \frac{\partial F}{\partial b}\right)\frac{db}{da} - \frac{\partial F}{\partial a} = 0,$$

with the complete primitive.

(Serret.)

Prove that the equations generally possess a singular solution of the first class.

Ex. 5. Obtain the various singular solutions that are possessed by the differential equations, of which

$$\left.\begin{aligned} y &= ax - 9b^2 + 18a^2b - 4a^4 \\ z &= bx + 12ab^2 - 4a^3b \end{aligned}\right\}$$

constitute the complete primitive.

(Serret.)

Note. From the general theory, as well as from the first of the examples considered, it appears that a system of a couple of equations may possess (though it does not necessarily possess) two classes of singular solutions derivable from the complete primitive. More generally, a system of n simultaneous equations of the first order in n dependent variables may, in addition to the complete primitive, possess n distinct classes of singular solutions derivable from the complete primitive—a result, as regards the kinds of solutions possessed, which is in accordance with the results of § 202. It is not my intention to discuss the inferences again, and to establish their relations with the primitive: this discussion will be found in a memoir by Serret*, and in the memoirs by Mayer, Goursat, and Dixon, already (§ 200) quoted. The last of these memoirs discusses some important applications to the theory of the bitangents of torses, and to cognate questions in geometry.

In connection with this part of the subject, reference should be made to the second part of the memoir by Hamburger (cited in § 207), where the functional relations of singular solutions and complete primitives of a single equation of the first order are investigated. So far as I am aware, there is no complete investigation, which deals with the corresponding relations, in the case of a system of two or more equations of the first order.

* *Liouville*, 1re Sér., t. xviii (1853), pp. 1—40.

CHAPTER XIV.

EQUATIONS OF THE SECOND ORDER AND THE FIRST DEGREE*.

REGULAR INTEGRALS OF AN EQUATION $w'' = f(w', w, z)$.

209. ONE of the simplest systems of two equations in two dependent variables is that, which is the equivalent of a single equation of the second order in a single dependent variable.

When the equation is linear in the derivative of the second order, the system is of the form

$$\left.\begin{aligned} \frac{dw'}{dz} &= f(w', w, z) \\ \frac{dw}{dz} &= w' \end{aligned}\right\}.$$

The general existence-theorem of § 10 can be applied to this system, with the following result:—

Let $z = c$, $w = \alpha$, $w' = \beta$ be an ordinary combination of values for the function f, so that f can be expanded as a regular function of $w' - \beta$, $w - \alpha$, $z - c$; let β be finite, so that the function, defined by the equations as the value of $\frac{dw}{dz}$, is regular; and let the domain of z round c, within which the function f is regular (with the corresponding ranges for w and w'), be given by $|z - c| \leqslant r$. Then there exists a regular function of z, which is a solution of the equation of the second order; at c, it acquires the value α, and its derivative acquires the value β; and so long as the variable is restricted to variation within the specified domain, the solution so

---

* Some references, more or less slight, to the subject of this chapter will be found in a few of the memoirs quoted in Chapters XII and XIII. Painlevé's Stockholm Lectures *Sur la théorie analytique des équations différentielles*, (1897), pp. 394—433, may also be consulted.

determined is the only solution, which satisfies the assigned conditions. The expression for the solution is

$$w - \alpha = \beta (z - c) + \tfrac{1}{2} (z - c)^2 f (\beta, \alpha, c) + \dots.$$

The sole limitation has been that $w' = \beta$, $w = \alpha$, $z = c$, constitute an ordinary combination of values for the function $f(w', w, z)$: in order that w' may be regular in the vicinity of the assigned initial value, β must be finite.

But as in the case of a single equation of the first order, it does not follow that the regular integral thus obtained, though unique as a regular integral, is the only solution of the equation, which satisfies the initial conditions. Any other such solution will be a non-regular function of the variable z; in order to obtain its full significance, it then would be necessary to consider variations of z outside the domain of c.

Thus if $w' - \beta$ is a factor of the function $f(w', w, z)$, the only regular solution of the equation is

$$w = \alpha + \beta (z - c).$$

If the initial value β' be zero, and if $w - \alpha$ be a factor of $f(w', w, z)$, the only regular solution of the equation is

$$w = \alpha.$$

If the initial value β be zero, and if $f(w', w, z)$ be expressible in the form

$$w' g (w', w, z) + (w - \alpha) h (w', w, z),$$

where g and h are regular functions in the vicinity of the assigned initial values, the only regular solution of the equation is

$$w = \alpha.$$

But if we admit into consideration the non-regular functions of z, some branches of which may at least satisfy the initial conditions, it is not difficult to construct examples, for which these special results of the general theorem are not valid. For instance, if $w = 0$, and $w' = 0$, when $z = 0$, be assigned as initial values for the equation

$$w'' = w' \frac{2w - 1}{z - 1},$$

a first integral satisfying* these conditions is

$$(z - 1) w' = w^2,$$

---

* A general first integral is $(z - 1) w' = w^2 + A^2$; the primitive is
$$w = A \tan \{B + A \log (z - 1)\}.$$

and a final integral is

$$w = \frac{1}{A - \log(z-1)}.$$

If that branch of the (infinitely-branched) logarithmic function be taken which can be represented by $\mathrm{Log}\,(z-1) + 2k\pi i$, where k is infinite, the corresponding branch of w is zero. As in the simpler case (§ 15) of a single equation of the first order, it happens that this branch coincides, within the domain $|z| < 1$, with the uniform branch $w = 0$; but when z is not restricted to lie within this domain, then the branches which vanish at $z = 0$ are only a portion of the function.

More generally, we have the result, analogous to Fuchs's theorem (§ 34) for the first order:

The integral of the equation

$$w'' = f(w', w, z),$$

determined as equal to u, a regular function of z, by the assigned initial conditions, is the sole integral of the equation determined by those conditions, only if the initial value of z is not a point of indeterminateness for the equation

$$\frac{d^2v}{dz^2} = f\left(\frac{dv}{dz} + \frac{du}{dz},\ v + u,\ z\right) - f\left(\frac{dv}{dz},\ v,\ z\right).$$

Thus for the equation

$$\frac{d^2w}{dz^2} = \gamma + (w' - \beta - \gamma z)^2\, F(w', w, z),$$

where F is a regular function of its arguments, if the assigned conditions are that $w = \beta c + \frac{1}{2}\gamma c^2$, $w' = \beta + \gamma c$, when $z = c$, the value $z = c$ is a point of indeterminateness.

Note. The existence-theorem, stated at the beginning of this section, is deduced as a special instance of Cauchy's general existence-theorem (§ 10) appertaining to a system of any number of equations each of the first order. The direct application of the analysis, which leads to the general theorem, is cumbrous for the present case; and it does not appear to be the obviously appropriate analysis. It is desirable to have a direct establishment of the theorem, appropriate to the case $n = 2$; the investigation is left to the student as an exercise.

INTEGRALS OF THE EQUATION $z^2w'' = f(zw', w, z)$.

210. Another existence-theorem for an equation, which does not satisfy the foregoing requirement as to form, is sufficiently important to be treated independently. The equation in question is

$$z^2w'' = f(zw', w, z),$$

where f is a regular function of its arguments. Integrals are required for values of z in the vicinity of the origin: it will be seen that, in certain circumstances, the equation possesses integrals, which are finite in that vicinity.

We assume

$$w = \alpha + \beta z + \gamma z^2 + uz,$$

so that

$$w' = \beta + 2\gamma z + u + zu';$$

the coefficients α, β, γ are constants; and γ is to be chosen so that, if possible, both u and u' are zero, when $z = 0$. Let f_{lmn} denote

$$\frac{\partial^{l+m+n}}{\partial t^l \partial w^m \partial z^n} f(t, w, z),$$

when, after differentiation, $t = 0$, $w = \alpha$, $z = 0$; then the equation for u is

$$2\gamma z^2 + z^3 u'' + 2z^2 u' - f(0, \alpha, 0)$$
$$= \Sigma\Sigma\Sigma \frac{1}{l!\,m!\,n!} z^n (\beta z + \gamma z^2 + uz)^m (\beta z + 2\gamma z^2 + zu + z^2 u')^l f_{lmn}$$

the summation being for integer values of l, m, n, such that $l + m + n \geqslant 1$. Making $z = 0$, we have

$$f(0, \alpha, 0) = 0:$$

so that a first condition is, that the initial value of w for $z = 0$ must be a root of the equation $f(0, \alpha, 0) = 0$. If this equation is an identity, then the initial value of w remains arbitrary.

Since $u' = 0$ when $z = 0$, we can regard the order of u' in powers of z, for sufficiently small values of $|z|$, as greater than zero, and the order of u as greater than unity. Thus there are no terms of the first order on the left-hand side: and therefore the terms of the first order on the right-hand side must vanish, so that

$$\beta f_{100} + \beta f_{010} + f_{001} = 0,$$

giving

$$\beta = -\frac{f_{001}}{f_{100} + f_{010}}.$$

Equating terms of the second order in powers of z on the two sides, we have

$$2\gamma = \tfrac{1}{2}f_{002} + \beta\,(f_{011} + f_{101}) + \tfrac{1}{2}\beta^2\,(f_{200} + 2f_{110} + f_{020}) + \gamma\,(f_{010} + 2f_{100}),$$

giving the value of γ. The remaining terms then constitute the equation

$$z^3 u'' + 2z^2 u' = z^2 u' f_{100} + zu\,(f_{100} + f_{010}) + z^3\kappa + F(z^2 u', zu, z),$$

where F is a regular function of its arguments: in F, the terms are all of order at least equal to four, the orders of u and u' being as explained above. Now let

$$u = vz,$$

so that

$$p = u' = z\frac{dv}{dz} + v,$$

the quantity v vanishing with z. Then

$$F(z^2 u', zu, z) = F(z^2 p, z^2 v, z),$$

so that F is divisible by z^2; the quotient, say Θ, is of order at least equal to two, because the orders of p and v are not less than zero, and, in general, are greater than zero. Thus, since $u'' = \dfrac{dp}{dz}$, the equations in p and v are

$$\left.\begin{aligned} z\frac{dp}{dz} &= \kappa z + (f_{100} - 2)\,p + (f_{100} + f_{010})\,v + \Theta \\[2mm] z\frac{dv}{dz} &= p - \phantom{(f_{100})}\, v \end{aligned}\right\},$$

where the lowest terms in Θ are of dimensions at least two in the variables p, v, z.

These equations are of the form already discussed in §§ 165 et seq.; the general characteristics of their integrals are known. The critical quadratic is

$$(\Omega + 2 - f_{100})\,(\Omega + 1) - (f_{100} + f_{010}) = 0,$$

that is,

$$\Omega^2 - \Omega\,(f_{100} - 3) + 2 - 2f_{100} - f_{010} = 0:$$

let its roots be Ω_1 and Ω_2.

If the integrals are to be regular, the value of β must not be infinite. Unless $f_{001} = 0$, this will be the case when $f_{100} + f_{010} = 0$; so that Ω may not be equal to -1, unless f_{001} certainly is zero: and even if this is so, it does not follow that there is a finite value for β. Also, the value of γ must not be infinite, and therefore $2 - 2f_{100} - f_{010}$ must not be zero: that is, Ω may not be equal to zero.

From the earlier investigations, we know that, if the equations have regular integrals, neither Ω_1 nor Ω_2 can be a positive integer, unless a certain condition is satisfied among the other coefficients; if therefore a root of the critical quadratic be an integer, it must be less than -1, unless a condition is satisfied.

If neither $\Omega_1 + 2$ nor $\Omega_2 + 2$ is a positive integer greater than zero, then there exists a pair-solution of the equations; the values for v and p vanish with z, and have the form

$$p = 2Az + \ldots,$$
$$v = Az + \ldots,$$

where

$$A(6 - 3f_{100} - f_{010}) = \kappa;$$

and these values are regular functions of z. If however the critical quadratic has an integer greater than -2 for a root, then this pair-solution (in the form of regular functions of z) does not exist, unless a certain condition among the coefficients of f be satisfied: when this ·condition is satisfied, there can exist an infinite number of regular solutions.

The further inferences, proved for the corresponding equations (§§ 173 *et seq.*), as regards the existence of non-regular functions of z, which vanish with z and satisfy the equations, need hardly be restated here. The existence and, if they exist, the character of such non-regular solutions depend upon the actual expressions of the roots Ω_1 and Ω_2 of the critical quadratic.

Assuming then that v denotes the most general value of v which, in association with the corresponding most general value of p, satisfies the equations for v and p, the value of w is

$$w = \alpha + \beta z + z^2(\gamma + v),$$

where α, β, γ have the values as obtained above. Hence it follows that *the equation*

$$z^2 w'' = f(zw', w, z)$$

possesses a solution, determined by the specified conditions, viz.

(i)   that it acquires a value α when $z = 0$;

(ii)   that β and γ are definite constants;

(iii) that its regularity or non-regularity, as a function of z, depends upon the roots of the critical quadratic

$$\Omega^2 - \Omega(f_{100} - 3) + 2 - 2f_{100} - f_{010} = 0.$$

Ex. 1.   Obtain the solutions (if any) of the equation

$$z^2 w'' = a \left(zw' - w\right)^2 + bz^2,$$

which are finite and have their first derivative finite, when $z = 0$.

Since w' is not infinite when $z = 0$, therefore zw'' is not infinite for $z = 0$: consequently $z^2 w''$ is zero for that value. Thus the equation $f(0, a, 0) = 0$ is

$$a^2 = 0,$$

that is, the integral must have an initial zero value. In accordance with the text, we take

$$w = \beta z + \gamma z^2 + zu,$$

so that

$$zw' - w = z^2 \left(\gamma + \frac{du}{dz}\right);$$

the equation becomes

$$\left(2\gamma + zu'' + 2u'\right) z^2 = az^4 \left(\gamma + \frac{du}{dz}\right)^2 + bz^2.$$

Hence $2\gamma = b$; and thus

$$zu'' + 2u' = az^2 \left(\tfrac{1}{2} b + \frac{du}{dz}\right)^2.$$

The equations are

$$\left. \begin{aligned} z \frac{dp}{dz} &= -2p + az^2 \left(\tfrac{1}{2} b + p\right)^2 \\ z \frac{dv}{dz} &= \quad p - v \end{aligned} \right\} :$$

the critical quadratic is

$$(\Omega + 1)(\Omega + 2) = 0,$$

both the roots of which are negative integers.

It therefore follows that there exists a function, which vanishes with z and satisfies the equations. When it is substituted in the expression for w, it provides a solution of the original equation ; this solution is a regular function of z. Also, because the roots of the critical quadratic are negative integers, this regular function is the only solution of the equation which vanishes with z ; the quantity β, being the initial value of w', is arbitrary.

This result may be verified by shewing, through direct quadrature, that the complete primitive of the original equation is

$$\frac{w}{z} = \beta - \frac{1}{a} \int \frac{1}{z\eta} \frac{d\eta}{dz} \, dz,$$

where

$$\eta = A J_0 (abz) + B Y_0 (abz),$$

β being an arbitrary constant, and J_0, Y_0 denoting Bessel's functions of order zero. In order that the imposed conditions may be satisfied, it is necessary that $B = 0$; the integral w then becomes a regular function of z.

Ex. 2.   Obtain the integrals of the equation

$$z^2 w'' = 2zw' - 2w (1 - w) - 2 \frac{z^2 w'^2}{1 - w},$$

which vanish and have a derivative that is not infinite, when $z = 0$.

(The complete primitive is

$$w = \frac{\beta z + \gamma z^2}{1 + \beta z + \gamma z^2}.)$$

Ex. 3.   Consider the equation

$$z^2 w'' = (m + n - 1) z w' - mnw + 2kz + (a, b, c, f, g, h \rangle z w', w, z)^2 + l z^3 + \dots,$$

so as to obtain uniform integrals, which are finite and have finite first and second derivatives, when $z = 0$.

If such an integral can exist, and if a be its value when $z = 0$, then we have

$$0 = - mna + ca^2,$$

so that either $a = 0$ or $a = - mn/c$.   The general expression in the vicinity of $z = 0$ must be of the form

$$w - a = \theta_1 z + \theta_2 z^2 + \dots.$$

For this value, we have

$$z^2 w'' - (m + n - 1) z w' + mnw = mna + \underset{p=1}{\Sigma} (p - m)(p - n) \theta_p z^p ;$$

and accordingly,

$$(p - m)(p - n) \theta_p = \text{coefficient of } z^p \text{ in } 2kz + (a, b, c, f, g, h \rangle z w', w, z)^2 + \dots$$

$$= \Theta_p,$$

say, the foregoing expression being substituted for w.

The Ω-equation is easily obtained : for

$$f_{100} = m + n - 1, \quad f_{010} = - mn,$$

and therefore

$$\Omega^2 - \Omega (m + n - 4) + 4 - 2m - 2n + mn = 0,$$

that is,

$$\{\Omega - (m - 2)\} \{\Omega - (n - 2)\} = 0.$$

Hence, if m and n be positive integers, the equation does not unconditionally possess regular integrals of the required type.

As a special instance, let $m = 1$, $n = 2$.   Then taking the value 0 of a, the condition for the possession of regular integrals is

$$k = 0.$$

Similarly for the value $- 2/c$ of a.

Again, let $m = 2$, $n = 3$.   Then taking the value 0 of a, the condition for the possession of regular integrals is

$$(a + 2h + b) k^2 + 2 (f + g) k + c = 0 ;$$

the value of θ_2 is given by

$$(4a + 3h + 2b + 4g + 2f) \theta_2 + l = 0 ;$$

the value of θ_3 is arbitrary, and the subsequent coefficients are expressible in terms of θ_3.   There thus is an infinitude of regular integrals, if the condition is satisfied.   Similarly for the value $-6/c$ of a.

INTEGRALS OF AN EQUATION $z^m w^{(m)} = f(z^{m-1} w^{(m-1)}, \ldots, zw', w, z)$.

211. A like result holds for an equation, which is of any order m and of similar form, say

$$z^m \frac{d^m w}{dz^m} = f\left(z^{m-1} \frac{d^{m-1} w}{dz^{m-1}},\ z^{m-2} \frac{d^{m-2} w}{dz^{m-2}},\ \ldots,\ z \frac{dw}{dz},\ w,\ z\right).$$

Let α denote a root of

$$f(0, 0, \ldots, 0, \alpha, 0) = 0,$$

in which all the arguments of f, except w, have been made zero; and let f_r denote the value of

$$\frac{\partial}{\partial\left(z^{m-r} \dfrac{d^{m-r} w}{dz^{m-r}}\right)} f\left(z^{m-1} \frac{d^{m-1} w}{dz^{m-1}},\ \ldots,\ z \frac{dw}{dz},\ w,\ z\right),$$

when, after differentiation, w is made equal to α, and all the other arguments are made zero: r having the values $1, 2, \ldots, m$. Also, let $E_m(\Omega)$ denote

$\Omega + m - f_1$,	$(m-1)f_1 + (m-2)f_2$,	$(m-2)f_2 + (m-3)f_3$,	$\ldots$,	$2f_{m-2} + f_{m-1}$,	$f_{m-1} + f_m$
1 ,	$\Omega + 1$,	0	, ...,	0 ,	0
0 ,	1 ,	$\Omega + 2$	, ...,	0 ,	0
0 ,	0 ,	1	, ...,	0 ,	0
....	....	....	....	....	....
0 ,	0 ,	0	, ...,	1 ,	$\Omega + m - 1$

where the last constituent in the first row differs in form from the preceding $m - 2$ constituents in that row. The expanded form of $E_m(\Omega)$ is easily obtained ; thus

$$E_3(\Omega) = \{(\Omega + 3 - f_1)(\Omega + 1) - (2f_1 + f_2)\}(\Omega + 2) + (f_1 + f_2),$$

$$E_4(\Omega) = [\{(\Omega + 4 - f_1)(\Omega + 1) - (3f_1 + 2f_2)\}(\Omega + 2)$$
$$+ (2f_2 + f_3)](\Omega + 3) - (f_2 + f_3),$$

and so on. Lastly, let $\Omega_1', \Omega_2', \ldots, \Omega_m'$ denote the m roots of the algebraical equation $E_m(\Omega) = 0$; and let $\Omega_i = \Omega_i' + m$, for $i = 1, \ldots, m$. Then we have the following theorems :—

If no one of the quantities $\Omega_1, \ldots, \Omega_m$ is a positive integer, then a solution of the equation exists, which acquires the value α, when $z = 0$, and is expressible in the form of a regular function of z : but if one, or more than one, of those quantities is a positive integer, then, unless a condition is satisfied among the coefficients of f, there is no such regular solution.

If the roots Ω_1', ..., Ω_κ' have their real parts positive, though they are not themselves positive integers, and no two of them be equal, and if $\Omega'_{\kappa+1}$, ..., Ω'_m have their real parts negative, (but, if actually integers, then each $< -m$), then there is a κ-tuple infinitude of integrals, which acquire a value α when $z = 0$, and are expressible as regular functions of $z^{\Omega_1'}$, ..., $z^{\Omega_\kappa'}$, z. If of the roots having their real parts positive two be equal, say Ω_1' and Ω_2', then the arguments $z^{\Omega_1'}$ and $z^{\Omega_2'}$ of the regular functions, that express the integrals, are to be replaced by $z^{\Omega_1'}$ and $z^{\Omega_1'} \log z$, and the infinitude of integrals is only $(\kappa - 1)$-tuple; and likewise if more than two of those roots be equal.

If a root Ω' is a positive integer, and if the condition for the existence of the (infinitude of) regular functions of z as integrals of the equation is not satisfied, then there exists an infinitude of integrals, which are expressible as regular functions of z and $z \log z$, and acquire a value α when z is zero. Similarly, if two or more roots are positive integers.

In fact, propositions similar to those in § 210 hold for the differential equation; they depend upon the character of the root-system of the critical equation

$$E_m(\Omega) = 0.$$

Note. The character of an integral being recognised in accordance with these propositions, its formal expression is obtainable (as usual) by substituting an expression in the form of a regular function, which possesses that character, and by determining the coefficients so as to make the equations satisfied identically.

INTEGRALS OF AN EQUATION $zw'' = f(w', w, z)$.

212.   Another class of equations, represented by

$$zw'' = f(w', w, z),$$

where f is a regular function of its arguments, can be treated in a similar manner, the integrals (if any) being required to be finite in the vicinity of $z = 0$. Let α be an initial value assigned to w, β to w'; then the initial value of w'' will be infinite unless

$$f(\beta, \alpha, 0) = 0,$$

13

a relation which accordingly will be assumed to be satisfied. Now take

$$w = \alpha + \beta z + y, \quad w' = \beta + y', \quad w'' = y'';$$

then

$$zy'' = f(\beta + y', \alpha + \beta z + y, z)$$
$$= y'f_1 + yf_2 + (\beta f_2 + f_3) z + \text{terms of higher orders,}$$

the quantities y and y' vanishing with z. Hence the equations may be taken in the form

$$\left.\begin{aligned}
z\frac{dy'}{dz} &= y'f_1 + yf_2 + (\beta f_2 + f_3) z + \dots \\
z\frac{dy}{dz} &= \qquad\qquad\qquad zy'
\end{aligned}\right\},$$

similar to the form discussed in § 210. The critical quadratic is

$$(\Omega - f_1)\,\Omega = 0,$$

the roots of which are f_1, 0; so that the character of the integral depends upon the quantity f_1, which is the value of

$$\frac{\partial f(w', w, z)}{\partial w'},$$

when, after differentiation, we take $w' = \beta$, $w = \alpha$, $z = 0$.

If f_1 is not a positive integer, then the equation possesses a solution expressible as a regular function of z, such that y vanishes with z: and the solution is uniquely determinate. If the real part of f_1 is then negative, this solution is the only integral of the equation determined by the initial values α and β; but if the real part of f_1 is positive, though f_1 is not itself a positive integer, then an infinitude of integrals is determined by the initial values α and β, these integrals being expressible as regular functions of z and z^{f_1}.

If f_1 is a positive integer, then in general the equation possesses no solution, which is expressible in the form of a regular function of z that vanishes with z: but it possesses an infinitude of solutions determined by the initial values α and β, these solutions being expressible as regular functions of z and $z \log z$. But if a particular condition among the coefficients be satisfied, there is an infinitude of solutions, expressible in the form of regular functions of z and determined by the initial values α and β: in that case, such solutions are the only integrals of the equation determined by those initial values.

Such integrals are, of course, only a particular class of integrals; and if the initial values α and β be such that $f(\beta, \alpha, 0)$ is not zero, then the subsequent inferences are no longer valid. The point $z = 0$ then usually is an essential singularity.

Note. There is a corresponding result for an equation

$$z \frac{d^m w}{dz^m} = f\left(\frac{d^{m-1} w}{dz^{m-1}}, \ldots, w, z\right)$$

of order m; the critical equation has $m - 1$ roots equal to zero and one root equal to f_1, where f_1 is the value of

$$\frac{\partial}{\partial\left(\frac{d^{m-1} w}{dz^{m-1}}\right)} f\left(\frac{d^{m-1} w}{dz^{m-1}}, \ldots, w, z\right),$$

when the initial values are substituted after differentiation. The results are similar to those for the equation of order 2, and depend upon the value of f_1 in the same manner as for that lower order[*].

Ex. Consider the equation

$$a^2 z w'' = (z w' - w)^2 - c^2.$$

If the integral, supposed to be finite when $z = 0$, has an initial value a, we have $a^2 = c^2$, so that $a = \pm c$: and β remains arbitrary. With the substitutions of the text, we have

$$a^2 z y'' = (z y' - y)\{-2a + (z y' - y)\};$$

and therefore, since $\frac{d}{dz}(z y' - y) = z y''$, the only regular solution of the equation is

$$z y' - y = 0,$$

that is, with the condition $y' = 0$ when $z = 0$, we have $y = 0$ as the solution, and therefore the integral of the original equation is

$$w = \pm c + \beta z,$$

subject to the condition assigned.

As the equation can be integrated by quadratures, it is possible to compare this result with the complete primitive. Let

$$z w' - w = -\frac{a^2}{\eta} \frac{d\eta}{dz};$$

after substitution, the equation for η is

$$\frac{d^2 \eta}{dz^2} - \eta \frac{c^2}{a^4} = 0,$$

so that

$$\eta = A e^{\frac{cz}{a^2}} + B e^{-\frac{cz}{a^2}},$$

---

[*] See also Königsberger, *Theorie d. Differentialgleichungen*, pp. 411—416.

and therefore

$$zw' - w = -c\,\frac{Ae^{\frac{2cz}{a^2}} - B}{Ae^{\frac{2cz}{a^2}} + B}.$$

Consequently,

$$\frac{w}{z} = C - c\int \frac{Ae^{\frac{2cz}{a^2}} - B}{Ae^{\frac{2cz}{a^2}} + B}\frac{dz}{z^2},$$

which is the complete primitive; it contains two arbitrary constants, viz.
C and the fraction B/A. Denoting this fraction by k, we have

$$\frac{w}{z} = C - c\int \left\{\frac{1-k}{1+k} + \frac{4k}{(1+k)^2}\frac{cz}{a^2} + \dots\right\}\frac{dz}{z^2},$$

so that

$$w = c\frac{1-k}{1+k} - \frac{4k}{(1+k)^2}\frac{c^2}{a^2} z\log z + \text{a regular function of } z,$$

this regular function vanishing with z.

If initially $w = c$, then $k = 0$ and the term in $z\log z$ disappears; if initially
$w = -c$, then $k = \infty$ and the term in $z\log z$ again disappears; for any other
assigned initial value of w, the term in $z\log z$ remains.

NON-REGULAR INTEGRALS OF $w'' = f(w', w, z)$.

213. The simplest deviation from the previous conditions
arises, when the initial value assigned to w' is infinite. In that
case, we make z the dependent variable and w the independent
variable, so that

$$z' = \frac{1}{w'}, \quad w'' = -\frac{1}{z'^3}z'',$$

and the equations are

$$\begin{aligned}\frac{dz'}{dw} &= -z'^3 f\left(\frac{1}{z'}, w, z\right) = g(z', z, w)\\[2mm]\frac{dz}{dw} &= z'\end{aligned}\Biggr\}.$$

The initial conditions now are that $z = c$, and $z' = 0$, when $w = a$.

If $z' = 0$, $z = c$, $w = a$, constitute an ordinary combination for
the function g, the preceding existence-theorem applies: and the
solution is

$$z - c = \tfrac{1}{2}(w - a)^2 g(0, c, a) + \dots,$$

so that, if $g(0, c, a)$ is different from zero, we have

$$w - a = (z - c)^{\frac{1}{2}} P\{(z - c)^{\frac{1}{2}}\},$$

where P is a regular function of its argument, and does not vanish when $z = c$: and if $g(0, c, a)$ is zero, we have a result of the form

$$w - a = (z - c)^{\frac{1}{n}} R\{(z - c)^{\frac{1}{n}}\},$$

where $n > 2$, and R is a regular function of its argument, which does not vanish when $z = c$. But when z' is an integral factor of g, so that $\frac{1}{z'} g(z', z, w)$ is regular in the vicinity of $z' = 0$, $z = c$, $w = a$, then we have

$$z - c = 0$$

everywhere in the domain considered: so that no solution of the original equation exists, which can express w as a function of z, unless the combination of values is a point of indeterminateness for the equation.

If, however, $z' = 0$, $z = c$, $w = a$, constitute a non-ordinary combination for the function g, then, instead of making the transformation which leaves w as the independent variable, it is better and simpler to investigate the existence of the integrals (if any) by the methods, which are used in connection with a combination of values that constitute an accidental singularity of the first kind or of the second kind, taking as the sole initial condition that w shall acquire a value a when $z = c$.

Ex. Determine a solution of the equation

$$w^2 \frac{d^2 w}{dz^2} = (cw - az)\left(\frac{dw}{dz}\right)^3,$$

such that $w = a$ and $\frac{dw}{dz} = \infty$ when $z = c$.

214. Deviations (other than that just considered) from the initial conditions, which uniquely determine the existence of an integral of the equation expressible as a regular function of z, may occur in various ways.

It is unnecessary to take any detailed account of two of the simplest cases. One of these arises when the initial value of z is infinite: we take a new variable $z'\left(= \frac{1}{z}, \text{ as usual}\right)$, and we have zero as the new initial value. The other arises when the initial value of w is infinite: a similar transformation leads to zero as the initial value of the new variable.

As regards the possibility of an infinite initial value of w', sometimes it is a condition that may be imposed upon the equation (and the equation is then treated as in § 213): sometimes it is a condition evolved by the form of the equation itself. Accordingly, in many cases, the assignment of an initial value to w' will be left open.

The most important class of cases is constituted by those, in which the assigned initial values constitute a non-ordinary combination for the function $f(w', w, z)$ on account of its form. To the discussion of some of these we now proceed.

ACCIDENTAL SINGULARITY OF THE FIRST KIND.

215. First, suppose that the values $w' = \beta$, $w = \alpha$, $z = c$ constitute an accidental singularity of the first kind of the function f, so that (Note, Chapter I) f can be represented in the form

$$\frac{1}{f(w', w, z)} = P(w' - \beta, w - \alpha, z - c),$$

where P is a regular function of w', w, z, which vanishes at β, α, c. In order to discuss the solution, we retain w' as a variable, making it now the independent variable; so that we have

$$\left.\begin{aligned}
\frac{dz}{dw'} &= P(w' - \beta, w - \alpha, z - c) \\
\frac{dw}{dw'} &= w' P(w' - \beta, w - \alpha, z - c) \\
&= \beta P + (w' - \beta) P
\end{aligned}\right\}.$$

Suppose that terms occur in P, independent of $w - \alpha$ and $z - c$; and let the lowest power of $w' - \beta$ in P occurring alone be $A(w' - \beta)^n$; then we have, as in other cases with similar expressions,

$$\left.\begin{aligned}
z - a &= \frac{A}{n+1}(w' - \beta)^{n+1} R(w' - \beta) \\
w - \alpha &= \frac{A\beta}{n+1}(w' - \beta)^{n+1} S(w' - \beta)
\end{aligned}\right\},$$

where R and S denote regular functions of their arguments, each of them becoming equal to unity when $w' = \beta$. Thus

$$w' - \beta = \left\{\frac{n+1}{A}(z - a)\right\}^{\frac{1}{n+1}} R_1\{(z - a)^{\frac{1}{n+1}}\},$$

where R_1 denotes a regular function of its argument, which becomes equal to unity when $z = a$; and therefore

$$w - \alpha = \beta (z - a) + (z - a)^{1 + \frac{1}{n+1}} R_2 \{(z - a)^{\frac{1}{n+1}}\},$$

where R_2 denotes a regular function of its argument, which does not vanish when $z = a$. The point $z = a$ is thus a branch-point for $n + 1$ branches of the integral of the equation, which circulate round the point a; they are equal to one another at the point, and have their first derivatives equal to one another at the point.

If however no terms occur in P which involve $w' - \beta$ alone, the preceding inference is not justified: the only regular solution of the equations is the solution $w = \alpha$, which requires that β should vanish: so that no regular solution of any kind exists, unless $\beta = 0$. But non-regular solutions may exist satisfying the conditions: the characteristics of these will depend upon the form of P.

To examine the possible cases, we effect a slight transformation on the equation. Let

$$z - c = x,$$
$$w - \alpha = \beta x + y;$$

so that

$$w' - \beta = y', \quad w'' = y''.$$

The equation in the new dependent variable is subject to the conditions that $y' = 0$, and $y = 0$, when $x = 0$. Also, let

$$P (w' - \beta, w - \alpha, z - c) = P (y', \beta x + y, x)$$
$$= Q (y', y, x),$$

so that as, by hypothesis, there is no term in P involving $w' - \beta$ alone, there is no term in Q involving y' alone; the equation now is

$$y'' = \frac{1}{Q (y', y, x)},$$

where Q is a regular function of its arguments and vanishes with them.

Note. Sometimes, however, it appears that a solution w may acquire a value α, when $z = c$, but that no definite precedent assignment of β is possible: that, in fact, the equation possesses a solution, only if the value of w' (for $z = c$) is determined by the form of the equation and not by external added conditions. In that case, we take $z - c = x$, $w - \alpha = y$, $w' = y'$: and the initial conditions then are that $y = 0$, when $x = 0$.

216. It may happen that $Q(y', y, x)$ has a factor, which is a function of y and x alone: let $y - f(x)$ be a linear factor of this function. If $f(0)$ is not zero, then $y - f(x)$ does not vanish for the assigned conditions: and therefore the linear factor does not contribute to the vanishing of Q, so that it ceases to be important for the determination of the leading term in y. If $f(0)$ is zero, then we take

$$y - f(x) = v,$$

(unless $f(x)$ is zero for all values of x, and then no change of variable is necessary): v vanishes when $x = 0$. The other factors of Q may, or may not, together vanish as a product, when $v = 0$, $x = 0$.

To determine the leading term in v, we assume that, for sufficiently small values of $|x|$, we may take

$$v = \rho x^\mu,$$

where $\mu > 0$. For this substitution, let $k\mu + l$ be the index of the lowest power of x occurring in Q, where $k \geqslant 1$. If $k > 1$, l may be negative. Then since

$$y''Q = 1,$$

and if the lowest term in y'' is of the same order as the lowest term in v'', we have

$$\mu - 2 + k\mu + l$$

as the index of the lowest term on the left-hand side: hence

$$\mu - 2 + k\mu + l = 0,$$

which gives rise to a value of μ. If μ be negative, we infer that no integral of the specified kind exists. If μ be positive but less than unity, then v' is infinite, when $x = 0$; so that, if only finite values of β are to be retained, again no integral of the specified kind exists. If μ be positive and greater than unity, then v' is 0, when $x = 0$; so that $f'(0)$ is the value of y', when $x = 0$; and, if this is the value of β, an integral does exist. And so on, for each such linear factor in succession.

More generally, if, in the case under consideration when $Q(y', 0, 0)$ vanishes whatever value be assigned to y', there be no separable factor dependent upon y and x alone, it is possible to use Weierstrass's theorem for the expression of Q. Instead, however, of adopting this process, it is sufficient to take the general form of Q. Adopting a substitution

$$y = cx^\mu,$$

which is the leading term in y for sufficiently small values of $|x|$ and has μ positive, we determine the lowest index of powers of x in

$$y''Q(y', y, x)$$

for this substitution. This lowest term is made equal to unity, so that there is one relation to determine μ, and there is one equation to determine c. (Sometimes the relation determining μ is satisfied identically; and the equation for c may then leave c arbitrary, if μ is properly determined.) If the value of μ be greater than unity, then all the assumptions made are satisfied: if the value of μ be positive but not greater than unity, then y' is not zero with x, and the initial assumption as to finiteness in value of β is not justified; if μ be zero or negative, the assumptions are not justified, and no solution of the required type exists.

217.   *Ex.* 1.   Consider the equation

$$w'' = \frac{1}{w - zw' - cw'^2},$$

as regards solutions (if any) defined by the initial conditions $w=0$, $w'=0$, when $z=0$.

The reciprocal of the expression for w'' is regular in the vicinity of these initial zero values. We have

$$\frac{dz}{dw'} = -cw'^2 - zw' + w,$$

$$\frac{dw}{dw'} = -cw'^3 - zw'^2 + ww';$$

so that

$$z = -\tfrac{1}{3}cw'^3 + \dots,$$

$$w = -\tfrac{1}{4}cw'^4 + \dots,$$

and therefore

$$w = -\frac{3\sqrt[3]{3}}{4c^{\frac{1}{3}}}z^{\frac{4}{3}} + \dots,$$

the point $z=0$ being accordingly a branch-point round which three branches of the integral circulate.

As a matter of fact, the integral can be obtained by quadratures. The equations are

$$\frac{dz}{dw'} + zw' = w - cw'^2,$$

$$\frac{dw}{dw'} = w'\frac{dz}{dw'}:$$

hence

$$\frac{d^2z}{dw'^2} + z + w'\frac{dz}{dw'} = \frac{dw}{dw'} - 2cw'$$

$$= w'\frac{dz}{dw'} - 2cw',$$

and therefore

$$\frac{d^2z}{dw'^2} + z = -2cw',$$

so that

$$z = -2cw' + A\sin w' + B\cos w'.$$

Therefore

$$w = cw'^2 + zw' + \frac{dz}{dw'}$$

$$= -2c - cw'^2 + B(w'\cos w' - \sin w') + A(\cos w' + w'\sin w').$$

Now $w' = 0$, $w = 0$, $z = 0$ are the initial values: hence

$$B = 0,$$

$$-2c + A = 0,$$

and consequently

$$z = -2c(w' - \sin w'),$$

$$w = -2c(1 - \cos w' - w'\sin w') - cw'^2,$$

and their expressions as regular functions of w' are

$$z = 2c\sum_{n=1}^{\infty}\frac{(-1)^n}{(2n+1)!}w'^{2n+1},$$

$$w = 2c\sum_{n=2}^{\infty}\frac{(-1)^{n-1}(2n-1)}{2n!}w'^{2n},$$

the summation being for integer values of n. In this form, w' can be regarded as an intermediate variable.

Ex. 2. Required the solution (if any) of the equation

$$w'' = -\frac{1}{w^3},$$

such that $w = 0$ when $z = 0$.

That no integral exists, unless β (as the initial value of w') is infinite, can easily be seen. For taking sufficiently small values of $|z|$, the most important term in a function that vanishes with z is of the form $w = cz^n$, where $n > 0$; then $w'' = n(n-1)cz^{n-2} + ...$, so that we must have

$$n - 2 = -3n,$$

that is, $n = \frac{1}{2}$. In this case, the most important term in w' is $\frac{1}{2}cz^{-\frac{1}{2}}$, which is infinite when $z = 0$: so that β must be infinite.

The complete primitive can be obtained explicitly in the form

$$a^2(a^2 + w^2) = (z + b)^2,$$

where a and b are arbitrary constants. If $w = 0$ when $z = 0$, then $b^2 = a^4$: and we have

$$aw = (2a^2z + z^2)^{\frac{1}{2}},$$

or

$$w = (2z + \kappa z^2)^{\frac{1}{2}},$$

where $\kappa = \dfrac{1}{a^2}$. If β be infinite, all the conditions are satisfied: and thus there is an unlimited number of integrals (because κ is arbitrary) satisfying the conditions.

Ex. 3. Discuss similarly the equation

$$4w^2 w'' = ww'^2 - 1,$$

with the condition that $w = 0$, when $z = 0$. Prove that no solution exists unless the initial value of w' is infinite: and that, if this initial value is infinite, a solution can be obtained involving an arbitrary constant.

Ex. 4. Consider the equation

$$w'' = \frac{1}{G(zw', w, z)},$$

where G is a polynomial function of its arguments, which vanishes when zw', w, z, vanish. To obtain the solutions (if any) which vanish when $z = 0$, assume as a leading term

$$w = \rho z^{\mu},$$

so that, as $\mu - 2$ is the order of w'', the lowest index in G after substitution is made must be equal to $2 - \mu$. This index being $k\mu + l$, we have

$$k\mu + l = 2 - \mu,$$

so that

$$\mu = \frac{2 - l}{1 + k};$$

as μ is to be positive, and k is positive because of the form of G, we must have $l < 2$. When μ is reduced to its lowest terms, let it be $\dfrac{p}{q}$; we then take

$$z = x^q, \quad w = \rho x^p + \ldots,$$

and w will then be expressible as a function of x, the exact form of which depends upon G.

As a special illustration, let the equation be

$$w'' = \frac{1}{aw^2 + bzw'w + cz^2},$$

where a, b, c are constants. For the substitution $w = \rho z^{\mu}$, the lowest terms in G have index 2μ or 2, equal to the smaller of these values: clearly

$$\mu - 2 + 2\mu = 0,$$

so that $\mu = \frac{2}{3}$. It proves possible to substitute

$$z^2 = x^3,$$

and still to keep the terms in the resulting equation uniform in expression: and we take a new dependent variable u in the form

$$w = xu.$$

Then on substitution, the equation for u is found to be

$$x^2 u'' = \tfrac{1}{2}u - \tfrac{3}{2}xu' + \frac{\tfrac{9}{4}}{(a + \tfrac{2}{3}b)u^2 + \tfrac{2}{3}bxu'u + cx},$$

where $u'=\dfrac{du}{dx}$, $u''=\dfrac{d^2u}{dx^2}$; any solution of this equation, which is not infinite when $x=0$, leads to a solution of the original equation. The form is a particular case of that considered in § 210. We have, with the notation there used,

$$f(\xi,\ \eta,\ \zeta)=\tfrac{1}{2}\eta-\tfrac{3}{2}\xi+\frac{\tfrac{\Omega}{4}}{(a+\tfrac{2}{3}b)\,\eta^2+\tfrac{2}{3}b\xi\eta+c\zeta}.$$

The initial value a, of u, is given by $f(0,\ a,\ 0)=0$, that is,

$$a=-3\,(4b+6a)^{-\frac{1}{3}}\ ;$$

and then

$$f_{100}=-\frac{9a+4b}{3a+2b},$$

$$f_{010}=\tfrac{3}{2},$$

$$f_{001}=-c\,(4b+6a)^{-\frac{2}{3}}.$$

Thus when $x=0$, the value of u', being $-\dfrac{f_{001}}{f_{100}+f_{010}}$, is

$$-\frac{(4b+6a)^{\frac{1}{3}}}{9a+2b},$$

which remains finite unless $b=-\tfrac{3}{2}a$.

The critical quadratic determining the character of the solution is

$$\Omega^2+\frac{18a+10b}{3a+2b}\,\Omega+\frac{39a+18b}{6a+4b}=0:$$

the roots evidently depending upon the value of $\dfrac{b}{a}$. If neither root be a positive integer, then a solution u exists which is finite, when $x=0$, and is a regular function of x: that is, there exists a solution of the original equation, which is a regular function of $z^{\frac{2}{3}}$. If the roots Ω_1 and Ω_2 have their real parts negative, then this solution is the only solution, which vanishes with z; but if their real parts are positive, though they are not themselves positive integers, then there exists a double infinitude of solutions of the original equation, which vanish with z, and are expressible as regular functions of $z^{\frac{2}{3}}$, $z^{\frac{2}{3}\Omega_1}$, $z^{\frac{2}{3}\Omega_2}$.

Knowing the character of the actual solutions, their form can be obtained by postulating a series of the appropriate character with unknown coefficients. When this is substituted, so that the equation must be satisfied identically, the comparison of coefficients leads to relations that determine the unknown coefficients in the postulated series.

Ex. 5.   Discuss the integrals (if any) of the equation

$$z^2w''=\frac{1}{(a,\ b,\ c,\ f,\ g,\ h\rangle zw',\ w,\ z)^2},$$

which are finite when $z=0$. Consider, in particular, the case when $a=0$, $b=0$, $c=0$.

Ex. 6.   Consider the equation

$$w''=\frac{1+2z+2z^2}{z^2\,(1+2z)}\,w':$$

solutions (if any) are required such that $w=0$, $w'=\beta$, when $z=0$.

Direct quadrature of the equation leads to the result

$$w = B + Az^2 e^{-1/z},$$

where A and B are constants, arbitrary so far as concerns the quadrature. If the solution required is limited by the condition, that the integral is to be regular, then A must be zero: and then $w' = 0$, so that, if β is different from zero, no regular solution exists subject to the assigned conditions.

We have, for the more general solution,

$$w' = A (1 + 2z) e^{-1/z}.$$

Now the origin is an essential singularity of the integral: so that $e^{-1/z}$, by making z approach the origin, can be made to have any value there. Let it approach the origin, so that $e^{-1/z}$ in the limit acquires the value β; then $A = 1$. With this path, we have $w = B$ when $z = 0$, so that $B = 0$: and thus

$$w = z^2 e^{-1/z}$$

is a solution, which satisfies the conditions, when z approaches the origin in the specified way.

Ex. 7.   Is there any solution of the equation

$$w'' = a \frac{(w - zw')^2}{z^2 (w - z)},$$

such that $w = a$ when $z = a$, where a is a constant and a is arbitrary?

Taking

$$w - z = y,$$

we have

$$w - zw' = y - zy',$$
$$w'' = y'' :$$

so that the equation is

$$y'' = a \frac{(y - zy')^2}{z^2 y}.$$

If for sufficiently small values of $|z - a|$, the most important term in y is $c (z - a)^\kappa$, where κ should have its real part positive if a solution of the desired type exists, and c is a constant, then the most important term in $y - zy'$ is $-a\kappa c (z - a)^{\kappa - 1}$, and the most important term in y'' is $\kappa (\kappa - 1) c (z - a)^{\kappa - 2}$. Selecting the corresponding terms from

$$z^2 yy'' = a (y - zy')^2,$$

we have

$$a^2 c^2 \kappa (\kappa - 1) (z - a)^{2\kappa - 2} = a a^2 \kappa^2 c^2 (z - a)^{2\kappa - 2},$$

which is satisfied if only

$$\kappa (\kappa - 1) = a\kappa^2.$$

Now κ is not zero: thus the only admissible value is

$$\kappa = \frac{1}{1 - a}.$$

If therefore the real part of a is less than $+1$, the necessary condition is satisfied.

The most important term in y' is $\kappa c\,(z-a)^{\kappa-1}$, that is, we have

$$w'_{z=a} = 1 + y'_{z=a}$$

$$= 1 + \kappa c \operatorname*{Lim}_{z=a} (z-a)^{\frac{a}{1-a}},$$

the value of which depends upon the form of a. If, for instance, $a = \tfrac{1}{2}(1+i)$, then

$$\frac{a}{1-a} = i, \qquad \kappa = \frac{1}{1-a} = 1+i;$$

and

$$|w'-1|_{z=a} = 2|c|e^{-\theta},$$

where θ is the argument of $z-a$ as it vanishes, and $|c|$ is so far arbitrary. Hence any initial value can be assigned to w', and the variable z can be made to approach a, so that the assigned initial value is acquired at $z=a$: even the value $w'=1$ can be assigned, but then θ must be infinitely great and positive.

It proves possible to integrate the equation by quadratures. Let

$$y = zv, \qquad z = \frac{1}{\zeta};$$

then

$$y - zy' = -z^2v' = \frac{dv}{d\zeta},$$

$$-zy'' = \frac{d}{dz}\left(\frac{dv}{d\zeta}\right) = -\zeta^2 \frac{d^2v}{d\zeta^2}:$$

and therefore

$$v\frac{d^2v}{d\zeta^2} = a\left(\frac{dv}{d\zeta}\right)^2.$$

We find

$$v = (A\zeta + B)^{\frac{1}{1-a}},$$

and therefore

$$y = z^{\frac{a}{1-a}}(A + Bz)^{\frac{1}{1-a}};$$

the primitive of the original equation thus is

$$w = z + z^{-\frac{a}{1-a}}(A + Bz)^{\frac{1}{1-a}}.$$

If the particular value of a, viz. $\tfrac{1}{2}(1+i)$, be adopted, the primitive can be taken in the form

$$w = z + Cz^{-i}(z-a)^{1+i},$$

where C is arbitrary, and the condition that $w=a$ when $z=a$ has been used.

Ex. 8.　Discuss in the same manner the possibility of the existence of solutions of the equation

$$w'' = a\,\frac{(w-zw')^n}{z^2\,(w-z)},$$

when w is required to assume a value a at the point $z=a$.

Ex. 9.　Consider the equations

$$y'' = \frac{a}{(y-xy')^n},$$

with the condition that $y=0$ when $x=0$, for various values of the positive integer n.

To obtain the leading term we assume, as explained,

$$y = \rho x^{\mu},$$

so that, for this substitution, we have

$$y - xy' = \rho (1-\mu) x^{\mu},$$

$$y'' = \mu (\mu-1) \rho x^{\mu-2};$$

and therefore

$$y'' (y - xy')^n = -\rho^{n+1} (1-\mu)^{n+1} \mu x^{\mu-2+n\mu}.$$

Accordingly, we take

$$\mu - 2 + n\mu = 0,$$

so that

$$\mu = \frac{2}{1+n};$$

and then

$$\rho (1-\mu) = \left(-\frac{a}{\mu} \right)^{\frac{1}{n+1}} = \left(-\frac{n+1}{2} a \right)^{\frac{1}{n+1}},$$

which determines ρ: the determination being effective unless $n=1$, when ρ becomes infinite.

Hence when $n>1$, a value of y is obtained equal to zero when $x=0$: it is such that y' is infinite when $x=0$. But, if $n=1$, the value of ρ is infinite on the foregoing determination.

The equation can be integrated by quadratures. Since

$$\frac{d}{dx} (y - xy') = -xy'',$$

we have

$$(y - xy')^{n+1} = -\tfrac{1}{2}(n+1) ax^2,$$

no constant being added, because y and xy' vanish when $x=0$. Thus

$$\frac{d}{dx} \left(\frac{y}{x} \right) = -\frac{xy'-y}{x^2}$$

$$= -\left\{ -\frac{a}{2}(n+1) \right\}^{\frac{1}{n+1}} x^{-\frac{2n}{n+1}},$$

and therefore

$$y = Ax - \frac{1+n}{1-n} \left\{ -\frac{a}{2}(n+1) \right\}^{\frac{1}{n+1}} x^{\frac{2}{1+n}},$$

where A is arbitrary.

If $n>1$, the value of y' when $x=0$ is infinite.

If $n=1$, the form obtained ceases to be effective. It is easy to shew that a solution of

$$y'' = \frac{a}{y - xy'},$$

vanishing when $x=0$, is

$$y = Ax - (-a)^{\frac{1}{2}} x \log x,$$

where A is arbitrary; evidently y' is infinite when $x=0$.

ACCIDENTAL SINGULARITY OF THE SECOND KIND FOR
$$w'' = f(w', w, z).$$

218. Next, suppose that the values $w' = \beta$, $w = \alpha$, $z = c$ constitute an accidental singularity of the second kind (Note, Chapter I) for the function $f(w', w, z)$, so that the equation takes the form

$$w'' = \frac{P_1(w', w, z)}{Q_1(w', w, z)},$$

where P_1 and Q_1 are regular functions of their arguments, and vanish when $w' = \beta$, $w = \alpha$, $z = c$. Write

$$z = c + x,$$
$$w = \alpha + \beta x + y,$$
$$w' = \beta + y', \quad w'' = y'';$$

then the equation is

$$y'' = \frac{P(y', y, x)}{Q(y', y, x)},$$

where P and Q are regular functions of their arguments, and vanish when $x = 0$, $y = 0$, $y' = 0$. As initial conditions, we require that y shall vanish with x: frequently integrals exist, though of such a nature that y' does not vanish with x: accordingly, the condition as regards y' will be left open, so that those cases may be included, which make y' finite (but not zero) or even infinite when $x = 0$: but where an initial condition can be imposed, it will be used for the fuller determination of the integral. (The alternative, in effect, deals with the possibility that β' may be infinite: and the corresponding substitutions would be $z - c = x$, $w = \alpha + y$, $w' = y'$.)

Without taking, in successive detail, all the more important of the forms that thus may arise, consider the most general of them all; let

$$P(y', y, x) = \Sigma\Sigma\Sigma \, a_{lmn} y'^l y^m x^n$$
$$Q(y', y, x) = \Sigma\Sigma\Sigma \, c_{\lambda\mu\nu} y'^\lambda y^\mu x^\nu$$

where $l + m + n \geqslant 1$ for P, and $\lambda + \mu + \nu \geqslant 1$ for Q. Since y is to vanish with x, let its leading term for sufficiently small values of $|x|$ be denoted by

$$y = ax^\theta,$$

where θ is a positive quantity; if y' also vanishes when $x = 0$, then θ will be greater than unity. When such a form is substituted for y in P and Q, let the terms of lowest index in powers of x be selected in both the functions. Let the term in P, which has a_{lmn} for its coefficient, be one of the terms of lowest index thus selected: the aggregate of terms of that index will be

$$x^{(l+m)\,\theta+n-l} \, \Sigma\Sigma\Sigma \, a_{lmn} \alpha^{l+m} \theta^l,$$

the summation extending over all the terms which have this common lowest index. Similarly, in Q, let the corresponding aggregate of terms with lowest index be

$$x^{(\lambda+\mu)\,\theta+\nu-\lambda} \, \Sigma\Sigma\Sigma \, c_{\lambda\mu\nu} \alpha^{\lambda+\mu} \theta^\lambda.$$

Now

$$y'' Q(y', y, x) = P(y', y, x);$$

hence, equating the indices of the lowest powers on the two sides, we have

$$\theta - 2 + (\lambda+\mu)\,\theta + \nu - \lambda = (l+m)\,\theta + n - l;$$

and equating their coefficients, we have*

$$\alpha\theta\,(\theta-1)\, \Sigma\Sigma\Sigma \, c_{\lambda\mu\nu} \alpha^{\lambda+\mu} \theta^\lambda = \Sigma\Sigma\Sigma \, a_{lmn} \alpha^{l+m} \theta^l.$$

The index-equation gives

$$\theta = \frac{2 + n - l - (\nu - \lambda)}{\lambda + \mu + 1 - (l+m)}.$$

In order to obtain the suitable values of θ as given by this fraction, we use a Puiseux diagram as in other cases (§§ 39—42). Referred to a pair of rectangular axes OX, OY, all the points $l+m$, $2+n-l$, connected with terms in the function P, are to be marked: likewise all the points $\lambda+\mu+1$, $\nu-\lambda$, connected with terms in the function Q. Through a point of algebraically smallest abscissa a line is drawn parallel to YO, and is made to turn in the counterclockwise sense until it meets one or more points; then it turns about the last of these in the same sense until it meets others; and so on, until it is parallel to OX. As before, θ is the tangent of the inclination, to the direction XO, of the line joining the point $(l+m, 2+n-l)$ to the point $(\lambda+\mu+1, \nu-\lambda)$; the orders of the corresponding terms in the equation $y''Q = P$, in powers of x, are

$$(\lambda+\mu+1)\,\theta + \nu - \lambda - 2, \quad (l+m)\,\theta + 2 + n - l - 2;$$

---

* The preceding equation assumes that θ is not unity, by taking $\theta - 2$ for the index of x in y'': the necessary correction, when θ is unity, is provided in the succeeding equation by the factor $1 - \theta$. So in other cases.

14

in virtue of the value of θ, these are equal to one another. On account of the mode of obtaining the portion of broken line, which determines this quantity θ, all other indices of powers of x, after the substitution $y \propto x^\theta$ is made, are greater than those selected.

Now θ, unless it is indeterminate, is the quotient of two numbers; when this fraction is in its lowest terms, let its expression be $\dfrac{p}{q}$. (When θ is indeterminate, the cause may be solely owing to the fact that particular points on the line give rise to $\dfrac{0}{0}$ as the expression, and other points on the line will (as in § 41) give a determinate expression for θ: or the cause may be that the portion of line is indeterminate, and this will be dealt with separately: see § 225.) Equating the two expressions for θ, we deduce the relation

$$p\,(l+m)+q\,(n-l)=p\,(1+\lambda+\mu)+q\,(\nu-\lambda-2)$$
$$=N,$$

say.

219. The variables now are changed according to the substitutions

$$x = t^q, \qquad y = ut^p,$$

where u is different from zero when t (and x) vanish. Then

$$y' = \frac{1}{q}\,t^{p-q}\left(t\frac{du}{dt}+pu\right),$$

$$y'' = \frac{1}{q^2}\,t^{p-2q}\left\{t^2\frac{d^2u}{dt^2}+(2p-q+1)\,t\frac{du}{dt}+p\,(p-q)\,u\right\};$$

inserting all these values in the equation

$$y''Q\,(y',\,y,\,x) = P\,(y',\,y,\,x),$$

we have

$$\frac{t^{p-2q}}{q^2}\left\{t^2\frac{d^2u}{dt^2}+(2p-q+1)\,t\frac{du}{dt}+p\,(p-q)\,u\right\}$$

$$\times\left[\Sigma\Sigma\Sigma\frac{c_{\lambda\mu\nu}}{q^\lambda}\left(t\frac{du}{dt}+pu\right)^\lambda u^\mu t^{\nu q+\mu p+\lambda(p-q)}+\ldots\right]$$

$$=\Sigma\Sigma\Sigma\frac{a_{lmn}}{q^l}\left(t\frac{du}{dt}+pu\right)^l u^m t^{nq+mp+l(p-q)}+\ldots,$$

where the unexpressed terms on the two sides involve powers of t that are higher than those in the expressed terms; and the aggregates of the unexpressed terms are regular functions of u and $pu + t\dfrac{du}{dt}$, besides involving these higher powers of t. Now the index of the lowest power of t on the left-hand side in the expressed terms is

$$= p - 2q + \nu q + \mu p + \lambda (p - q)$$
$$= p (1 + \lambda + \mu) + q (\nu - \lambda - 2) = N,$$

say; that on the right similarly is

$$= nq + mp + l (p - q)$$
$$= p (l + m) + q (n - l) = N;$$

and every other index of a power of t in succeeding terms is greater than N. Dividing out by t^N, we have

$$\left[\Sigma\Sigma\Sigma \frac{c_{\lambda\mu\nu}}{q^\lambda} \left(t \frac{du}{dt} + pu \right)^\lambda w^\mu + t R_1 \left(t \frac{du}{dt}, u, t \right) \right]$$
$$\times \left\{ t^2 \frac{d^2 u}{dt^2} + (2p - q + 1) t \frac{du}{dt} + p (p - q) u \right\}$$
$$= q^2 \Sigma\Sigma\Sigma \frac{a_{lmn}}{q^l} \left(t \frac{du}{dt} + pu \right)^l w^m + t R_2 \left(t \frac{du}{dt}, u, t \right),$$

where R_1 and R_2 are regular functions of their arguments. In the summations on the two sides, the included terms correspond to points lying upon that portion of the broken line, which determines the fraction $\dfrac{p}{q}$ and thus gives rise to the transformation of variables adopted.

220. As u is to be different from zero when $t = 0$, let its value be α: also let

$$F_1 (\alpha) = \Sigma\Sigma\Sigma \, c_{\lambda\mu\nu} \left(\frac{p}{q} \right)^\lambda p (p - q) \alpha^{\lambda + \mu + 1}$$
$$F_2 (\alpha) = q^2 \Sigma\Sigma\Sigma \, a_{lmn} \left(\frac{p}{q} \right)^l \alpha^{l + m}$$
$$F (\alpha) = F_1 (\alpha) - F_2 (\alpha)$$

Then α must be a (non-zero) root of the equation

$$F (\alpha) = 0.$$

First, let α not be a root of $F_1 (\alpha) = 0$; and let

$$u = \alpha + v,$$

so that v must vanish with t. The coefficient of

$$t^2 \frac{d^2v}{dt^2} + (2p - q + 1)\, t \frac{dv}{dt}$$

is

$$F_1(\alpha) + S\left(t\frac{dv}{dt}, v, t\right),$$

where S is a regular function of its arguments and vanishes when they vanish. Let all the other terms be transferred to the right-hand side: their aggregate is a regular function of t, v, $t\frac{dv}{dt}$, which vanishes when they vanish, say $T\left(t\frac{dv}{dt}, v, t\right)$, so that

$$t^2 \frac{d^2v}{dt^2} + (2p - q + 1)\, t \frac{dv}{dt} = \frac{T\left(t\dfrac{dv}{dt}, v, t\right)}{F_1(\alpha) + S\left(t\dfrac{dv}{dt}, v, t\right)}.$$

Since $F_1(\alpha)$ is not zero, the fraction can be expanded as a regular function of $t\frac{dv}{dt}$, v, t which, owing to the form of T, vanishes when those quantities vanish: and we manifestly may incorporate the term $(2p - q + 1)\, t\frac{dv}{dt}$ from the left-hand side. Thus the final form is

$$t^2 \frac{d^2v}{dt^2} = f\left(t\frac{dv}{dt}, v, t\right),$$

where f is a regular function of its arguments and vanishes when they vanish.

Secondly, let α be a root of $F_1(\alpha) = 0$ as well as of $F(\alpha) = 0$: since $F_1(\alpha) = 0$, the expansion that leads to the last form is not possible. It is clear that the equation can be taken in the form

$$t^2 \frac{d^2v}{dt^2} = \frac{g\left(t\dfrac{dv}{dt}, v, t\right)}{h\left(t\dfrac{dv}{dt}, v, t\right)},$$

where g and h are regular functions of their arguments, and vanish when their arguments vanish.

221. Further, it has been assumed that the terms, which, in arising out of the equation

$$y''Q(y', y, x) = P(y', y, x),$$

contribute (after the selected substitution for y) to the lowest power of x, are drawn from both sides of the equation. But a reference to the tableau of points for the line, by which the index of that substitution is obtained, shews that the line may be determined by points associated with P only, or by points associated with Q only. If this should happen, then the terms contributing to the lowest power of x are drawn from one side of the equation only: so that the preceding reduced form is no longer obtained.

Thus suppose that a particular substitution is determined solely by a particular group of terms in P, and that all the terms in $y''Q$ are higher in index for that substitution than those terms in P. Then the corresponding transformation gives

$$
t^s \left\{ t^2 \frac{d^2 u}{dt^2} + (2p - q + 1) t \frac{du}{dt} + p(p-q) u \right\}
$$
$$
= \frac{q^2 \, \Sigma\Sigma\Sigma \dfrac{a_{lmn}}{q^l} \left(t \dfrac{du}{dt} + pu \right)^l u^m + t R_2 \left(t \dfrac{du}{dt}, u, t \right)}{\Sigma\Sigma\Sigma \dfrac{c_{\lambda\mu\nu}}{q^\lambda} \left(t \dfrac{du}{dt} + pu \right)^\lambda u^\mu + t R_1 \left(t \dfrac{du}{dt}, u, t \right)},
$$

where s is a positive integer $\geqslant 1$. Now u is to have a value, say α, different from zero when $t = 0$; hence we must have

$$
F_2(\alpha) = q^2 \, \Sigma\Sigma\Sigma \, a_{lmn} \left(\frac{p}{q} \right)^l \alpha^{l+m} = 0,
$$

and we choose a non-zero root of $F_2(\alpha) = 0$.

We accordingly take

$$
u = \alpha + v.
$$

The numerator becomes

$$
F_2(\alpha) + T \left(t \frac{dv}{dt}, v, t \right),
$$

where T is a regular function of its arguments and vanishes when they all vanish. The denominator becomes

$$
F_1(\alpha) + S \left(t \frac{dv}{dt}, v, t \right).
$$

If the root of $F_2(\alpha) = 0$ does not make $F_1(\alpha)$ vanish, the fraction can be expanded as a regular function of $t \dfrac{dv}{dt}$, v, t, which vanishes when they vanish. Transferring the terms

$$
t^s \left\{ (2p - q + 1) t \frac{dv}{dt} + p(p-q)(v + \alpha) \right\}
$$

from the left to the right side, and absorbing them in the regular function, we do not alter the general character of the latter; and thus the final form of the equation is

$$t^{s+2}\frac{d^2v}{dt^2} = f\left(t\frac{dv}{dt}, v, t\right),$$

where s is a positive integer $\geqslant 1$, and f is a regular function which vanishes when all its arguments vanish.

If, however, the root of $F_2(\alpha) = 0$ is also a root of $F_1(\alpha) = 0$, the preceding expansion is not possible: and the resulting equation for v can be taken in the form

$$t^{s+2}\frac{d^2v}{dt^2} = \frac{g\left(t\dfrac{dv}{dt}, v, t\right)}{h\left(t\dfrac{dv}{dt}, v, t\right)},$$

where g and h are regular functions, both of which vanish with their arguments.

If on the other hand a substitution is determined by a group of terms in Q, so that all the lowest terms in P are of an order in t higher than those in $y''Q$, the equation becomes

$$t^2\frac{d^2u}{dt^2} + (2p - q + 1)\,t\,\frac{du}{dt} + p\,(p-q)\,u$$

$$= t^r\frac{q^2\,\Sigma\Sigma\Sigma\,\dfrac{a_{lmn}}{q^l}\left(t\dfrac{du}{dt} + pu\right)^l u^m + tR_2\left(t\dfrac{du}{dt}, u, t\right)}{\Sigma\Sigma\Sigma\,\dfrac{c_{\lambda\mu\nu}}{q^\lambda}\left(t\dfrac{du}{dt} + pu\right)^\lambda w^\mu + tR_1\left(t\dfrac{du}{dt}, u, t\right)} :$$

and u is to be different from zero when $t = 0$. If then α be the value of u when $t = 0$, we must have

$$p\,(p-q)\,\Sigma\Sigma\Sigma\,\frac{c_{\lambda\mu\nu}}{q^\lambda}\,p^\lambda\alpha^{\lambda+\mu+1} = 0,$$

and we choose a non-zero root of this equation. (If $p = q$, then α thus far remains arbitrary: in that case, we can remove a factor t from both sides of the equation.) We take

$$u = \alpha + v,$$

where v vanishes when $t = 0$: the dimensions of v in powers of t, and the form of the equation in v, depend upon the forms of the regular functions R_1 and R_2.

222. As regards the form

$$t^2 \frac{d^2 v}{dt^2} = f\left(t \frac{dv}{dt}, v, t\right),$$

where v is to vanish with t, the tests have already been obtained in § 210; the first of them being satisfied, viz. that $v = 0$ when $z = 0$, because $f(0, 0, 0) = 0$. With the notation of that investigation, the critical quadratic is

$$\Omega^2 - \Omega \left(f_{100} - 3\right) + 2 - 2f_{100} - f_{010} = 0.$$

Let its roots be Ω_1 and Ω_2.

If neither Ω_1 nor Ω_2 is an integer, or if, when either is an integer, it is less than -1, then there exists a solution of the foregoing equation, which is a regular function of t; it vanishes when $t = 0$, and it has

$$-\frac{f_{001}}{f_{100} + f_{010}}$$

(which is finite or zero) as the value of its first derivative when $t = 0$. Consequently, in such cases, there exists a solution of the original equation, which acquires a value α when $z = c$; it is expressible in the form of a regular function of $x^{\frac{1}{q}}$, that is, the point $z = c$ is a branch-point for the integral of the equation. The integral is composed of a number of groups of branches: each group consists of a single circulating set, the branches of each of the groups being equal to one another at $z = c$: the number of groups is the number of non-zero distinct roots of the coefficient-equation for u.

If Ω_1 and Ω_2, though not positive integers, have their real parts positive, then there is a double infinitude of solutions of the v-equation, which vanish with t and are expressible as regular functions of t, t^{Ω_1}, t^{Ω_2}; so that the original equation then possesses, in the vicinity of $z = c$, corresponding solutions, which at $z = c$ acquire a common value, and are expressible as regular functions of $x^{\frac{1}{q}}$, $x^{\frac{\Omega_1}{q}}$, $x^{\frac{\Omega_2}{q}}$.

If either Ω_1, or Ω_2, or both Ω_1 and Ω_2, be an integer $\geqslant -1$, then the v-equation has no regular solution, which vanishes with t, unless a condition is satisfied: but it has an infinitude of solutions, which vanish with t and are expressible in the form of regular functions of t and $t \log t$. The original equation then has an infinitude of solutions, which acquire a common value at $z = c$

and are expressible as regular functions of $x^{\frac{1}{q}}$ and $x^{\frac{1}{q}} \log x$. If however the condition is satisfied, the equation possesses no solutions of this non-regular class: but it then possesses an infinitude of solutions, which acquire a common value at $z = c$, and are expressible as regular functions of $x^{\frac{1}{q}}$.

Ex. 1. Consider the equation

$$w'' = \frac{aw' + bw + cz}{aw' + \beta w + \gamma z},$$

so as to obtain integrals (if any) such that $w = 0$, $w' = 0$, when $z = 0$.

When the tableau of points is constructed, the points for the numerator are 1, 1 : 1, 2 : 0, 3 ; those for the denominator are 2, −1 : 2, 0 : 1, 1. It is at once clear that there is only one value of θ, viz. $\theta = 2$; and so we take

$$w = z^2 u,$$

where u is not to vanish with z. Then

$$z^2 u'' + 4zu' + 2u = \frac{a(zu' + 2u) + bzu + c}{a(zu' + 2u) + \beta zu + \gamma} ;$$

so that, if ξ denote the value of u when $z = 0$, we have

$$2\xi = \frac{2a\xi + c}{2a\xi + \gamma},$$

and therefore

$$4a\xi = a - \gamma \pm \{(a - \gamma)^2 + 4ca\}^{\frac{1}{2}}.$$

Now let

$$u = \xi + v,$$

so that v vanishes with u; we have

$$z^2 v'' + 4zv' + 2v = \frac{(2a - 4a\xi)v + (a - 2a\xi)zv' + (b\xi - 2\beta\xi^2)z + (b - 2\beta\xi)zv}{\gamma + 2a\xi + 2av + azv' + \beta\xi z + \beta zv},$$

and therefore, unless $\gamma + 2a\xi = 0$, we have

$$z^2 v'' = f(zv', v, z),$$

where f is a regular function of its arguments and vanishes with them.

Solutions of this equation are required which vanish with z. Because $f(0, 0, 0)$ vanishes, the first test of § 210 is satisfied. The quantities there denoted by f_{100}, f_{010}, are

$$f_{100} = -4 + \frac{a - 2a\xi}{\gamma + 2a\xi} = -4 + \rho, \text{ say,}$$

$$f_{010} = -2 + \frac{2a - 4a\xi}{\gamma + 2a\xi} = -2 + 2\rho ;$$

so that the critical quadratic, being

$$\Omega^2 - \Omega(f_{100} - 3) - 2f_{100} - f_{010} + 2 = 0,$$

is

$$\Omega^2 - \Omega(\rho - 7) + 12 - 4\rho = 0,$$

the roots of which are

$$\Omega = -4, \quad \Omega = \rho - 3.$$

One of these roots is a negative integer. As regards the other, let

$$p = 4\,\frac{a\gamma - ca}{(a+\gamma)^2};$$

then

$$\rho = \frac{1-(1-p)^{\frac{1}{2}}}{1+(1-p)^{\frac{1}{2}}},$$

where either sign of the radical can be taken : so that the character of the root $\rho - 3$ depends upon the value of p.

If for either value of ρ, the real part of $\rho - 3$ is negative, then the equation possesses one and only one solution (for that value of ρ) which vanishes with z ; it is a regular function of z.

If for either value of ρ, the real part of $\rho - 1$ $(=\rho - 3 + 2,$ § 210) is positive, though $\rho - 1$ is not a positive integer, the equation possesses an infinitude of solutions (for that value of ρ) which vanish with z ; they are regular functions of z and $z^{\rho-1}$. If $\rho - 1$ is a positive integer, then the equation possesses an infinitude of solutions, which vanish with z, and are regular functions of z and $z \log z$; but for particular cases, when the appropriate relation among the coefficients is satisfied (§ 210), the solutions, which vanish with z, are infinite in number (i.e. the solution contains an arbitrary constant), and they are regular functions of z.

To obtain the actual expression of the regular functions, which are solutions of the equation, let

$$w = \xi z^2 + k_3 z^3 + k_4 z^4 + \dots;$$

then

$$aw' + bw + cz = z\left[c + 2a\xi + \sum_{m=3} (mak_m + bk_{m-1})\,z^{m-2}\right],$$

$$aw' + \beta w + \gamma z = z\left[\gamma + 2a\xi + \sum_{m=3} (mak_m + \beta k_{m-1})\,z^{m-2}\right];$$

and therefore

$$\left[2\xi + \sum_{m=3} m\,(m-1)\,k_m z^{m-2}\right]\left[\gamma + 2a\xi + \sum_{m=3} (mak_m + \beta k_{m-1})\,z^{m-2}\right]$$
$$= c + 2a\xi + \sum_{m=3} (mak_m + bk_{m-1})\,z^{m-2}.$$

The terms independent of z on the two sides are equal ; by equating the coefficients of z^{m-2} on the two sides, we have

$$m\,(m-1)\,k_m\,(\gamma + 2a\xi) + 2\xi\,(mak_m + \beta k_{m-1}) + K'_{m-1} = mak_m + bk_{m-1},$$

where K'_{m-1} is an aggregate of coefficients k with suffix less than m, and therefore

$$mk_m\left\{(\gamma + 2a\xi)\,(m-1) + 2a\xi - a\right\} = - K'_{m-1} + bk_{m-1} - 2\beta\xi k_{m-1}$$
$$= K_{m-1},$$

say. Taking account of the expression for ρ, this gives

$$m\,(\gamma + 2a\xi)\,(m - 1 - \rho)\,k_m = K_{m-1},$$

and the smallest value of m for which this holds is $m = 3$.

Now if $\rho - 1$ is not a positive integer, ρ cannot have any one of the values 2, 3, ... ; hence the foregoing equations, for the successive values of m, determine the coefficients k_3, k_4, ... uniquely in terms of ξ. Thus there are two regular integrals of the required type, corresponding to the two values of ξ.

If $\rho - 1$ be a positive integer, then the coefficient of $k_{\rho+1}$ vanishes; hence $k_{\rho+1}$ is infinite, unless

$$K_\rho = 0,$$

and this is the condition specified. If this condition be not satisfied, then the corresponding integral as a regular function of z does not exist. If the condition be satisfied, then k_ρ is arbitrary; the remaining equations determine the succeeding coefficients k, which involve k_ρ: and then there is an infinitude of regular integrals for this value of ρ. (It should be noted that, as the product of the two values of ρ is unity, the eventuality considered can arise for only one of them at the utmost.) In particular, if $\rho = 2$, then $\xi = \dfrac{a - 2\gamma}{6a}$, and

$$K_2 = b\xi - 2\beta\xi^2$$

$$= \frac{1}{18a^2}(a - 2\gamma)(3ab - a\beta + 2\beta\gamma),$$

so that, if $K_2 = 0$, then either

$$a = 2\gamma, \text{ or } a = 2\gamma + \frac{3ab}{\beta}.$$

If a has neither of these values, there is no integral in the form of a regular function, such that $w = 0$, $w' = 0$, when $z = 0$; if it has either of them, then k_3 is arbitrary, and there is an infinitude of regular integrals.

If when $\rho = 2$, K_2 is not zero, then the equation has an infinitude of integrals, which vanish with z, and are expressible as regular functions of z and $z \log z$. The form is

$$w = \frac{a - 2\gamma}{6a} z^2 + \theta z^3 \log z + \Sigma\Sigma c_{mn} z^m (z \log z)^n,$$

the double summation extending over values of m and n, such that

$$m \geqslant 2, \quad m + n \geqslant 4;$$

the coefficient θ is arbitrary and enters into the expression of the coefficients c_{mn}: and it must be remembered that, because $\rho = 2$, the relation

$$2(a + \gamma)^2 = 9(a\gamma - ca)$$

is satisfied.

Similarly for higher integer values of ρ.

Ex. 2.   Discuss in a similar manner the equation

$$aw'' = \frac{w' - \beta z}{w'^2 - 2\beta^2 z^2},$$

so as to obtain the integrals w (if any) such that w and w' vanish with z.

Ex. 3.   Consider the equation

$$w'' = \frac{azw' + bw + cz}{azw' + \beta w + \gamma z},$$

to obtain integrals which vanish when $z = 0$.

The points to be marked in the tableau as belonging to the numerator are 1, 2 : 1, 2 : 0, 3 ; and those to be marked as belonging to the denominator are 2, 0 : 2, 0 : 1, 1. There are seen to be two possible values of θ, viz. 2, 1.

First, let $\theta = 2$, and write

$$w = z^2 u :$$

then the equation for u is

$$z^2 u'' + 4zu' + 2u = \frac{c + (b + 2a)\, zu + az^2 u'}{\gamma + (\beta + 2a)\, zu + az^2 u'}.$$

Let

$$u = \frac{1}{2} \frac{c}{\gamma} + v ;$$

then the equation for v is

$$z^2 v'' + 4zv' + 2v = \frac{\left\{ b + 2a - \dfrac{c}{\gamma}(\beta + 2a) \right\} z \left(\dfrac{c}{2\gamma} + v \right) + \left(a - \dfrac{c}{\gamma} a \right) z^2 v'}{\gamma + (\beta + 2a)\dfrac{cz}{2\gamma} + (\beta + 2a)\, zv + az^2 v'},$$

so that

$$z^2 v'' = f(zv', v, z),$$

where f is a regular function of its arguments and vanishes with them. The quantities denoted by f_{100}, f_{010} are

$$f_{100} = -4, \quad f_{010} = -2 ;$$

so that the critical quadratic is

$$\Omega^2 + 7\Omega + 12 = 0.$$

Each of the roots is a negative integer less than -1 ; hence there is one integral of the equation in the form of a regular function of z, which vanishes with z ; and this solution is the only solution (for the value $\theta = 2$), which vanishes with z. Accordingly, knowing that a regular solution exists of which the leading term involves z^2, we take

$$w = k_2 z^2 + k_3 z^3 + k_4 z^4 + \dots ;$$

substituting in the original equation and multiplying up by the denominator, we find

$$(2k_2 + 6k_3 z + 12k_4 z^2 + \dots)\{ \gamma + \Sigma\, (\beta + ma)\, k_m z^{m-1} \} = c + \Sigma\, (b + ma)\, k_m z^{m-1},$$

which must be an identity. Hence

$$c = \gamma\, 2k_2,$$
$$(b + 2a)\, k_2 = \gamma\, 6k_3 + (\beta + 2a)\, k_2 .\ 2k_2,$$
$$(b + 3a)\, k_3 = \gamma 12 k_4 + (\beta + 2a)\, k_2 .\ 6k_3 + (\beta + 3a)\, k_3 . 2k_2,$$
$$(b + 4a)\, k_4 = \gamma 20 k_5 + (\beta + 2a)\, k_2 . 12 k_4 + (\beta + 3a)\, k_3 . 6k_3 + (\beta + 4a)\, k_4 .\ 2k_2,$$

and so on : these equations determine the coefficients k in succession.

Secondly, let $\theta = 1$, and write

$$w = zu :$$

then the equation for u is

$$zu'' + 2u' = \frac{c + (a + b)\, u + azu'}{\gamma + (a + \beta)\, u + azu'}.$$

Let

$$u = p + qz + rz^2 + vz,$$

choosing p, q, r, so that vz is of the third order at least, that is, so that v and v' vanish when $z=0$. Accordingly, we substitute and equate terms of lowest order on the two sides : we find, from terms of order zero, that

$$2q = \frac{c+(a+b)\,p}{\gamma+(a+\beta)\,p} :$$

from terms of order unity, that

$$3r = q^2 \left\{ \frac{2a+b}{c+(a+b)\,p} - \frac{2a+\beta}{\gamma+(a+\beta)\,p} \right\} :$$

and then the rest of the equation is

$$z^2 v'' + 4zv' + 2v = \text{regular function of } zv', \, v, \, z,$$

the regular function containing no term in zv' alone, or in v alone, or in z alone. The critical equation therefore is

$$\Omega^2 + 7\Omega + 12 = 0,$$

the same as before. Thus there is only one solution of the equation in v, which vanishes with z and the derivative of which vanishes with z : and thus there is only one solution (for the value $\theta = 1$) of the original equation which vanishes with z, this solution being regular. Moreover, p is arbitrary, and this is the initial value of w' ; hence the integral is made determinate uniquely by assigning an initial value to w'. Knowing that such a solution exists, we can obtain its formal expression as in the former case for $\theta = 2$.

REMAINING REDUCED FORMS OF § 221.

223.   As regards the form

$$z^{s+2}w'' = f(zw', \, w, \, z),$$

where s is a positive integer greater than zero, it is easy to see, by taking one of even the simplest forms of the equation, that *the point $z = 0$ is an essential singularity of the complete primitive, and the equation does not unconditionally possess an integral, which is a regular function of z and vanishes with z.* Thus consider the simple form

$$z^3 w'' = czw' - 2cw + P(z),$$

where $P(z)$ is a regular function of z, say in the form

$$P(z) = a_1 z^3 + 2a_2 z^4 + 3a_3 z^5 + \ldots,$$

the power-series being supposed to converge.   We have

$$\int_0^z \frac{P(z)}{z^3}\, dz = a_1 z + a_2 z^2 + a_3 z^3 + \ldots$$
$$= Q(z),$$

say, so that $Q(z)$ converges. The equation is

$$e^{-\frac{c}{z}}\left\{\frac{d^2}{dz^2}(we^{\frac{c}{z}}) + \frac{c}{z^2}\frac{d}{dz}(we^{\frac{c}{z}})\right\} = \frac{P(z)}{z^3},$$

that is,

$$\frac{d}{dz}\left\{e^{-\frac{c}{z}}\frac{d}{dz}(we^{\frac{c}{z}})\right\} = \frac{P(z)}{z^3},$$

and therefore

$$e^{-\frac{c}{z}}\frac{d}{dz}(we^{\frac{c}{z}}) = A + Q(z),$$

where A is an arbitrary constant. Hence

$$w = Be^{-\frac{c}{z}} + Ae^{-\frac{c}{z}}\int e^{\frac{c}{z}}dz + e^{-\frac{c}{z}}\int e^{\frac{c}{z}}Q(z)dz.$$

If B be different from zero, the point $z = 0$ is an essential singularity of the integral. Also $z = 0$ is easily seen to be an essential singularity of the function, which is represented by the integral

$$e^{-\frac{c}{z}}\int e^{\frac{c}{z}}dz.$$

If therefore the equation is to possess a regular integral, we must have $A = 0$, $B = 0$.

Further, if $R(z) = z^2 Q(z)$, the integral, on the assumption that A and B are zero, is

$$w = e^{-\frac{c}{z}}\int e^{\frac{c}{z}}z^{-2}R(z)dz$$

$$= -\frac{1}{c}\left\{R(z) + \left(\frac{z^2}{c}\frac{d}{dz}\right)R(z) + \left(\frac{z^2}{c}\frac{d}{dz}\right)^2 R(z) + \ldots\right\},$$

choosing the constant of integration so that $v = 0$ (if it exist) when $z = 0$. But it is known (§ 82) that this series does not converge, unless c is a zero of a special function, closely allied with $R(z)$. That this may be the case, we must have

$$\frac{a_1}{2!} + \frac{a_2 c}{3!} + \frac{a_3 c^2}{4!} + \ldots = 0,$$

a condition (§ 82) which is necessary and sufficient to secure the convergence of the expression of w in powers of z, that is, for the existence of an integral expressible as a regular function of z vanishing with z.

Hence, in even a simple case, the equation does not unconditionally possess a regular integral that vanishes with z.

Ex. Discuss the possible integrals of the equations

$$\text{(i)} \quad z^3 w'' = \frac{z+2}{z+1}(w - zw'),$$

$$\text{(ii)} \quad z^3 w'' = aw(w - zw'),$$

in the vicinity of $z = 0$.

224. As regards the other forms which have not been fully reduced, such as e.g.

$$t^2 \frac{d^2v}{dt^2} = \frac{g\left(t\,\dfrac{dv}{dt},\, v,\, t\right)}{h\left(t\,\dfrac{dv}{dt},\, v,\, t\right)},$$

where g and h are regular functions of their arguments, which vanish when t, v, $t\dfrac{dv}{dt}$ all are zero, we adopt the same method of the Puiseux diagram as for the first stage in the reduction which led to this form: and for the purpose, we should associate the factor t^2 with the denominator. The following examples will suffice to give an indication of the detailed process.

Ex. Discuss the equations

$$\text{(i)} \quad z^2 w'' = \frac{a\,(zw' - pw)}{b\,(zw' - qw)},$$

(where p and q are integers):

$$\text{(ii)} \quad z^2 w'' = \frac{(a,\, b,\, c,\, f,\, g,\, h \rangle zw',\, w,\, z)^2}{(a,\, \beta,\, \gamma \rangle zw',\, w,\, z)}:$$

$$\text{(iii)} \quad z^2 w'' = \frac{(a,\, \beta,\, \gamma \rangle zw',\, w,\, z)}{(a,\, b,\, c,\, f,\, g,\, h \rangle zw',\, w,\, z)^2}:$$

so as to obtain the integrals (if any), which vanish and have their first derivative not infinite when $z = 0$.

DISCUSSION OF THE INDETERMINATE CASE OF § 218.

225. In the initial investigation, one case was deferred (§ 218), viz. that in which the deduced value of θ was indeterminate, because there was no determinate piece of line. This can happen only when several points in the tableau coincide, and all other points (if any) in the tableau are such as, with that condensed point, to give negative values for θ.

As regards the several points which coincide, being those which give rise to the common lowest order in powers of x, the index of x is of the form $k\theta + k'$; and the index-equation is merely the identity

$$k\theta + k' = k\theta + k'.$$

Now the terms on the left-hand side of the equation

$$y''Q(y', y, x) = P(y', y, x),$$

are such that $\lambda + \mu + 1 = k$, and those on the right-hand side are such that $l + m = k$. The coefficient-equation, determining the initial coefficient α, is

$$\alpha\theta(\theta - 1)\,\Sigma\Sigma\Sigma\,c_{\lambda\mu\nu}\alpha^{\lambda+\mu}\theta^\lambda = \Sigma\Sigma\Sigma\,a_{lmn}\alpha^{l+m}\theta^l\,;$$

so that, since $\lambda + \mu + 1 = k$ and $l + m = k$ for each of the terms included in either of the summations, and because zero-values for α are to be rejected, the quantity α^k can be removed from the two sides, and we have

$$\theta(\theta - 1)\,\Sigma\Sigma\Sigma\,c_{\lambda\mu\nu}\theta^\lambda = \Sigma\Sigma\Sigma\,a_{lmn}\theta^l.$$

The coefficient-equation thus becomes an index-equation: and the leading coefficient α remains arbitrary. Only such values of θ, being roots of this equation, are to be retained as have their real parts positive.

Ex. 1. The simplest case in which the index-equation becomes evanescent, because the diagram furnishes no line that determines a quantity θ, arises when the numerator and the denominator in the fractional expression for y'' contain only one term each. If, in this form, the equation is

$$\kappa y'' = \frac{y'^l y^m z^n}{y'^\lambda y^\mu z^\nu},$$

and the index-equation is evanescent, then

$$l + m = \lambda + \mu + 1,$$
$$2 + n - l = \nu - \lambda,$$

so that

$$l - \lambda = \mu - m + 1 = 2 + n - \nu$$
$$= q + 2,$$

say, where q is an integer; thus the form of equation is

$$\kappa y'' = \frac{y'^{q+2} z^q}{y^{q+1}}.$$

The index-equation is known to be evanescent. Substituting $y = az^\theta$, the coefficient-equation reduces to

$$\kappa\theta(\theta - 1) = \theta^{q+2},$$

or, removing the zero-root $\theta = 0$ which is to be excluded, we have

$$\theta^{q+1} - \kappa(\theta - 1) = 0\,;$$

only those roots will be retained as admissible, which have their real parts positive.

We shall discuss only the case $q > 0$. In connection with such a root, substitute

$$y = uz^\theta,$$

where u is not to vanish when $z = 0$; we have

$$\kappa \{z^2 u'' + 2\theta z u' + \theta(\theta - 1)u\} = \frac{1}{u^{q+1}}(zu' + \theta u)^{q+2}$$

$$= \theta^{q+2} u + (q+2)\theta^{q+1} z u' + z^2 u'^2 P,$$

where P is the quotient, by u^{q+1}, of a regular function of zu' and κ, that does not vanish with zu', if u have a value different from zero. On account of the equation satisfied by θ, the terms involving u alone on the two sides cancel; bringing the term in zu' from the left to the right-hand side, the coefficient

$$= (q+2)\theta^{q+1} - 2\kappa\theta$$
$$= \kappa(q+2)(\theta-1) - 2\kappa\theta$$
$$= \kappa(q\theta - q - 2)$$
$$= \kappa p,$$

say. Then the equation, on division by κz, becomes

$$zu'' = pu' + zu'^2 \frac{P}{\kappa};$$

and the character of the solution depends (§ 212) upon the value of p.

Clearly an arbitrary initial value a, different from zero, can be assigned to u; and the initial value of u' is zero. First, if p be not an integer, then it has been proved (§ 212) that the equation possesses a solution, which is a regular function of z, and acquires the value a when $z = 0$: it is not difficult, taking account of the form of P, to prove that this solution is $u = a$; and then the corresponding solution of the original equation is

$$y = ax^\theta.$$

If, in addition, the real part of p should be positive, there is an infinitude of integrals, which acquire the value a when $z = 0$, and are regular functions of z and z^p. This is not the case, if the real part of p is negative.

If p be a positive integer, then in general the only solution of the equation is $u = a$, leading to the same integral of the original equation as before.

Note. When $q = 0$, the equation is

$$\kappa y'' = \frac{y'^2}{y},$$

and the primitive is

$$y = (az + b)^{\frac{\kappa}{\kappa - 1}},$$

a and b being arbitrary: $b = 0$, when y is to vanish with z.

When $q = -1$, the equation is

$$\kappa y'' = \frac{y'}{z},$$

and the primitive is

$$y = az^{1+\frac{1}{\kappa}} + b,$$

a and b being arbitrary: $b = 0$, when y is to vanish with z.

Ex. 2. Discuss the equation for negative values of q.

Ex. 3. Required the integrals (if any) of the equation

$$y'' = \frac{2zy'y - z^2y'^2 - y^2}{z^2y},$$

which vanish with z.

The points to be marked in the tableau, corresponding to the three terms in the numerator, are the point 2, 2 for each of them : and the point, corresponding to the term in the denominator, is also the point 2, 2. There thus is no line derivable from the tableau : and the index-equation for the lowest terms is evanescent.

The coefficient-equation for the lowest terms, on the substitution $y = az^\theta$ where a is not zero, becomes

$$\theta(\theta - 1) = 2\theta - \theta^2 - 1,$$

that is, $\theta = 1$ or $\theta = \frac{1}{2}$.

First, let $\theta = 1$: and write

$$y = uz,$$

where u is neither zero nor infinite when $z = 0$; the equation for u is

$$zu'' + 2u' = -z\frac{u'^2}{u},$$

a solution of which manifestly is given by $u = $ constant. As a matter of fact, the index-equation for the lowest term of u (taken in the form Bz^ϕ) is evanescent ; the coefficient-equation becomes

$$\phi(\phi - 1) + 2\phi = -\phi^2,$$

that is,

$$2\phi^2 + \phi = 0.$$

The root $\phi = -\frac{1}{2}$ would make u infinite when $z = 0$: it is therefore rejected. The root $\phi = 0$ makes $u = B$, when $z = 0$: the solution thus obtainable is

$$y = Bz,$$

a solution which vanishes when $z = 0$, and the derivative of which acquires a finite assigned value when $z = 0$.

Secondly, let $\theta = \frac{1}{2}$: and write

$$z = x^2, \quad y = ux,$$

where u is not to vanish when $x = 0$. The equation for y is

$$x^2\frac{d^2y}{dx^2} - x\frac{dy}{dx} = -\frac{1}{y}\left(x\frac{dy}{dx} - 2y\right)^2 ;$$

the equation for u is, on removal of a factor x,

$$x\frac{d^2u}{dx^2} \cdot \frac{du}{dx} = -\frac{x}{u}\left(\frac{du}{dx}\right)^2.$$

Let a (different from zero and not infinite) be the value of u when $x = 0$; then if β is the value of $\frac{du}{dx}$ when $x = 0$, we have, by § 212,

$$\beta = 0,$$

and a remains arbitrary. Thus we can write

$$u = a + v,$$

where v' is zero when $x=0$; and we have

$$x\frac{d^2v}{dx^2} = \frac{dv}{dx} - \frac{x}{a+v}\left(\frac{dv}{dx}\right)^2.$$

The right-hand side is a regular function of v, $\dfrac{dv}{dx}$, x. The quantity f_1, of § 212, is unity; so that, unless the appropriate condition is satisfied, the equation in v possesses no regular solution. But if that condition be satisfied, the equation possesses an unlimited number of regular solutions, which vanish with x; that is, there is a solution involving an arbitrary constant. That the latter is the case, can be seen at once : for, since $\beta=0$, a regular solution, if it exist, must be of the form

$$v = a_2 x^2 + a_3 x^3 + a_4 x^4 + \dots,$$

so that

$$x\frac{d^2v}{dx^2} - \frac{dv}{dx} = 3a_3 x^2 + 8a_4 x^3 + \dots,$$

$$x\left(\frac{dv}{dx}\right)^2 = 4a_2{}^2 x^3 + 12a_2 a_3 x^4 + \dots,$$

and therefore equating coefficients in

$$(a+v)\left(x\frac{d^2v}{dx^2} - \frac{dv}{dx}\right) = -x\left(\frac{dv}{dx}\right)^2,$$

the terms in x have disappeared (which is the condition for this case) : and then

$$3a_3 = 0,$$
$$8aa_4 = -4a_2{}^2,$$
$$\vdots$$

leaving a_2 arbitrary. It is easy to see that all the odd powers of x in v have zero coefficients.

Thus there exists an integral of the original equation in the form

$$y = z^{\frac12}(a + a_2 z + a_4 z^2 + \dots),$$

where a and a_2 are arbitrary constants. As the equation in v possesses an infinitude of regular solutions vanishing with x, there are no non-regular solutions vanishing with x; hence the foregoing integral of the original equation is (for $\theta=\frac12$) the only integral that vanishes with z.

The complete primitive of the original equation can be obtained by simple quadratures in the form

$$y = (Az + Bz^2)^{\frac12},$$

with which both the preceding integrals should be compared.

CHAPTER XV.

EQUATIONS OF THE SECOND ORDER AND DEGREE HIGHER THAN THE FIRST*.

226. IN the preceding chapter, the equations of the second order were taken to be of the first degree: so that, in every instance, the value of w'' could be expressed as a uniform function of w', w, z, though not always as a regular function of those quantities. For particular combinations of initial values, the expression for w'' might become infinite or might become indeterminate : but the expression, being uniform in the immediate

* For a very limited portion of the investigations contained in this chapter, the memoirs quoted at the beginning of Chapter XIII may be consulted. With hardly an exception, the present discussion is limited (as the title indicates) to equations of the second order. Some of the results obtained for the second order can be at once stated for an equation of the nth order by a mere generalisation of the symbols.

In particular, the novel method expounded in Chapter VIII, for the singular solutions of equations of the first order, finds its natural generalisation in the present chapter for equations of the second order ; and it will easily be seen that it is capable of extension to equations of higher order.

Moreover, as there is a geometrical interpretation for equations of the first order, associated with the theory of envelopes, there is a corresponding geometrical interpretation for equations of the second order, associated with the theory of osculants. A full development of this theory (which, so far as I know, has not been attempted, either from a geometrical or an analytical standpoint) might be expected to throw at least as much light upon the relation of singular integrals to complete primitives of equations of the second order, as has the theory of envelopes in the corresponding question of equations of the first order. Further, as the circle is the simplest curve, determined by three-point contact with a curve, (so that the values of y, y', y'' are the same for the curve and the circle), the geometrical interpretation of singular solutions naturally leads to the introduction of the osculating circle, and thence, as will be seen in this chapter, to the theory of (circular) osculants. Similarly, for the equations of the third order, the corresponding discussion could lead to the theory of parabolic osculants (that is, parabolas with four-point contact) and associated loci; and likewise for equations of the fourth order, the discussion could lead to the theory of conic osculants (that is, conics with five-point contact) and associated loci.

vicinity of such values, could be used to indicate the character of the integral.

We shall now consider equations of the second order, which are of a degree higher than the first; they will be taken algebraical in w'', w', and generally also in w, but merely analytical in z. The first purpose is the determination of the number and the character of those integrals of the equation, which satisfy the condition of acquiring an assigned value for an assigned value of z. The corresponding condition, as regards their first derivative also, will sometimes (but not universally) be exacted; though it may happen that the equation and the initially assigned value of w may combine to determine a value or values of w' for the assigned value of z. The equation will be denoted by

$$F(z, w, w', w'') = 0,$$

of degree n in w'' in the most general case.

Manifestly it is possible, if it be desired, to make the discussion depend upon that in Chapter XIII, which deals with the integrals of two simultaneous equations of the first order. For this purpose, we take w and $(w' =) v$ as the two dependent variables: the equations are

$$\begin{aligned} F &= F(z, w, v, V) = 0 \\ G &= W - v = 0 \end{aligned},$$

where $V = \dfrac{dv}{dz}$, $W = \dfrac{dw}{dz}$. The critical function is the Jacobian of F and G with regard to V and W: in this case it is

$$\frac{\partial F}{\partial V},$$

which accordingly must play a critical part in the properties of the integrals determined by the equation. The characteristics of the integrals are better investigated independently: and associated geometrical interpretations (for real values of the variables) will be connected with curvature relations of plane curves, instead of with tangential relations of tortuous curves.

As regards the equation

$$F(z, w, w', w'') = 0,$$

let $z = c$, $w = \alpha$, $w' = \beta$ be an initial set of values: then if γ denote an initial value of w'', the whole set of initial values satisfy

$$F(c, \alpha, \beta, \gamma) = 0,$$

which accordingly is an equation of order n to determine γ.

Simple roots γ.

227. First, let γ be a root of this equation, which is finite and unrepeated. In connection with the original equation, assume

$$z - c = x,$$

$$w - a = \beta x + \tfrac{1}{2}\gamma x^2 + y,$$

$$w' - \beta = \gamma x + y',$$

$$w'' - \gamma = y'',$$

so that $y'' = 0$, when $x = 0$; thus y' will be of order higher than 1 in powers of x, and y will be of order higher than 2 in powers of x, for sufficiently small values of $|x|$. Substitute these values in the original equation, so that

$$0 = F(c + x, \ a + \beta x + \tfrac{1}{2}\gamma x^2 + y, \ \beta + \gamma x + y', \ \gamma + y'')$$

$$= x\frac{\partial F}{\partial c} + (\beta x + \tfrac{1}{2}\gamma x^2 + y)\frac{\partial F}{\partial a} + (\gamma x + y')\frac{\partial F}{\partial \beta} + y''\frac{\partial F}{\partial \gamma} + \ldots,$$

where $\dfrac{\partial F}{\partial c}, \ldots$ are the values of $\dfrac{\partial F}{\partial z}, \ldots$ when, after differentiation, the values c, a, β, γ are substituted for z, w, w', w''. Now γ is an unrepeated root, so that $\dfrac{\partial F}{\partial \gamma}$ is not zero: the equation therefore may be solved, giving y'' explicitly as a regular function of x, y, y', in the form

$$y'' = f(x, y, y'),$$

where the regular function f vanishes when $x = 0$, $y = 0$, $y' = 0$. Hence this equation is a differential equation to determine y, the initial conditions being that $y = 0$, $y' = 0$, when $x = 0$. By § 209, we know that there exists a solution, expressible in the form of a regular function of x, subject to the initial conditions; also we know that this is the only regular solution.

It therefore follows that a root γ of the equation

$$F(c, a, \beta, \gamma) = 0,$$

which is finite and unrepeated, determines an integral of the equation which, in the immediate vicinity of $z = c$, is a regular function of z satisfying the initial conditions. This holds for each simple finite root: so that there are as many such regular integrals as there are simple finite roots.

Multiple roots γ.

228.   Secondly, let γ be a root of the equation, which is finite and of multiplicity m; then

$$\frac{\partial^r F}{\partial \gamma^r} = 0, \qquad\qquad (r = 1, \ldots, m-1).$$

The lowest power of y'' alone, that occurs in the transformed equation after effecting the same substitution as before, is y''^m.

If $\dfrac{\partial F}{\partial c} + \beta \dfrac{\partial F}{\partial \alpha} + \gamma \dfrac{\partial F}{\partial \beta}$ is not zero, then a term involving the first power of x alone also occurs. The other terms involve y, or y'. or higher powers of x, or combinations of y, y', and x, or powers of y'' with index $< m$ in combination with x, y, y', or powers of y'' with index $> m$, and so on.   With the hypothesis adopted, take

$$y = x^2 u, \quad y' = xv,$$

so that u and v vanish with x; then the equation becomes

$$y''^m R\,(y'') + x S\,(v,\,u,\,x,\,y'') = 0,$$

where R and S are regular functions of their arguments and do not vanish when their arguments vanish.   This equation can be solved for y'', with the result

$$y'' = x^{\frac{1}{m}}\, g\,(v,\,u,\,x^{\frac{1}{m}}),$$

where g is a regular function of its arguments, which does not vanish when $v = 0$, $u = 0$, $x = 0$.   To obtain the integral, let

$$x = t^m;$$

then the equations are

$$\left.\begin{aligned}
t\frac{dv}{dt} + mv &= my'' = mtg\,(v,\,u,\,t) \\[2mm]
t\frac{du}{dt} + 2mu &= m\frac{y'}{x} = mv
\end{aligned}\right\},$$

the initial conditions being that $v = 0$, $u = 0$, when $t = 0$.   The critical quadratic is

$$(\Omega + m)\,(\Omega + 2m) = 0,$$

so that, as m is an integer $\geqslant 2$, both the roots of this quadratic are negative integers less than -1.   Accordingly, the equations for u and v have solutions, which are expressible as regular functions of t and vanish with t: and they constitute the only solutions determined by the condition of vanishing with t. Hence there is a solution of the original equation, determined by the initial conditions, and expressible as a regular function of $(z-c)^{\frac{1}{m}}$: that is, corresponding to the m-tuple finite root, there is a cyclical group of m branches of the integral which, at $z = c$, have a common value α and, for their first derivatives, have a common value β. To obtain the form of the expression, let $g(0, 0, 0) = \rho$; then

$$v = \frac{m}{m+1} \rho t Q(t), \quad u = \frac{m^2}{(m+1)(2m+1)} \rho t P(t),$$

where $Q(t)$ and $P(t)$ are regular functions of t such that $P(0) = 1$, $Q(0) = 1$. Hence the cyclical group of branches of the integral is given by

$$w - \alpha = \beta(z - c) + \tfrac{1}{2}\gamma(z - c)^2$$
$$+ \frac{\rho}{\left(1 + \dfrac{1}{m}\right)\left(2 + \dfrac{1}{m}\right)} (z - c)^{2 + \frac{1}{m}} P\{(z - c)^{\frac{1}{m}}\},$$

where γ is the root of multiplicity m, and ρ is the value of $g(0, 0, 0)$.

It has been assumed that the term involving the first power of x exists in the transformed equation. It is necessary to take the alternative possibility into account, and consequently to suppose that

$$\frac{\partial F}{\partial c} + \beta \frac{\partial F}{\partial \alpha} + \gamma \frac{\partial F}{\partial \beta} = 0.$$

Various cases arise for consideration according as $\dfrac{\partial F}{\partial \beta}$ is not zero, or $\dfrac{\partial F}{\partial \beta} = 0$ and $\dfrac{\partial F}{\partial \alpha}$ not zero, and so on. The generally effective method for such cases is of even wider application than to these immediately: and therefore the discussion of the cases will be deferred until that method is discussed (see § 230, Exx. 2, 3, 4). It will be seen that the integral is composed of a number of branches, equal to the multiplicity of the root γ of the equation

$$F(c, \alpha, \beta, \gamma) = 0:$$

these branches are such that they acquire a common value at $z = c$, their first derivatives acquire a common value there, and their second derivatives acquire a common value there; but the third derivatives are not, in general, all equal to one another, though they may, in sets, have a value common to a set.

Evanescent form of $F = 0$.

229. Next, it might happen that, for the initial values assigned to w and w' when $z = c$, the equation $F(c, a, \beta, \gamma) = 0$ is evanescent, so that no inference as to the value of γ can be drawn. We then take

$$z - c = x, \quad w - a = y,$$

so that $w' = y'$ and $w'' = y''$, using this form so as to admit of possibly infinite values for β and γ; and we assume that the equation takes the form

$$F = \Sigma\Sigma\Sigma\Sigma \, a_{klmn} y''^k y'^l y^m x^n = 0,$$

which now will be supposed an evanescent equation when $x = 0$, $y = 0$. (If β is known to be finite, the substitution

$$w - a = \beta x + y$$

leads to an equation of the same form.) To determine the integrals (if any) in the immediate vicinity of $x = 0$, which vanish with x, we assume

$$y = \rho x^\mu,$$

for sufficiently small values of $|x|$: and we equate to zero the coefficient of the lowest power of x that arises after substitution. To this lowest power, let the terms

$$a_{klmn} y''^k y'^l y^m x^n, \quad a_{k'l'm'n'} y''^{k'} y'^{l'} y^{m'} x^{n'},$$

contribute; since the index is the same for both, we have

$$k(\mu - 2) + l(\mu - 1) + m\mu + n = k'(\mu - 2) + l'(\mu - 1) + m'\mu + n',$$

so that

$$\mu = \frac{n' - 2k' - l' - (n - 2k - l)}{k + l + m - (k' + l' + m')}.$$

We mark, in a tableau referred to rectangular axes $O\xi$ and $O\eta$, all the points $k + l + m$, $n - 2k - l$, corresponding to the terms in the transformed equation; and with these we construct a Puiseux diagram in the usual manner (§§ 39—42). If any portion of the broken line thus obtained gives a positive value for μ—being the tangent of the inclination of that portion to the axis ξO—such value gives an appropriate substitution.

The coefficient of this lowest term is to be made zero: we thence have a relation that determines one or more non-zero values of ρ.

Let μ, the ratio of two integers, be $\dfrac{p}{q}$, when expressed in its lowest terms; and take

$$x = t^q, \quad y = ut^p,$$

where u is to acquire the (non-zero) value ρ when $t = 0$. Then on removing the lowest power of t, the equation takes the form

$$\Sigma\Sigma\Sigma\Sigma\, a_{klmn}\, Y''^k Y'^l u^m + tR\,(t^2u'',\, tu',\, u,\, t) = 0,$$

where R is a regular function of its arguments, Y' denotes

$$\frac{1}{q}\,(tu' + pu),$$

Y'' denotes

$$\frac{1}{q^2}\,\{t^2u'' + (2p - q + 1)\,tu' + p\,(p - q)\,u\},$$

u' and u'' are the first and the second derivatives of u with regard to t: and the summation extends over all the terms, that give rise to the lowest power of t. Manifestly, the equation determining ρ is

$$\Sigma\Sigma\Sigma\Sigma\, a_{klmn} \left\{\frac{p}{q}\left(\frac{p}{q} - 1\right)\right\}^k \left(\frac{p}{q}\right)^l \rho^{k+l+m} = 0:$$

and only non-zero roots are to be retained.

Let ρ_1 be such a root: and write

$$u = \rho_1 + v,$$

$$u + \frac{1}{p}\,tu' = \rho_1 + v + \frac{1}{p}\,tv',$$

$$t^2u'' + (2p - q + 1)\,tu' + p\,(p - q)\,u = p\,(p - q)(\rho_1 + J);$$

then on substituting in R the value for t^2u'' given by the last equation, we have

$$\Sigma\Sigma\Sigma\Sigma\, a_{klmn} \left(\frac{p}{q}\right)^l \left\{\frac{p}{q}\left(\frac{p}{q} - 1\right)\right\}^k (\rho_1 + J)^k \left(\rho_1 + v + \frac{1}{p}\,tv'\right)^l (\rho_1 + v)^m$$

$$+ tR_1\,(tv',\, v,\, t,\, J) = 0,$$

where R_1 is a regular function of tv', v, t, J. This is the equation to determine J; we thus obtain the various equations (if there be more than one) satisfied by the quantity u associated with ρ_1.

If ρ_1 be such that the quantity

$$\Sigma\Sigma\Sigma\Sigma\, a_{klmn} \left\{\frac{p}{q}\left(\frac{p}{q} - 1\right)\right\}^k \left(\frac{p}{q}\right)^l k\rho_1^{k+l+m-1}$$

is not zero, that is, if, for the initial values considered, the equation

$$\frac{\partial F}{\partial y''} = 0$$

is not satisfied, then in the foregoing equation for J, there is a term involving the first power of J alone; and thus it follows that J is expressible as a regular function of tv', v, t, which vanishes with its arguments. We then have

$$t^2 v'' + (2p - q + 1) tv' + p (p - q) v = p (p - q) J (tv', v, t);$$

the character of the quantity v (if any) then determined depends (§ 210) upon the coefficients of tv' and of v on the right-hand side.

If however ρ_1 is such that the coefficient of the first power of J vanishes, that is, if the equation

$$\frac{\partial F}{\partial y''} = 0$$

is satisfied for the initial values, then the preceding determination of J is not effective. We then are, in fact, in a position much the same as at first, though it is simpler: as will appear, we must begin with the equation which determines J_1, again applying the method which so far has been applied in this section to $F = 0$.

Note. It should, moreover, be remarked that the application of the Puiseux diagram, while effective for the case when the original equation becomes evanescent for the initial values, can be made to cases where the original equation, though not evanescent, does not readily lend itself to any algebraical solution for y''. It then is frequently convenient to know the leading term, say ρx^μ, in y, because of the assistance thereby rendered towards that algebraical solution; and accordingly the method is then adopted for the purpose, particularly in connection with some of the cases left over for consideration.

Some examples will serve to illustrate in detail the process indicated.

230. *Ex.* 1. Determine the integrals (if any) of the equation

$$a^2 z w''^2 + w'' w'^2 w = b^2 z,$$

which vanish with z.

Insertion of initial values makes the equation evanescent. Constructing the Puiseux diagram, we mark the points 2, -3: 4, -4: 0, 1; and thus there are two values of θ, viz. $\theta = \frac{1}{2}$, $\theta = 2$.

First, let $\theta = \frac{1}{2}$; and write
$$z = t^2, \quad w = ut,$$
so that
$$w' = \frac{1}{2t}\frac{dw}{dt} = \frac{1}{2t}(tu' + u),$$
$$w'' = \frac{1}{4t^2}\frac{d^2w}{dt^2} - \frac{1}{4t^3}\frac{dw}{dt}$$
$$= \frac{1}{4t^3}(t^2u'' + tu' - u),$$

where u' and u'' are derivatives of u with regard to t. Substituting, we find
$$a^2(t^2u'' + tu' - u)^2 + (t^2u'' + tu' - u)(tu' + u)^2u = 16b^2t^6.$$

Now u is to be finite and different from zero when t is 0: if therefore its value be ρ, we have
$$\rho^2 = a^2.$$

Solving the quadratic for $t^2u'' + tu' - u$, we have
$$2a^2(t^2u'' + tu' - u) + (tu' + u)^2u = \{u^2(u + tu')^4 + 64a^2b^2t^6\}^{\frac{1}{2}}.$$

Let $u = \rho + v$; so that
$$u^2(u + tu')^4 + 64a^2b^2t^6 = u^2(u + tu')^4\left\{1 + \frac{64a^2b^2t^6}{\rho^6}P\right\};$$

and therefore
$$\{u^2(u + tu')^4 + 64a^2b^2t^6\} = \pm u(u + tu')^2(1 + t^6Q),$$

where P and Q are regular functions of t, v, and tv'.

With the positive sign, we have
$$t^2u'' + tu' - u = \frac{t^6}{2a^2}u(u + tu')^2 Q;$$

but $\rho = 0$ is the only value given by this equation when $t = 0$: and we therefore neglect it.

With the negative sign, we have
$$t^2u'' + tu' - u = -\frac{1}{a^2}u(u + tu')^2(1 + \tfrac{1}{2}t^6 Q),$$

or taking $u = \rho + v$, we have
$$t^2v'' + tv' - v = -3v - 2tv' + S(tv', v, t),$$

where S is a regular function, the terms of which are of at least the second order in their arguments: and v is to vanish with t. The quantities (§ 210) denoted by $f_{100}, f_{010}, f_{001}$, are
$$f_{100} = -3, \quad f_{010} = -2, \quad f_{001} = 0;$$

so that $\beta = 0$ (or v' vanishes with t), and the critical quadratic is
$$\Omega^2 + 6\Omega + 10 = 0,$$

having $\Omega = -3 \pm i$ for its roots. The real parts of both roots are negative, the roots themselves being complex: hence there is a solution in the form of a regular function of t, which vanishes with t. It is the only solution which vanishes with t.

This holds for each of the two values of ρ, viz. $\rho = \pm a$: it is manifest from the original equation that the two values of w, satisfying the equation, have their sum zero.

Secondly, let $\theta = 2$; and write

$$w = uz^2.$$

Then, after removal of a factor z, the equation for u is

$$a^2 (z^2u'' + 4zu' + 2u)^2 + (z^2u'' + 4u'z + 2u)(zu' + 2u)^2 uz^3 = b^2;$$

so that, solving for z^2u'', we have

$$z^2u'' + 4zu' + 2u = \pm \frac{b}{a} + z^3R_1 (zu', u, z).$$

Clearly

$$u = \pm \frac{1}{2}\frac{b}{a} + v,$$

where v vanishes with z; and then

$$z^2v'' + 4zv' + 2v = z^3 R (zv', v, z).$$

The quantities $f_{100}, f_{010}, f_{001}$ of § 210 are

$$f_{100} = -4, \quad f_{010} = -2, \quad f_{001} = 0,$$

so that $\beta = 0$ (or v' vanishes with z): and the critical quadratic is

$$\Omega^2 + 7\Omega + 12 = 0.$$

As the two roots of this are negative integers less than -1, there is a solution of the equation, which is expressible as a regular function of z, and vanishes with z: it is the only solution, $\left(\text{for this value of } \theta \text{ and with either sign of } \dfrac{b}{a}\right)$, which vanishes with z.

This holds for each of the values of u, when $z = 0$: it is again manifest from the original equation that the two values of w, satisfying the equation, have their sum zero.

Thus the solutions of the equation, which vanish with z, are

$$\left.\begin{array}{l} w = \pm az^{\frac{1}{2}} P (z^{\frac{1}{2}}) \\ w = \pm \dfrac{1}{2}\dfrac{b}{a} z^2 Q (z) \end{array}\right\},$$

where $P(z)$ and $Q(z)$ are regular functions of z, each of which is equal to unity when $z = 0$.

Ex. 2.　Find the integrals vanishing with x (if any) of the equation

$$fy''^2 + gy' + hx^2 + 2kxy'' + lx^3 = 0,$$

where f, g, h, k, l are constants.

The points to be marked in the tableau are $2, -4$ for the first term: $1, -1$ for the second: $0, 2$ for the third: $1, -1$ for the fourth: $0, 3$ for the fifth. They determine a single line for which $\mu = 3$, the first four lying upon it: and so we take

$$y = ux^3, \quad y' = vx^2,$$

where u and v are not zero when $x = 0$. We have

$$vx^2 = y' = x^2 (xu' + 3u),$$

so that

$$xu' + 3u = v;$$

and

$$y'' = x (xv' + 2v),$$

so that substituting in the original equation, we have

$$f(xv' + 2v)^2 + gv + h + 2k(xv' + 2v) + lx = 0.$$

This is merely a quadratic in v', and so it could easily be solved explicitly : we adopt, however, a different method, which is suitable for equations that cannot so easily be solved. Let ρ and σ be the values of u and v, when $x=0$: and let

$$u = \rho + U, \quad v = \sigma + V,$$

where U and V vanish when $x=0$: then

$$3\rho = \sigma,$$

$$xU' + 3U = V,$$

are the deductions from the first equation. From the second, we have

$$f(2\sigma)^2 + g\sigma + h + 2k(2\sigma) = 0 ;$$

σ will be used to denote any root of it. Now let ξ denote $xv' + 2v$, so that 2σ is the initial value of ξ : then

$$f\xi^2 + gv + h + 2k\xi + lx = 0.$$

Hence

$$f\{\xi^2 - (2\sigma)^2\} + 2k(\xi - 2\sigma) + g(v - \sigma) + lx = 0,$$

and therefore

$$\xi - 2\sigma = -\frac{g(v - \sigma) + lx}{2k + f(\xi + 2\sigma)}.$$

But

$$\xi - 2\sigma = xV' + 2V, \quad v - \sigma = V :$$

thus

$$xV' + 2V = -\frac{gV + lx}{2k + 4f\sigma + f(xV' + 2V)}$$

$$= -\frac{g}{2k + 4f\sigma} V - \frac{l}{2k + 4f\sigma} x + \dots,$$

the right-hand side being a regular function of x, V, and xV', and the unexpressed terms ($= T$, say) being of at least the second order in x, V, and xV'. Writing

$$\epsilon = 2 + \frac{g}{2k + 4f\sigma}, \quad l' = \frac{-l}{2k + 4f\sigma},$$

the equations are

$$\left.\begin{array}{l} xU' + 3U = V \\ xV' + \epsilon V = l'x + T(x, V, xV') \end{array}\right\} :$$

and the critical quadratic is

$$(\Omega + 3)(\Omega + \epsilon) = 0.$$

One of the roots, -3, is a negative integer < -1 : the other is $-\epsilon$, upon the character of which the solution depends.

If ϵ be a negative integer, then in general (Chap. XII) there is no solution of the equation, which is expressible as a regular function of x and vanishes with x ; though there may be an unlimited number, if the appropriate relation of §§ 179, 180 is satisfied. There is a solution which vanishes with x, contains an arbitrary constant, and is expressible as a regular function of x and $x \log x$; but if the relation of §§ 179, 180 appropriate to the existence of regular integrals is satisfied, the non-regular integral does not exist.

If ϵ, though not a negative integer, has its real part negative, then the equation possesses a solution, which vanishes with x and is a regular function of x: it also possesses a solution, which vanishes with x, contains an arbitrary constant, and is expressible as a regular function of x and $x^{-\epsilon}$.

If ϵ has its real part positive, then the equation possesses a solution, which vanishes with x and is expressible as a regular function of x: it is the only solution, which vanishes with x.

In general, there are two roots of the equation

$$f(2\sigma)^2 + g\sigma + h + 2k(2\sigma) = 0,$$

and each root determines a corresponding quantity ϵ: the above tests must be applied in connection with each of the two quantities.

The solution is, however, ineffective if, for either of the roots (it cannot in general hold for both),

$$k + 2f\sigma = 0,$$

which is the condition that, for the initial values,

$$\frac{\partial F}{\partial y''} = 0,$$

when the irrelevant factor x is rejected. We then have

$$f(xV' + 2V)^2 = -(gV + lx)$$

for the one root, the earlier form being valid for the other root. By actual substitution and comparison of coefficients, which will be found to lead to diverging series, it can be proved that, when l is different from zero, there is no solution of this equation which vanishes with x. Hence, for this case, there is no corresponding solution of the original equation : and it is only in connection with the other root of the quadratic that a solution exists. It is easy to see that the case contemplated is a special case, which can arise only if the relation

$$k^2 + \tfrac{1}{2}kg = fh$$

among the coefficients is satisfied.

The general value of y is

$$y = (\tfrac{1}{3}\sigma + U)x^3,$$

in connection with the respective values of σ.

Note. One of the cases, left undecided in § 228, can be made to depend upon this result. Let

$$\frac{\partial F}{\partial c} + \beta\frac{\partial F}{\partial a} + \gamma\frac{\partial F}{\partial \beta} = 0,$$

and suppose that $\dfrac{\partial^2 F}{\partial \gamma^2}$ is not zero : then the equation is

$$0 = fy''^2 + gy' + hx^2 + 2kxy'' + \text{other terms of higher order},$$

where

$$f = \tfrac{1}{2}\frac{\partial^2 F}{\partial \gamma^2}, \quad g = \frac{\partial F}{\partial \beta},$$

$$h = \tfrac{1}{2}\left(\frac{\partial}{\partial c} + \beta\frac{\partial}{\partial a} + \gamma\frac{\partial}{\partial \beta}\right)\left(\frac{\partial F}{\partial c} + \beta\frac{\partial F}{\partial a} + \gamma\frac{\partial F}{\partial \beta}\right),$$

$$k = \tfrac{1}{2}\left(\frac{\partial}{\partial c} + \beta\frac{\partial}{\partial a} + \gamma\frac{\partial}{\partial \beta}\right)\frac{\partial F}{\partial \gamma}.$$

All the preceding analysis applies except that, in the equation

$$x V' + \epsilon V = l' x + T(x, V, x V'),$$

the regular function T is less simple than in the particular result, though its general character is the same.

The quantity ϵ that determines the character of the solution is

$$\epsilon = 2 + \frac{g}{2k + 4f\sigma},$$

provided the condition, which makes it possible to have $k + 2f\sigma = 0$, is not satisfied. According to the form of ϵ, we have the various inferences. In every case, corresponding to a solution y, there is a value of w given by

$$w = a + \beta(z - c) + \tfrac{1}{2}\gamma(z - c)^2 + y,$$

so that the different branches of the integral, that correspond to the (repeated) value γ, are equal to one another at $z = c$, have their first derivatives equal to one another, and their second derivatives equal to one another: but, in general, their third derivatives are not equal to one another.

We assume for the present purpose that g and k do not simultaneously vanish; otherwise the term

$$y\, \frac{\partial F}{\partial a}$$

would arise for consideration. (This term would affect the characteristic form of the solution, only if the terms in x^2 and x^3 were explicitly absent from the equation.)

If g vanishes but not k, that is, if $\dfrac{\partial F}{\partial \beta} = 0$ without other conditions, then $\epsilon = 2$: we have the third of the alternatives as giving, for each of the two values of σ, the sole solutions of the required type that exist.

Ex. 3. Consider the equation

$$f y'^3 + g y' + h x^2 + k x y'' + l y = 0,$$

to obtain integrals (if any) such that $y = 0$, $y' = 0$, when $x = 0$: the coefficients f, g, h, k, l being constants.

The points to be marked in the tableau are

$$3, -6: 1, -1: 0, 2: 1, -1: 1, 0:$$

for the terms in succession; there are two values of μ, viz. $\mu = \tfrac{5}{2}$, $\mu = 3$.

First, let $\mu = \tfrac{5}{2}$: and write

$$x = t^2, \quad y = u t^5, \quad y' = v t^3,$$

where u and v are different from zero when $t = 0$. Then

$$v t^3 = y' = \frac{1}{2t}(t^5 u' + 5 t^4 u),$$

so that

$$t u' + 5 u = 2 v.$$

Also

$$y'' = \tfrac{1}{2} t (t v' + 3 v),$$

so that the equation, after substitution and removal of a factor t^3, becomes

$$\tfrac{1}{8} f (t v' + 3v)^3 + g v + h t + \tfrac{1}{2} k (t v' + 3v) + l u t^2 = 0.$$

Let $u=\rho$, $v=\sigma$, when $t=0$; then
$$5\rho=2\sigma,$$
and
$$\tfrac{1}{8}f(3\sigma)^3+g\sigma+\tfrac{1}{2}k\,3\sigma=0.$$
Now $\sigma\neq0$, so that there are two roots of this equation suitable, being such that
$$\tfrac{27}{8}\sigma^2f+g+\tfrac{3}{2}k=0.$$
We have
$$\tfrac{1}{8}f\{(tv'+3v)^3-(3\sigma)^3\}+g\,(v-\sigma)+ht+\tfrac{1}{2}k\,\{(tv'+3v)-3\sigma\}+lut^2=0,$$
and therefore
$$tv'+3v-3\sigma=-\frac{g\,(v-\sigma)+ht+lut^2}{\tfrac{1}{8}f\{(tv'+3v)^2+3\sigma\,(tv'+3v)+9\sigma^2\}+\tfrac{1}{2}k}.$$
Now let
$$u=\rho+U,\quad v=\sigma+V,$$
where U and V vanish when $t=0$; then
$$tU'+5U=2V,$$
$$tV'+3V=-\frac{g\,V+ht+l\rho t^2+lUt^2}{\tfrac{1}{8}f\{27\sigma^2+9\sigma\,(tV'+3V)+(tV'+3V)^2\}+\tfrac{1}{2}k}$$
$$=-\frac{g}{\tfrac{27}{8}f\sigma^2+\tfrac{1}{2}k}V+\frac{h}{\tfrac{27}{8}f\sigma^2+\tfrac{1}{2}k}t+T(t,\ V,\ tv',\ U),$$
where T is a regular function of its arguments.

The critical quadratic for these equations is
$$(\Omega+5)(\Omega+\epsilon)=0,$$
where
$$\epsilon=3+\frac{g}{\tfrac{27}{8}f\sigma^2+\tfrac{1}{2}k}:$$
the conclusions as regards the character of the solutions are similar to those in Ex. 2.

Secondly, let $\mu=3$; we write
$$y=ux^3,\quad y'=vx^2.$$
We have
$$xu'+3u=v,$$
and
$$y''=x\,(xv'+2v),$$
so that, substituting and removing a factor x^2, we find
$$fx\,(xv'+2v)^3+gv+h+k\,(xv'+2v)+lux=0.$$
Let ρ and σ (both different from zero) be the values of u and v when $t=0$, so that
$$3\rho=\sigma,$$
$$g\sigma+h+2k\sigma=0\ ;$$
there thus is one value for σ. (It is assumed that h is not zero : for the term hx^2 in the equation helps to determine the value $\mu=3$.) Let
$$u=\rho+U,\quad v=\sigma+V,$$
so that we have
$$xU'+3U=V,$$
$$xV'+(2+g)\,U=\text{regular function of }x,\ v,\ xv'.$$

The critical quadratic is

$$(\Omega+3)(\Omega+2+g)=0;$$

the conclusions, as regards the possible integrals, depend upon the form of $2+g$, and are similar to those in Ex. 2.

Thus in general there are three kinds of solutions for this equation : the first two are such that $y=0$, $y'=0$, $y''=0$, $y'''=\infty$, when $x=0$; the third is such that $y=0$, $y'=0$, $y''=0$, $y'''=-\dfrac{2h}{g+2k}$, when $x=0$.

Note. Another of the cases, left undecided in § 228, can be made to depend upon this result. Let

$$\frac{\partial F}{\partial c}+\beta\frac{\partial F}{\partial a}+\gamma\frac{\partial F}{\partial \beta}=0:$$

and suppose that $\dfrac{\partial^2 F}{\partial\gamma^2}$ as well as $\dfrac{\partial F}{\partial\gamma}$ is zero: but not $\dfrac{\partial^3 F}{\partial\gamma^3}$. The equation is

$$0=fy''^3+gy'+hx^2+2kxy''+ly+\text{higher terms},$$

where

$$f=\frac{1}{6}\frac{\partial^3 F}{\partial\gamma^3}, \qquad l=\frac{\partial F}{\partial a}, \qquad g=\frac{\partial F}{\partial\beta},$$

and h, k are as before (Ex. 2, Note). All the analysis of the preceding example applies, except that, in the equation

$$tV'+\epsilon V=\frac{h}{2^{\frac{2}{7}}f\sigma^2+\frac{1}{2}k}t+T(t,\,V,\,tV',\,U),$$

the regular function T is less simple than in the example, though its general character is the same.

We thus have the various inferences as regards the possible solutions y. The corresponding values of w are given by

$$w=a+\beta(z-c)+\tfrac{1}{2}\gamma(z-c)^2+y,$$

so that the three branches of the integral, corresponding to the triple root γ, have a common value at c, have their first derivatives equal to one another, and their second derivatives equal to one another, at that point; but of the third derivatives, two are infinite in value and the third is finite at the point.

Ex. 4. Obtain the integrals of the equation

$$fy''^m+gy'+hx^2+kxy''+ly=0,$$

(where m is an integer >3) such that $y=0$, $y'=0$, when $x=0$: and discuss the relations of the m branches to one another.

Apply the results to the case, left undecided in § 228, when

$$\frac{\partial F}{\partial c}+\beta\frac{\partial F}{\partial a}+\gamma\frac{\partial F}{\partial\beta}=0,$$

and

$$\frac{\partial^r F}{\partial\gamma^r}=0,$$

for $r=1$, 2, ..., $m-1$, but not for $r=m$: and indicate the relations of the m branches of the solution of the original equation to one another.

Ex. 5.   Obtain the integrals of the equation

$$(z^2 w w'' - z^2 w'^2 + w^2)^2 = 4zw\,(zw' - w)^3,$$

which are such that $w = 0$, $w' = \beta$, when $z = 0$.

[The primitive is

$$w = \beta z e^{-\frac{z}{c(z-c)}},$$

β, c being arbitrary constants.]

Infinite values of γ.

231.   Lastly, it may happen that, for the initially assigned values of z, w, w', the equation

$$F(c,\,\alpha,\,\beta,\,\gamma) = 0$$

determines, not m values (equal or unequal in sets or singly), but only $m - n$ values, of γ: in fact, that it is only of degree $m - n$ in γ, though the original differential equation is of degree m. The inference is that, for those initial values, n of the values of γ are infinite.   In order to take account of the branches of the integral (if any) associable with such values, we write

$$z - c = x, \quad w - \alpha = \beta x + y, \quad w' = \beta + y', \quad w'' = y'',$$

in the original equation; and we consider those values of $\dfrac{1}{y''}$, which are zero when x, y, y' vanish, given by the transformed equation of the form

$$u_0 y''^m + u_1 y''^{m-1} + u_2 y''^{m-2} + \ldots + u_m = 0,$$

so that u_0, u_1, ..., u_{n-1} vanish with x, y, y'.

There are two ways of proceeding.   The simplest case occurs when $n = 1$, so that u_0 is the only coefficient that vanishes, when x, y, y' vanish.   Then one value of y'' is infinite for these initial values: the corresponding root of the algebraical equation is

$$\frac{1}{y''} = -\frac{u_0}{u_1} - \frac{u_2 u_0^2}{u_1^3} - \frac{2u_2^2 + u_1 u_3}{u_1^5} u_0^3 + \ldots$$

$$= -\frac{u_0}{u_1} P,$$

where P is a regular function of x, y, y', which is equal to unity when its arguments vanish.   Hence

$$y'' = -\frac{1}{u_0} \frac{u_1}{P},$$

where u_1/P can be expressed as a regular function of x, y, y'. This regular function u_1/P is finite when x, y, y' vanish, because P then is unity; and it is not zero when they vanish, because u_1 is not zero. Hence

$$y'' = -\frac{q(x, y, y')}{p(x, y, y')},$$

where p and q are regular functions of their arguments, of which q does not vanish and p does vanish when $x = 0$, $y = 0$, $y' = 0$. The combination of initial values is therefore an accidental singularity of the first kind for the quantity y'' thus determined: the character of such a combination, in relation to the corresponding integral, has already been discussed (§ 215). It is known that, in general, the point c is an algebraical branch-point; and the integral consists of a number of branches which have a common value there, the first derivatives of which have a common value there, and the second derivatives of which become infinite there.

The other method of proceeding applies to the case just considered: but it applies also to other and more complex cases, when the infinite value of y'' is multiple, and when the explicit determination of y'' in terms of x, y, y' may be difficult or practically unattainable. We assume

$$y = \rho x^\mu,$$

as the leading term of an integral for sufficiently small values of $|x|$: and we construct the customary Puiseux diagram. When the broken line in the diagram has been obtained, those portions of it, for which $1 < \mu < 2$, correspond to the present case: and the method of proceeding is precisely similar to that in preceding cases.

It may be added that this latter method is also suited for the discussion of what is practically an omitted case, owing to a tacit assumption. The form of the equation may require that integrals, which acquire an assigned value for an initial value of the variable, have all their derivatives infinite for that initial value: so that β would be infinite. We should then assume

$$w - \alpha = y, \quad w' = y', \quad w'' = y'', \quad z - c = x:$$

and for the transformed equation construct the Puiseux diagram in those cases, where it proves impracticable to obtain the expression of y'' in terms of x, y, y'. The portions of the broken line, for

16—2

which $0 < \mu < 1$, correspond to this case: the further development is as before.

In the several cases, the point may be an ordinary point, or it may be a branch-point for the integrals: the integrals determined have a common value at the point, a common value (which may be infinite) for their first derivatives, and a common value (which may be infinite) for their second derivatives.

It is unnecessary to add illustrative examples since, in each distinct set of conditions, the method of proceeding has coincided with one of the earlier methods; and each of them in its turn has been illustrated by special examples.

Ex. Obtain the integrals of the equation

$$(2z^2w'' - 5zw' + 6w)(2z^2w'' - 7zw' + 10w) + z^2(4z^2w'' - 12zw' + 15w) = 0,$$

which vanish with z.

[The primitive is

$$w = az^{\frac{5}{2}} + abz^2 + bz^{\frac{3}{2}},$$

where a and b are arbitrary constants.]

Summary.

232. The results, which have been obtained, may be summarised as follows for the equation

$$F(z, w, w', w'') = 0,$$

taken to be of degree m in w''.

When a root of the equation

$$F(c, \alpha, \beta, \gamma) = 0,$$

regarded as an algebraical equation in γ, is finite and distinct from every other, an integral of the equation thus determined is a regular function of z, which acquires an assigned value α at $z = c$, and the first derivative of which acquires an assigned value β there: the second derivative acquires the value γ.

When a root γ of the equation is finite, but of multiplicity n (where $n > 1$), then a corresponding integral of the equation is composed of n functions, which acquire a common value α at $z = c$, the first derivatives of which acquire a common value β at $z = c$, and the second derivatives of which acquire the common value γ at $z = c$: their third derivatives are, in general, not all equal to one another, but may be equal in sets, and all of them may be unequal to one another. The functions, which compose the integral, may be regular functions or multiform functions of z.

When a root γ of the equation is infinite and of multiplicity p, (where $p \geqslant 1$), a corresponding integral of the equation is composed of p functions, which acquire a common value at $z = c$; their first derivatives are the same there, and likewise their second derivatives (being infinite): their third derivatives likewise are infinite, but the infinity is not, in general, the same for all. The functions, which compose the integral, are generally multiform functions of z.

Combining these results for all the roots, we have the composite integral of the equation as determined by assigned initial values.

The test of multiplicity of a root γ is that it should also satisfy the equation

$$\frac{\partial F}{\partial \gamma} = 0:$$

so that it is a root common to the two equations

$$F(c, \alpha, \beta, \gamma) = 0, \quad \frac{\partial F}{\partial \gamma} = 0.$$

Let $\Delta(c, \alpha, \beta)$ be the eliminant of F and $\dfrac{\partial F}{\partial \gamma}$; then if the relation $\Delta = 0$ is not satisfied, the roots γ are distinct: if it is satisfied, at least one of the roots γ is of multiple occurrence.

GEOMETRICAL INTERPRETATION.

233. As in the case of an equation of the first order, so also in the case of the equation of the second order, it is convenient to associate a geometrical interpretation with the analytical results. The variables w and z are restricted to have only real values: and the integral relation between w and z, involving a couple of arbitrary constants when it is the primitive, then represents a doubly-infinite system of curves. The quantity w' at any point determines the direction of the tangent to the curve: and the quantity w'', in conjunction with w', determines the curvature at the point.

Denoting any arbitrary point z, w in the plane by c, a, we have a limited number of directions β through the point, determined by roots of the equation

$$\Delta (c, a, \beta) = 0 ;$$

let these be denoted by $\beta_1, \ldots, \beta_\kappa$, among which $\beta = \infty$ will be considered to be included: and let $w' = \xi$ denote any other direction through the point, ξ being a parametric quantity which can range between $+ \infty$ and $- \infty$. Then assigning a value ξ to w', the differential equation of degree m and order two defines m integral curves, which pass through the point and have a common tangent there: but the curvatures of the m curves are distinct from one another. As the direction ξ swings round the point c, a, the m curves touching one another move with it; they keep their curvatures unequal at the point, so long as ξ does not coincide with one of the quantities β. But when ξ is equal to any one of these quantities β, the m curves through the point, which touch one another, do not have all their curvatures unequal: one group of curves, containing two or more members, will certainly have the curvatures of all members of that group equal to one another, so that the curves osculate at the point: and there may be more than one of such groups, the common curvature of a group not being the same as that of another group. Those of the m curves, which do not belong to such a group, have each of them its own curvature, distinct from all the other curvatures of the curves touching one another. When the swinging tangent passes through a critical direction β, and resumes a parametric direction ξ, the m curvatures of the touching curves again become distinct.

In general, the point c, a is a critical point (such as a cusp of appropriate order) for those curves, which have equal curvature, and the critical direction β for their common tangent: it is an ordinary point for the other curves, each with its own isolated curvature, which have that critical direction for its tangent: and it is an ordinary point for curves through it, which have a parametric (non-critical) direction. Not every critical point, however, is a cusp: thus it can happen that the curvature is infinite (the circle of curvature being a point), and the curve may be continuous through the point: such a case is provided by the curve

$$w^3 = z^5,$$

when, at the origin, we assign a parametric direction coinciding with the axis of z.

Similarly, a direction through $z = c$, provided by $\beta = \infty$, does not necessarily provide a critical point on the integral curve, though the value $z = c$ may be a branch-point of the integral function. For instance, let

$$w^2 = zG(z),$$

where $G(z)$ is a regular function of z, such that $G(0) = 1$; then $z = 0$ is an ordinary point of the curve, the direction coinciding with the axis of w. The discrimination in such a case would be possible, by making z the dependent variable and w the independent variable: it would be effected, as for the other critical values if necessary, by consideration of the properties of the differential equation itself.

Taking then the point c, α as a moving point, and a parametric direction ξ as accompanying the point, the integral curves move with them; they change their curvature, alike with the variation of ξ through the moving point, and with the motion of the point.

THE DISCRIMINANT OF $F = 0$: SINGULAR SOLUTIONS.

234. Instead of considering the discriminant of $F(c, \alpha, \beta, \gamma)$ with regard to γ, consider the discriminant of $F(z, w, w', w'')$ with regard to w''; and let

$$\Delta(z, w, w') = 0$$

be the eliminant of

$$F(z, w, w', w'') = 0, \quad \frac{\partial}{\partial w''} F(z, w, w', w'') = 0.$$

The functional relation of w' and w'' to w does not arise in the course of the elimination: and, in effect, all that is done is to eliminate q between the equations

$$F(z, w, p, q) = 0, \quad \frac{\partial}{\partial q} F(z, w, p, q) = 0,$$

with the result

$$\Delta(z, w, p) = 0.$$

Now when q is determined from $F = 0$ in terms of z, w, p, and is substituted in the other equation, the result (being rationalised) is $\Delta = 0$. When the value of q is substituted in $F = 0$, the result is an identity, so that

$$\frac{\partial F}{\partial z} + \frac{\partial F}{\partial w}\frac{dw}{dz} + \frac{\partial F}{\partial p}\frac{dp}{dz} = 0,$$

the term in $\dfrac{dq}{dz}$ disappearing on account of $\dfrac{\partial F}{\partial q} = 0$. Let Θ denote

$$\frac{\partial F}{\partial z} + \frac{\partial F}{\partial w}p + \frac{\partial F}{\partial p}q,$$

so that

$$\Theta = \frac{\partial F}{\partial w}\left(p - \frac{dw}{dz}\right) + \frac{\partial F}{\partial p}\left(q - \frac{dp}{dz}\right).$$

Moreover, the value of p satisfies

$$\Delta(z, w, p) = 0;$$

suppose that $p = \dfrac{dw}{dz}$, so that $\Delta = 0$ becomes

$$\Delta(z, w, w') = 0,$$

an equation of the first order. We now have

$$\Theta = \frac{\partial F}{\partial p}\left(q - \frac{dp}{dz}\right)$$

$$= \frac{\partial F}{\partial p}\left(q - \frac{d^2w}{dz^2}\right);$$

consequently, if $q = w''$, we must have $\Theta = 0$, that is, *if a solution of the equation*

$$F(z, w, w', w'') = 0$$

is such as to satisfy

$$\frac{\partial F(z, w, w', w'')}{\partial w''} = 0,$$

then it must also satisfy

$$\frac{\partial F}{\partial z} + \frac{\partial F}{\partial w}w' + \frac{\partial F}{\partial w'}w'' = 0;$$

and if a value of w'' given by $F = 0$ satisfies the other two equations, the corresponding value of w' is given by the equation

$$\Delta(z, w, w') = 0,$$

where Δ is the eliminant of F and $\dfrac{\partial F}{\partial w''}$.

It is clear that, in general, the equation

$$\frac{\partial F}{\partial z} + p\frac{\partial F}{\partial w} + q\frac{\partial F}{\partial p} = 0$$

is not satisfied solely as a consequence of

$$F(z, w, p, q) = 0, \quad \frac{\partial F}{\partial q} = 0.$$

If this were so, the three equations would not determine p, q, w, as functions of z, but would determine only two of them; that is to say, regarded as involving p, q, w, there would be a functional relation between F_z, F_q, F, where

$$F_z = \frac{\partial F}{\partial z} + p \frac{\partial F}{\partial w} + q \frac{\partial F}{\partial p}, \quad F_q = \frac{\partial F}{\partial q};$$

and therefore the relation

$$J \left(\frac{F, F_q, F_z}{p, q, w} \right) = 0$$

would be satisfied. This is an equation affecting the form of F, which is supposed quite general; consequently, it is not satisfied by any assigned function F.

Hence, in general, it is possible to eliminate only two of the quantities, say p and q, between $F = 0$, $F_z = 0$, $F_q = 0$: let the result be denoted by

$$E = E(z, w) = 0.$$

235. Suppose that, in a particular case, it is the fact that one of the equations

$$F = 0, \quad F_q = 0, \quad F_z = 0,$$

say the third, is satisfied in virtue of the other two. To form Δ, one method would be to solve $F_q = 0$ for q, substitute the various roots in F in turn, and take the product of the substituted results, say

$$\Delta = \Pi(F).$$

Let F' denote the factor F which vanishes (and secures the vanishing of F) in virtue of that value of q, which satisfies the three equations: and suppose, for simplicity, that no one of the other factors of Δ vanishes. Thus we can take

$$\Delta = F' \Pi'(F).$$

Then

$$\Delta_z = F_z' \Pi'(F) + F' [\Pi'(F)]_z.$$

But $F_z' = 0$, $F' = 0$; hence $\Delta_z = 0$, that is,

$$\frac{\partial \Delta}{\partial z} + \frac{\partial \Delta}{\partial w} w' + \frac{\partial \Delta}{\partial w'} w'' = 0,$$

in virtue of the value of $q = w''$, which satisfies the three equations. Consequently, if the three equations are definitely satisfied by the one value of w'' in virtue of $\Delta = 0$, then also the equation

$$\frac{\partial \Delta}{\partial z} + \frac{\partial \Delta}{\partial w} w' + \frac{\partial \Delta}{\partial w'} w'' = 0$$

is satisfied.

Clearly, if the three equations are not definitely satisfied by the one value of w'' in virtue of $\Delta = 0$, then

$$\frac{\partial \Delta}{\partial z} + \frac{\partial \Delta}{\partial w} w' + \frac{\partial \Delta}{\partial w'} w''$$

does not vanish for that value.

236. Now

$$\Delta (z, w, w') = 0$$

is an equation of the first order: if it be of degree κ in w', there are in general κ solutions, such that $w = \alpha$ when $z = c$; and they acquire $\beta_1, ..., \beta_\kappa$ as the values of their first derivatives respectively at $z = c$, these quantities being the roots of

$$\Delta (c, \alpha, \beta) = 0.$$

Consider integrals of

$$F(z, w, w', w'') = 0,$$

such that $w = \alpha$, $w' = \beta$, when $z = c$; then the values of w'' at c are given by

$$F(c, \alpha, \beta, \gamma) = 0,$$

some of which are equal to one another (while the rest of them may be unequal) because $\Delta (c, \alpha, \beta) = 0$. These values determine integrals. Each value γ', whose occurrence is simple, determines a regular function of $z - c$, which has the common value α at c, the first derivative of which has the common value β, and the second derivative of which has an isolated value γ' there: also γ' is such that $\dfrac{\partial F}{\partial \gamma}$ is not zero, when that value of γ is substituted.

A value γ'', whose occurrence is multiple, determines a number of functions, which have the common value α at c, the first derivatives of which have the common value β there, and the second derivatives of which are equal to γ'' there: also γ'' is such that $\dfrac{\partial F}{\partial \gamma}$ is zero, when that value of γ is substituted.

Since γ' is such that $\left[\dfrac{\partial F}{\partial \gamma}\right]_{\gamma=\gamma'}$ is not zero, there is no relation between the second derivative of the solution of $F = 0$, and the second derivative of the solution of $\Delta = 0$, for the values c, α, β.

Next, γ'' is such that $\left[\dfrac{\partial F}{\partial \gamma}\right]_{\gamma=\gamma''}$ is zero. If γ'' be such that

$$\Delta_{z'} = \frac{\partial \Delta}{\partial z} + w' \frac{\partial \Delta}{\partial w} + w'' \frac{\partial \Delta}{\partial w'},$$

does not vanish for z, w, w', $w'' = c$, α, β, γ''—and this is the case in general—there is no relation between the second derivative of the solution of $F = 0$, and the second derivative of the solution of $\Delta = 0$, for the values c, α, β. If, however, γ'' be such that

$$\Delta_z = 0$$

is satisfied for z, w, w', $w'' = c$, α, β, γ'', then the second derivative of the solution of $F = 0$ is the same as the second derivative of the solution of $\Delta = 0$; as this holds for the solution of $\Delta = 0$ in general, this solution can be regarded as a solution of $F = 0$. It is a *singular solution*.

Now consider the equation

$$E(z, w) = 0;$$

let c, α denote simultaneous values of z and w, that satisfy the equation. Construct a solution of

$$\Delta(z, w, w') = 0,$$

such that $w = \alpha$, when $z = c$: let w', thus determined, have a value β, when $z = c$. Construct a solution of

$$F(z, w, w', w'') = 0,$$

such that $w = \alpha$, $w' = \beta$, when $z = c$; let w'' have a value γ at c, and suppose that this is one of the multiple roots of

$$F(c, \alpha, \beta, \gamma) = 0:$$

multiple roots of $F = 0$ exist, because $\Delta(c, \alpha, \beta) = 0$. Hence

$$\frac{\partial F}{\partial \gamma} = 0.$$

Now since E is the eliminant of F, $\dfrac{\partial F}{\partial \gamma}$, and $[F_z]_{z=c}$, and since E, F, $\dfrac{\partial F}{\partial \gamma}$ all vanish, it follows that

$$\frac{\partial F}{\partial c} + p \frac{\partial F}{\partial \alpha} + q \frac{\partial F}{\partial \beta} = 0;$$

and therefore, as in § 235, that

$$\frac{\partial \Delta}{\partial z} + w' \frac{\partial \Delta}{\partial w} + w'' \frac{\partial \Delta}{\partial w'} = 0,$$

when $z = c$. Hence the second derivative of the solution of $\Delta = 0$ is, at the particular point c, equal to the second derivative of the solution of $F = 0$: that is, the solutions of $F = 0$ and $\Delta = 0$ agree, at the point c, up to their second derivatives inclusive. The solution of $\Delta = 0$ is then a solution of $F = 0$ at the point; but this holds only at the point, and not for variation from the point.

237.   It has been seen that, if the equations

$$F = 0, \quad \frac{\partial F}{\partial w''} = 0, \quad \frac{\partial F}{\partial z} + w' \frac{\partial F}{\partial w} + w'' \frac{\partial F}{\partial w'} = 0,$$

are satisfied in virtue of the eliminant of F and $\frac{\partial F}{\partial w''}$, say

$$\Delta (z, w, w') = 0,$$

then an integral of $\Delta = 0$ can provide a singular solution of the equation $F = 0$. Now for the particular values of w' given by $\Delta = 0$, the equations $F = 0, \frac{\partial F}{\partial w''} = 0$ are simultaneously satisfied: but, in general, the third equation then is not satisfied. It is, however, possible to construct the form of F, for which it is identically satisfied: so that the corresponding equation will have a singular solution.   To obtain these, it is sufficient to regard

$$\frac{\partial F}{\partial z} + w' \frac{\partial F}{\partial w} + w'' \frac{\partial F}{\partial w'} = 0$$

as a partial differential equation satisfied by the form of F, a function involving z, w, w', w''; we take the most general integral of this formal equation.   The subsidiary system is

$$\frac{dz}{1} = \frac{dw}{w'} = \frac{dw'}{w''} = \frac{dw''}{0};$$

three independent integrals are

$$u_1 = w'',$$
$$u_2 = w''z - w',$$
$$u_3 = w''z^2 - 2w'z + 2w:$$

and therefore the most general form of F is

$$F = \Phi (w'', \; w''z - w', \; w''z^2 - 2w'z + 2w),$$

where Φ is any arbitrary function of its three arguments: say

$$F = \Phi\left(u_1, u_2, u_3\right).$$

To obtain the primitive, we have

$$\frac{dF}{dz} = \left(\frac{\partial\Phi}{\partial u_1} + z\frac{\partial\Phi}{\partial u_2} + z^2\frac{\partial\Phi}{\partial u_3}\right) w''',$$

so that, as $F = 0$ permanently, and therefore $\dfrac{dF}{dz} = 0$, we can have $w''' = 0$: that is,

$$w'' = \alpha,$$

where α is an arbitrary constant.   Hence, since

$$\frac{d}{dz}\left(w''z - w'\right) = zw''' = 0,$$

we have

$$zw'' - w' = \beta:$$

and similarly

$$z^2 w'' - 2w'z + 2w = \gamma,$$

so that

$$2w = \gamma - 2\beta z + \alpha z^2,$$

with the relation

$$\Phi\left(\gamma, \beta, \alpha\right) = 0,$$

among the arbitrary constants.

An equation of this form possesses singular solutions[*].

Ex.   Prove that the alternative equivalent of $\dfrac{dF}{dz} = 0$, given by

$$\frac{\partial\Phi}{\partial u_1} + z\frac{\partial\Phi}{\partial u_2} + z^2\frac{\partial\Phi}{\partial u_3} = 0,$$

leads to a singular solution of the original equation.

(Dixon.)

238.   Only very brief indications have been given as to the relation, borne to the original equation, by singular solutions of $\Delta = 0$, when they exist.   Such a solution provides a value of w' which, with the value of w, satisfies $\Delta = 0$; if it does not also provide a value of w'' which, with the values of w and w', satisfies $F = 0$, it is not an integral of that equation.   If, however, the condition be satisfied, it is a singular integral of the equation of the *second kind*.

[*] This class of equations was obtained, otherwise and (I believe) for the first time, by Dixon, *Phil. Trans.* (1895, A), pp. 563—564.   He regards it as an extension of Clairaut's form

$$\Phi\left(w', zw' - w\right) = 0,$$

of equations of the first order; and he obtains it for equations of order n.

Ex. Prove that an equation $F(z, w, w', w'') = 0$ possesses a singular integral of the second kind, if the equations

$$F = 0, \quad \frac{\partial F}{\partial w'} = 0, \quad \frac{\partial F}{\partial w''} = 0,$$

$$\frac{\partial F}{\partial z} + w' \frac{\partial F}{\partial w} = 0,$$

$$\begin{vmatrix} \dfrac{\partial^2 F}{\partial z^2} + \dfrac{\partial^2 F}{\partial z \partial w} w' & , & \dfrac{\partial F}{\partial w} + \dfrac{\partial^2 F}{\partial z \partial w'} & , & \dfrac{\partial^2 F}{\partial z \partial w''} \\[2mm] \dfrac{\partial^2 F}{\partial z \partial w'} + \dfrac{\partial^2 F}{\partial w' \partial w} w' & , & \dfrac{\partial^2 F}{\partial w'^2} & , & \dfrac{\partial^2 F}{\partial w' \partial w''} \\[2mm] \dfrac{\partial^2 F}{\partial z \partial w''} + \dfrac{\partial^2 F}{\partial w'' \partial w} w' & , & \dfrac{\partial^2 F}{\partial w'' \partial w'} & , & \dfrac{\partial^2 F}{\partial w''^2} \end{vmatrix} = 0,$$

are all satisfied by the same values of w' and w'', in virtue of a single relation $\Theta(z, w) = 0$: which relation is, in fact, the singular integral of the second kind.

Are these conditions necessary as well as sufficient?

239. To resume the geometrical interpretation, we can exhibit the relations between the equations

$$\left. \begin{aligned} F(x, y, y', y'') &= 0 \\ \Delta(x, y, y') &= 0 \\ E(x, y) &= 0 \end{aligned} \right\},$$

as follows.

Denote by $x = c$, $y = \alpha$, any point in the plane not lying on the curve $E = 0$. Through that point, there will in general pass a (finite) number of curves, which satisfy the equation $\Delta = 0$: let the tangents at the point be called the critical lines, and the curves be called the critical curves.

Through the point c, α, and touching any line that is not one of the critical lines, pass a finite number of curves satisfying the equation $F = 0$; these curves, which possess a common tangent, have their curvatures distinct from one another. Through the point c, α, and touching one of the critical lines, there pass a finite number of curves satisfying the equation $F = 0$; they all touch the critical curve which has the critical line for a tangent; two or more of them have the same curvature, and there may be more than one set, the members in a set having the same curvature, but except for the curves in such set or sets, the curvatures are distinct: and in general, no one of these curves is osculated by the critical curve. The condition for osculation is that

$$\frac{\partial \Delta}{\partial x} + \frac{\partial \Delta}{\partial y} y' + \frac{\partial \Delta}{\partial y'} y'' = 0$$

should be satisfied; since the point c, α is not on the curve $E = 0$, the condition is not satisfied.

When the curve $E = 0$ does not exist, the condition for osculation is satisfied identically; the equation of the osculating critical curve then is a singular integral of the original equation.

The critical curves may have an envelope, so that the envelope is a curve satisfying the equation $\Delta = 0$. Even if the critical curves provide a singular integral of the original equation, their envelope will not do so, unless it osculates them as well as envelopes them. Should the latter condition be satisfied, it provides a second singular integral of the original equation.

Denote by $x = c'$, $y = \alpha'$, any point in the plane lying on the curve $E = 0$. Through that point there will, in general, pass a (finite) number of curves satisfying the equation $\Delta = 0$; let the curves be called the sub-critical curves, and their tangents at the point be called the sub-critical lines. The sole difference from the case, when the point c, α lies off the curve $E = 0$, is in the case of curves, which satisfy the equation $F = 0$ and touch a sub-critical line, and when therefore also the sub-critical curve has that line for its tangent; the sub-critical curve osculates one of the set of curves, which have the same curvature at the point, though it does not osculate any of the remainder.

It thus is possible to construct instances of equations of the second order, which certainly possess singular solutions of one kind or of both kinds: though the following process* is not essentially necessary to the existence of such solutions. Take any curve A, other than a straight line in general (though this condition need not be observed universally). Construct a series of curves A_1, osculating the curve A and depending upon a parameter, which will give a family of osculating curves: also a series of curves B_1, touching but not osculating A and depending upon one parameter. Next, construct a series of curves A_2 osculating the curves A_1, so that their equation contains the parameter in A_1 and another independent parameter; the equation will thus contain two independent

---

* It is indicated by Dixon, *Phil. Trans.* (1895, A), p. 562, in connection with equations of order n, solely as regards those which possess singular solutions of all classes. An instance, in which the mode of construction does not obviously imply the existence of a singular integral, and which in fact does possess such an integral, is given by the first of the succeeding examples.

parameters. Construct a series of curves B_2 in the same relation to the curves B_1 as A_2 to A_1, and containing two independent parameters: likewise a series of curves touching (but not osculating) one or more of the curves B_1, so that their equation contains two independent parameters.

Then regarding the constants of the original equation as non-parametric constants, the equations of the respective families A_2, B_2, C_2 each contain two independent parameters. They therefore are integral equivalents of respective differential equations of the second order.

The differential equation of A_2 will certainly possess a singular integral of the first kind, being the equation of A_1: and it may possess other singular integrals of the first kind. It will also possess a singular integral of the second kind, being the equation of A; and it may possess other singular integrals of the second kind, if A_1 possess osculating envelopes other than A.

The differential equation of B_2 will certainly possess a singular integral of the first kind, being the equation of B_1; and it may possess other singular integrals of the first kind. It will not of necessity possess a singular integral of the second kind: this, however, can occur, if B_1 possesses an osculating envelope, but it is not the equation of A: and, in general, it is not the case.

The differential equation of C_2 will not, of necessity, possess a singular integral of the first kind: this, however, can occur if C_2 possesses an osculating envelope, but it is not the equation of B_1: and, in general, it is not the case. Similarly it will not, in general, possess a singular integral of the second kind.

Ex. 1. Consider the equation

$$a(a-2x)q^2 + 2aqp(1+p^2) + (1+p^2)^3 = 0,$$

where
$$p = \frac{dy}{dx}, \qquad q = \frac{d^2y}{dx^2}.$$

The primitive is
$$(y-a)^2 + (x-\beta)^2 = 2ax.$$

It can be interpreted as the doubly-infinite aggregate of circles having double contact with the singly-infinite aggregate of parabolas, which touch the axis of y, have their axis parallel to the axis of x, and are of latus rectum $2a$. In the first place, it is clear that a parabola, which is an envelope (but not an osculating envelope) of the single infinitude of circles for variations of β, is not part of a first singular integral; and also that the axis of y, which is an envelope (but not an osculating envelope) alike of the parabolas and the circles, is not part of a singular integral.

The eliminant of $F=0$ and $\dfrac{\partial F}{\partial q}=0$ is

$$\Delta=(1+p^2)^2\{a^2p^2-a(a-2x)(1+p^2)\}=0.$$

Evidently $p^2+1=0$ gives

$$p=\pm i,$$
$$q=0,$$

which satisfy the original equation. We have

$$y+ix=A,\quad y-ix=B,$$

two singular first integrals; there is no corresponding singular second integral for either of them.

Taking the alternative part of $\Delta=0$, we have

$$p^2=\frac{\tfrac12 a-x}{x};$$

the (repeated) value of q given by the original equation is

$$q=-\tfrac12\frac{p(1+p^2)}{\tfrac12 a-x}$$
$$=-\frac{\tfrac14 a}{\{x^3(\tfrac12 a-x)\}}.$$

But from the above value of p derived through $\Delta=0$, we have

$$\frac{dp}{dx}=\frac{-\tfrac14 a}{\{x^3(\tfrac12 a-x)\}^{\frac12}},$$

so that

$$\frac{dp}{dx}=q,$$

where the latter q is that given by the equation. Hence the equation

$$p^2=\frac{\tfrac12 a-x}{x}$$

provides a singular integral of the original equation: its primitive is obtained by the elimination of ϕ between

$$\left.\begin{array}{l}y-c=\tfrac14 a\,(\phi+\sin\phi)\\ x=\tfrac14 a\,(1-\cos\phi)\end{array}\right\}.$$

The equation $\Delta=0$ has no singular integral other than $x=0$; this does not provide any solution of the original equation.

Continuing the geometrical interpretation of the results, we see that the first singular integral is the (singly-infinite) family of cycloids, which touch the axis of y at their vertices, and have their cusps on a parallel to the axis of y given by $x=\tfrac12 a$. As regards the osculation of any circle

$$(y-a)^2+(x-\beta)^2=2ax$$

of the primitive at any point, by the cycloid passing through the point, we have

$$y - a = -\frac{p^2 + 1}{q}$$
$$x - \beta = a + \frac{p(p^2 + 1)}{q}$$

for the circle. Now p and q are to be the same for the circle and the cycloid, as of course x and y (being the coordinates of the common point): but for the cycloid

$$\frac{p(p^2 + 1)}{q} = -a + 2x,$$

and therefore

$$x = -\beta.$$

Thus

$$y - a = \{-2\beta(2\beta + a)\}^{\frac{1}{2}}.$$

Hence for real points of contact, we must have 2β negative (or zero); and if 2β be negative, then $a + 2\beta > 0$. In that case, we have

$$\cos \phi = 1 + \frac{4\beta}{a} < 1,$$
$$= -1 + \frac{2}{a}(a + 2\beta) > -1,$$

and so there is a real value for ϕ: also

$$y - a = \pm \tfrac{1}{2}a \sin \phi,$$

and therefore

$$a \pm \tfrac{1}{2}a \sin \phi - c = \tfrac{1}{4}a(\phi + \sin \phi),$$

so that

$$c = a - \tfrac{1}{4}a(\phi - \sin \phi), \text{ or } = a - \tfrac{1}{4}a(\phi + 3\sin \phi).$$

If β be zero, then $x = 0$, $y = a$, $\phi = 0$, $c = a$.

As regards the circles

$$(y - a)^2 + (x - \beta)^2 = 2ax,$$

we thus have three cases, in their relation to tangency and osculation by the cycloid

$$y - c = \tfrac{1}{4}a(\phi + \sin \phi)$$
$$x = \tfrac{1}{4}a(1 - \cos \phi)$$.

If $\beta > 0$, the point of osculation is imaginary.

If $\beta = 0$, the point of osculation is the vertex of the cycloid; $c = a$: and the circle then is the circle of curvature of the parabola $(y - a)^2 = 2ax$ at its vertex. The circle of curvature, the parabola, and the cycloid, have a common centre of curvature at

$$x = a, \quad y = a.$$

If $\beta < 0$ and $2\beta + a > 0$, the circle has imaginary double contact with the parabola $(y - a)^2 = 2ax$. The points of osculation are two, viz.

$$x = -\beta, \quad y = a \pm \{-2\beta(a + 2\beta)\}^{\frac{1}{2}},$$

on the circle ; the two cycloids are determined by

$$c - a + \tfrac{1}{4} a (\phi + \sin \phi) = \pm \tfrac{1}{2} a \sin \phi,$$

where

$$\cos \phi = 1 + \frac{4\beta}{a}.$$

If $\beta < 0$ and $2\beta + a < 0$, the circle is imaginary.

As regards the equation

$$p^2 = \frac{\tfrac{1}{2} a - x}{x},$$

the line $x = \tfrac{1}{2} a$ is a locus of cusps of the cycloids : it is not a solution of that equation. The line $x = 0$ is a singular solution, and provides the same value of p at a common point as the cycloid ; but the curvature of the line is zero and that of the cycloid is finite, so that the value of q is not the same as for the cycloid at the point, and therefore not the same as for the circle at the point. It does not provide a singular integral of the original equation.

Ex. 2. Discuss the equation

$$27cy'' (y'^2 + 1)^3 = 4 \{(x + c) y'' - y' (y'^2 + 1)\}^3,$$

as to the relation between its complete primitive and its singular solutions.

(The equation represents the circles of curvature of the family of parabolas in the preceding question.)

Ex. 3. Discuss the equation

$$y''^2 - 2xy'^3 y'' + y'^6 - y'^4 = 0,$$

as to the relation between its complete primitive and its singular solutions.

(The equation represents the doubly-infinite family of parabolas $(y - a)^2 = 2\beta x + \beta^2 + 1$. The β-envelope of this family osculates all the parabolas for parametric values of β : yet it is not a singular integral : explain this.)

Ex. 4. Shew that the equation

$$y''^2 - 2y'y'' + 1 = 0$$

has no singular solution ; and prove that the primitive of $\Delta = 0$ is the locus of the cusps of the integral curves belonging to the complete primitive.

(Goursat.)

Ex. 5. Shew that the equation

$$(1 + x^2) y''^2 - (2xy' + \tfrac{1}{2} x^2) y'' + y'^2 + xy' - y = 0$$

has a singular integral of the first kind : and that the discriminant-equation $\Delta = 0$ has a singular integral of its own, which is not an integral of the original equation.

Obtain the various integrals.

(Lagrange, Serret, Goursat.)

Ex. 6. Shew that the equation

$$(2yy'' - y'^2)^3 = 4y'' (y' - xy'')^3$$

has a singular integral of the first kind, and a singular integral of the second kind. Obtain their equations : and exhibit the geometrical relations to one another of the respective integrals.

(Dixon.)

Ex. 7. Let $F(w'', w', w)$ denote an irreducible homogeneous polynomial function of its arguments, the coefficients being constants ; a solution of the equation

$$F(w'', w', w) = 0$$

is given by

$$w = Ae^{\theta z},$$

where A is arbitrary, and θ is any root of the algebraical equation

$$F(\theta^2, \theta, 1) = 0.$$

Determine whether the integral so obtained is a particular form of the general primitive or is a singular integral.

(Appell.)

EQUATIONS OF ANY ORDER AND DEGREE.

240. It now is easy to state the results, obtainable by a corresponding line of proof, for a single equation of any order and degree. Denoting $\dfrac{d^r w}{dz^r}$ by w_r, we have the most general equation of order n in the form

$$F(z, w, w_1, \ldots, w_n) = 0 ;$$

and we denote by

$$F_1(z, w, w_1, \ldots, w_{n-1}) = 0$$

the eliminant of

$$F = 0, \quad \frac{\partial F}{\partial w_n} = 0.$$

Let the degree of $F = 0$ be m.

Then assigning values $\alpha, \alpha_1, \ldots, \alpha_{n-1}$ to $w, w_1, \ldots, w_{n-1}$, when $z = c$, the solution of the equation $F = 0$ determines m functions, which, at $z = c$, have a common value, and the first $n - 1$ derivatives of which have respective common values at that point. When

$$F_1(c, \alpha, \alpha_1, \ldots, \alpha_{n-1})$$

is different from zero, the nth derivatives of the m functions are distinct from one another at $z = c$. When

$$F_1(c, \alpha, \alpha_1, \ldots, \alpha_{n-1})$$

is zero, a set of two or more of the nth derivatives have a common value, and there may be more than one such set : the common values being different from set to set, and different from the single values for the remaining functions: and generally the $(n+1)$th derivatives are distinct from one another.

But though

$$F_1(z, w, w_1, \ldots, w_{n-1}) = 0,$$

at the point, it is not the case that, in general, either for a set of functions w or an isolated function w arising as a solution of $F = 0$, the value of the nth derivative is the same as the value of the nth derivative of a function, which satisfies

$$F_1(z, w, w_1, \ldots, w_{n-1}) = 0 ;$$

that is, in general, a solution of $F_1 = 0$ is not a solution of $F = 0$. But if

$$\frac{\partial F}{\partial z} + \frac{\partial F}{\partial w} w_1 + \sum_{r=1}^{n-1} \frac{\partial F}{\partial w_r} w_{r+1} = 0,$$

either identically or in virtue of

$$F = 0, \quad \frac{\partial F}{\partial w_n} = 0,$$

then a solution of $F_1 = 0$ such that the value of w_s is α_s, $(s = 0, 1, \ldots, n-1)$, when $z = c$, is a solution of $F = 0$: it provides a first singular integral. The general primitive of $F_1 = 0$ then is a singular integral of $F = 0$.

Further $F_1 = 0$, being an equation of order $n-1$, can have singular integrals: these usually have their derivatives, up to order $n - 1$ inclusive, equal to the derivatives, up to that order, of the complete primitive for a value $z = c$: but the nth derivatives are usually not the same. Unless the first singular integral of $F_1 = 0$ (should it possess a singular integral) has its derivatives of order n equal to that of the primitive, it cannot provide a second singular integral for $F = 0$.

And so on in succession. It is possible, under the proper respective aggregates of conditions, that an equation of order n and degree higher than unity can possess, in addition to its complete primitive, singular integrals of the first kind, the second kind, up to the nth kind. The complete primitive involves n arbitrary parameters; a singular integral of the rth kind involves $n - r$ arbitrary parameters, and it is a singular integral of each of the intermediate equations*.

---

* Reference, in this connection, should be made to the memoirs by Mayer, Goursat, and Dixon, quoted in § 200.

Ex. Let

$$u_s = w_n z^s - w_{n-1} s z^{s-1} + w_{n-2} s(s-1) z^{s-2} - \ldots + (-1)^s w_{n-s} s!,$$

for $s = 0, 1, \ldots, n$; shew that the equation

$$\Phi(u_0, u_1, u_2, \ldots, u_n) = 0,$$

where Φ is an arbitrary function of its arguments, possesses a first singular integral: and obtain the complete primitive.

(Dixon.)

ANALYTICAL RELATION OF SINGULAR SOLUTIONS TO THE PRIMITIVE.

241. The analytical relation of the primitive of the equation

$$F(z, w, w', w'') = 0$$

to the primitive of the equation

$$\Delta(z, w, w') = 0,$$

when at the particular point the value of w' is one of the critical values, can be exhibited in the vicinity of the point (and not merely at the point) as follows. Let $w = \eta$ be the primitive of the equation $\Delta = 0$, so that

$$\Delta(z, \eta, \eta') = 0;$$

then because Δ is the eliminant of $F = 0$ and $\dfrac{\partial F}{\partial w''} = 0$, it follows that, among the roots W of

$$F(z, \eta, \eta', W) = 0,$$

some are certainly equal to one another. Let ζ denote such a repeated root, of multiplicity κ. Now let

$$w = \eta + u, \quad w' = \eta' + u',$$

so that, at the initial point $z = c$, the values of u and u' are zero in relation to the critical values; and let

$$w'' = \zeta + v,$$

so that the equation is

$$F(z, \eta + u, \eta' + u', \zeta + v) = 0.$$

It is known that when $u = 0$, $u' = 0$, then $v = 0$ is a repeated root of $F = 0$, looked upon as an equation in v, of multiplicity κ: consequently· the equation must have the form

$$v^\kappa = G(z, u, u', v) + v^{\kappa+1} H(z, u, u', v),$$

where G is a regular function of z and of the arguments u, u', v, which vanishes when $u = 0$, $u' = 0$, and H is a regular function of its arguments; and therefore

$$u'' + \eta'' - \zeta = v$$
$$= \{G(z, u, u', v) + v^{\kappa+1} H(z, u, u', v)\}^{\frac{1}{\kappa}}.$$

242. First, suppose that η is not an integral of the equation, so that $\zeta - \eta''$ is not identically zero. It is a function of z; let c denote an ordinary point of this function (not being a zero), and let

$$(\zeta - \eta'')_{z=c} = C,$$

so that, in the immediate vicinity of $z = c$, we have

$$u'' = C + \{G(z, u, u', v) + \ldots\}^{\frac{1}{\kappa}}.$$

As regards the value of u thus determined, it is dependent upon the form of the function G. Without entering into full discussion, consider the case in which G contains a term au' by itself as well as other terms; then au' is the most important term in the radical. Manifestly C is the leading term in u'', and therefore $C(z - c)$ is the leading term in u'; accordingly we take

$$\left. \begin{aligned} z - c &= t^\kappa \\ u' &= t^\kappa (C + U) \\ u &= \tfrac{1}{2} t^{2\kappa} (C + V) \end{aligned} \right\},$$

so that U and V vanish with t. Then

$$\frac{1}{\kappa} t \frac{dU}{dt} + U + C = C + [t^\kappa \{a(C + U) + \ldots\}]^{\frac{1}{\kappa}}$$
$$= C + t(aC)^{\frac{1}{\kappa}} + t H_1(U, V, t),$$

where H_1 is a regular function of its arguments and vanishes with them: that is,

and

$$\left. \begin{aligned} t \frac{dU}{dt} + \kappa U &= \kappa (aC)^{\frac{1}{\kappa}} t + \kappa t H_1(U, V, t) \\ \\ t \frac{dV}{dt} + 2\kappa V &= 2\kappa U \end{aligned} \right\}.$$

The critical quadratic of this pair of equations is

$$(\Omega + \kappa)(\Omega + 2\kappa) = 0;$$

so that, as κ is a positive integer $\geqslant 2$, the only solutions of these equations, which vanish with t, are regular functions of t: and

$$V = \frac{2\kappa^2}{(\kappa+1)(2\kappa+1)} (aC)^{\frac{1}{\kappa}} t + \ldots$$
$$= tP(t),$$

where P is a regular function. Thus

$$w - \eta = u$$
$$= \tfrac{1}{2}C(z-c)^2 + \tfrac{1}{2}(z-c)^{2+\frac{1}{\kappa}} P\{(z-c)^{\frac{1}{\kappa}}\},$$

which is the relation between w, the solution of the differential equation, and η the solution of $\Delta(z, w, w') = 0$, the two functions having their first derivatives equal at the general point c. As c moves, w and η both change; their difference is of the foregoing form, c being regarded as parametric.

Now as c moves, it may pass through a zero of

$$\zeta - \eta'',$$

the function which is not identically zero. In the immediate vicinity of such a zero, let the leading term be

$$b(z-c)^\beta,$$

where $\beta > 0$, so that, on the present hypothesis, we have

$$u'' = b(z-c)^\beta + \ldots + \{au' + \ldots\}^{\frac{1}{\kappa}},$$

and therefore u'' vanishes, when $z = c$, as well as u and u'. In the immediate vicinity of $z - c$, let the leading term in u be of order μ, so that we may take

$$u = \alpha(z-c)^\mu,$$

for sufficiently small values of $|z-c|$: the indices to be compared are

$$\mu - 2, \quad \beta, \quad \frac{\mu-1}{\kappa}.$$

If $\beta \geqslant \dfrac{1}{\kappa-1}$, we take

$$\mu = \frac{2\kappa-1}{\kappa-1},$$

which makes the first and the third equal to one another, and neither of them greater than the second. If $\beta < \dfrac{1}{\kappa-1}$, we take

$$\mu = 2 + \beta,$$

which makes the first and the second equal to one another, and both of them less than the third.

In the former case, we take

$$z - c = t^{\kappa-1}$$
$$u' = t^{\kappa} (a + U)$$
$$u = t^{2\kappa-1} \left(\frac{\kappa - 1}{2\kappa - 1} a + V \right)$$

so that U and V vanish with t. Then a is given by the equation

$$\frac{\kappa}{\kappa - 1} a = (aa)^{\frac{1}{\kappa}},$$

or

$$\frac{\kappa}{\kappa - 1} a = b + (aa)^{\frac{1}{\kappa}},$$

according as $\beta > \dfrac{1}{\kappa - 1}$ or $\beta = \dfrac{1}{\kappa - 1}$. Proceeding as before, we have equations for U and V in the form

$$t \frac{dU}{dt} + \kappa U = \text{regular function of } t, U, V$$
$$t \frac{dV}{dt} + (2\kappa - 1) V - (\kappa - 1) U = 0$$

The critical quadratic is

$$(\Omega + \kappa)(\Omega + 2\kappa - 1) = 0;$$

so that, as $\kappa \geqslant 2$, the only solutions of the equation which vanish with t are regular functions of t, and then

$$w - \eta = u$$
$$= (z - c)^{\frac{2\kappa-1}{\kappa-1}} P \{(z-c)^{\frac{1}{\kappa-1}}\},$$

where P is a regular function of its argument that does not vanish when $z = c$.

In the latter case, β is less than $\dfrac{1}{\kappa - 1}$: we suppose it to be real and (when expressed in its lowest terms) of the form $\dfrac{p}{q}$. We then take

$$z - c = t^{\kappa q}$$
$$u' = t^{\kappa q + \kappa p} \left(\frac{bq}{p + q} + U \right)$$
$$u = t^{2\kappa q + \kappa p} \left\{ \frac{bq^2}{(p + q)(p + 2q)} + V \right\}$$

U and V vanishing with t; the equations that determine these new variables are

$$t\frac{dU}{dt} + \kappa(q+p)U = t^{q-(\kappa-1)p}\, G(t, U, V)$$
$$t\frac{dV}{dt} + \kappa(2q+p)V - \kappa qU = 0$$,

where G is a regular function of its arguments; also $q - (\kappa - 1)p$, which is equal to $q(\kappa - 1)\left(\dfrac{1}{\kappa - 1} - \beta\right)$, is a positive integer. As before, the only solutions, which vanish with t, express U and V as regular functions of t; and then

$$w - \eta = u$$
$$= (z - c)^{2+\beta}\, P\,\{(z - c)^{\frac{1}{\kappa q}}\},$$

where P is a regular function of its argument that does not vanish when $z = c$.

In both of these cases, c is a zero of the non-evanescent function $\zeta - \eta''$. The quantity η may be regarded as a solution of the equation at the point c, but not in the immediate vicinity of this point; that is, it is a solution only at a point.

243. Secondly, suppose that η is a solution of the equation, so that ζ is actually equal to η''; then $v = u''$, and the equation is

$$u'' = \{G(z, u, u', u'') + \ldots\}^{\frac{1}{\kappa}}.$$

In this case also the form of u depends upon the form of $G(z, u, u', u'')$, which vanishes with u and u': the leading term in u will manifestly be the leading term in the solution of the equation

$$u''^{\kappa} = G(z, u, u', u'').$$

Because u and u' vanish for a parametric value c of z, and because G vanishes with u and u', therefore u'' vanishes for that parametric value c of z; hence the leading term is of the form

$$\alpha(z - c)^{\theta},$$

where $\theta > 2$, and it can be obtained by the method of § 229. When θ, if fractional, is in its lowest terms, let its value be p/q: we then take

$$z - c = t^{q}$$
$$u' = t^{p-q}(\theta\alpha + U)$$
$$u = t^{p}(\alpha + V)$$.

With these values, and taking account of the way in which θ has been obtained, we have

$$G(z, u, u', u'') = t^{\kappa(p-2q)} G_1,$$

where G_1 is a regular function of its arguments not vanishing with t, U, V; and thus

$$\frac{1}{q} t \frac{dU}{dt} + \frac{p-q}{q}\left(U + \frac{p}{q}\alpha\right) = G_1^{\frac{1}{\kappa}}$$

$$= \frac{p(p-q)}{q^2}\alpha + \text{regular function of } U \text{ and } V,$$

$$t\frac{dV}{dt} + pV - qU = 0.$$

The form of the solution depends upon the coefficient of U in $G_1^{\frac{1}{\kappa}}$: unless it is such that the first equation becomes

$$t\frac{dU}{dt} - \gamma U = \lambda t + f(U, V, t),$$

where no term in f is of dimensions less than two in its arguments, and where γ is a positive integer, then a solution exists, which is a regular function of t and vanishes with t. Consequently

$$w - \eta = u$$
$$= (z-c)^{\frac{p}{q}} P\{(z-c)^{\frac{1}{q}}\},$$

where P is a regular function of its argument that does not vanish when $z = c$.

It may happen that the equation for u'' takes a form

$$u'' = u\{K(z, u, u', u'')\}^{\bar{\kappa}},$$

where K is a regular function not vanishing with u, u', u'': a solution is $u = 0$, and then

$$w - \eta = 0,$$

that is, the integral η then is a *particular* form of the primitive.

In the other cases, $w = \eta$ is a *singular* integral of the equation.

Ex. 1. Consider the equation

$$a^4 w''^2 - 2a^3 w' w'' + 9zw = 0.$$

We have

$$\Delta = a^2 w'^2 - 9zw,$$

and therefore

$$a^2 \eta'^2 - 9z\eta = 0,$$

so that

$$a^2 \eta = (z^{\frac{3}{2}} + \beta)^2.$$

Taking
$$w = u + \eta, \quad w' = u' + \eta', \quad w'' = v + \zeta,$$
where v is to vanish with u and u', we have
$$a^4 v^2 - 2a^3 u' v - 2a^3 u' \zeta + 9zu = 0,$$
and
$$a\zeta = \eta',$$
being the repeated root of
$$a^4 \zeta^2 - 2a^3 \eta' \zeta + 9z\eta = 0.$$
Hence
$$a^2 v - au' = \{2a^2 u' \eta' - 9zu + a^2 u'^2\}^{\frac{1}{2}}.$$
Now
$$v = -\zeta + w''$$
$$= u'' + \frac{1}{a^2}\left\{6z + \tfrac{3}{2}\beta z^{\frac{1}{2}} - 3\frac{z^2}{a} - 3\frac{\beta}{a}z^{\frac{1}{2}}\right\}.$$

Take any value $z = c$, which is not zero and not a root of
$$6z + \tfrac{3}{2}\beta z^{-\frac{1}{2}} - 3\frac{z^2}{a} - 3\frac{\beta}{a}z^{\frac{1}{2}} = 0,$$
and let $z - c = Z$; then
$$v = u'' + C_0 + C_1 Z + \dots .$$
Also
$$\eta' = \frac{3z^2 + 3\beta z^{\frac{1}{2}}}{a^2} = c_0 + c_1 Z + \dots ,$$
say, so that the equation is
$$a^2 (u'' + C_0 + C_1 Z + \dots) - au' = \{2a^2 u' (c_0 + c_1 Z + \dots) - 9zu + a^2 u'^2\}^{\frac{1}{2}}.$$

Now $u = 0$, $u' = 0$, when $Z = 0$; hence $u'' = -C_0$, when $Z = 0$, and therefore the leading term of u' is $-C_0 Z$. Accordingly, let
$$\left.\begin{array}{l} Z = t^2 \\ u' = t^2 (-C_0 + U) \\ u = t^4 (-\tfrac{1}{2}C_0 + V) \end{array}\right\};$$
then
$$u'' = -C_0 + U + \tfrac{1}{2}t\frac{dU}{dt},$$
and the equation is
$$t\frac{dU}{dt} + 2U = \left(-\frac{8C_0 c_0}{a^2}\right)^{\frac{1}{2}} t + \Theta_2(t, U, V),$$
where Θ_2 is a regular function containing no terms of dimensions lower than 2 in its arguments: also
$$t\frac{dV}{dt} + 4V - 4U = 0.$$
As the critical quadratic is
$$(\Omega + 2)(\Omega + 4) = 0,$$
the only solutions that vanish with t are the regular functions of t, being
$$U = \frac{2}{3}\left(-\frac{2C_0 c_0}{a^2}\right)^{\frac{1}{2}} t + \dots ,$$
$$V = \frac{8}{15}\left(-\frac{2C_0 c_0}{a^2}\right)^{\frac{1}{2}} t + \dots ;$$

and therefore

$$w - \eta = u$$

$$= (z-c)^2 \left\{ -\frac{1}{2} C_0 + \frac{8}{15} \left(-\frac{2C_0 c_0}{a^2} \right)^{\frac{1}{2}} (z-c)^{\frac{1}{2}} + \dots \right\}.$$

The values of w and η are equal at $z=c$; likewise those of w' and η'; but the values of w'' and η'' are unequal at $z=c$.

Ex. 2. Discuss the nature of the solution $a^2\eta = z^3$ of the equation $\Delta = 0$, in relation to the primitive of the equation just considered.

CLASSIFICATION OF INTEGRALS OF AN EQUATION OF THE SECOND ORDER, AS ASSOCIATED WITH THE PRIMITIVE.

244. The preceding results offer initial suggestions in connection with the classification of the integrals of a differential equation of the second order. The corresponding question for equations of the first order has received ample consideration from various writers: and, as is not infrequently the case in generalising theorems, not a few results can be written down for an equation of general order as soon as the step, next after the discussion of the simplest instance, viz. of an equation of the first order, has been taken to the discussion of one of the second order. As the relations between the various integrals are initially more of a formal than of a functional character, it will be convenient from the beginning to associate with the differential equation the customary geometrical interpretation in regard to plane curves.

It is known that, when an integral equation

$$g(x, y, c) = 0$$

satisfies a differential equation

$$\phi(x, y, p) = 0,$$

and when the integral curve has a proper envelope, which is included in the result of the elimination of c between

$$g = 0, \quad \frac{\partial g}{\partial c} = 0,$$

the equation of the envelope is an integral of the differential equation. The eliminant may include equations, which do not belong to the envelope: such an one is the node-locus, for example, and its equation is not an integral of the differential equation: so that, as is well known, the c-discriminant of the integral does not necessarily provide a solution of the differential equation.

As there is thus such a limitation upon the results achieved in connection with the equation of the first order, there will be corresponding limitations upon results, which are based merely upon broad processes, without fully detailed investigation, applied to equations of the second order. The succeeding discussion must therefore be regarded only as preliminary.

245. Let there be a doubly-infinite system of curves represented by

$$f(x, y, \alpha, \beta) = 0,$$

where α and β are two independent parameters: their equation can be regarded as a complete primitive of the differential equation of the second order, which results from the elimination of α and β between the three equations

$$0 = f,$$

$$0 = \frac{\partial f}{\partial x} + p \frac{\partial f}{\partial y},$$

$$0 = \frac{\partial^2 f}{\partial x^2} + 2p \frac{\partial^2 f}{\partial x \partial y} + p^2 \frac{\partial^2 f}{\partial y^2} + q \frac{\partial f}{\partial y}.$$

So far as regards elimination of α and β, it does not affect the result what their nature is, when these three equations are taken in this form. If therefore it is possible to determine them in ways other than as parameters, the final differential equation will be the same, if the three equations keep the same form. Accordingly, we assume that both α and β are functions of x to be determined (if possible) so that the equations have the same form; where convenient, this assumption will be replaced by the equivalent assumption that α is a function of x, and β is a function of α, both functions initially being unknown.

No change in the form of the first equation is thus caused. The second equation remains unchanged, provided

$$\frac{\partial f}{\partial \alpha} \frac{d\alpha}{dx} + \frac{\partial f}{\partial \beta} \frac{d\beta}{dx} = 0 :$$

and therefore, if any form of α and β other than parametric constants be possible, it must be included in the relation

$$P = \frac{\partial f}{\partial \alpha} + \frac{\partial f}{\partial \beta} \beta' = 0,$$

where $\beta' = d\beta/d\alpha$. The third equation remains unchanged, provided

$$\left(\frac{\partial^2 f}{\partial x \partial \alpha} + p \frac{\partial^2 f}{\partial y \partial \alpha}\right) \frac{d\alpha}{dx} + \left(\frac{\partial^2 f}{\partial x \partial \beta} + p \frac{\partial^2 f}{\partial y \partial \beta}\right) \frac{d\beta}{dx} = 0,$$

or therefore values, that are not merely parametric, must be included in the relation

$$\frac{\partial^2 f}{\partial x \partial \alpha} + \beta' \frac{\partial^2 f}{\partial x \partial \beta} + p \left(\frac{\partial^2 f}{\partial y \partial \alpha} + \frac{\partial^2 f}{\partial y \partial \beta} \beta'\right) = 0.$$

The form of p is to be the same, for purposes of elimination, whether α be parametric or variable: taking its value from the second of the three equations, we have

$$Q = \frac{\partial f}{\partial y}\left(\frac{\partial^2 f}{\partial x \partial \alpha} + \beta' \frac{\partial^2 f}{\partial x \partial \beta}\right) - \frac{\partial f}{\partial x}\left(\frac{\partial^2 f}{\partial y \partial \alpha} + \frac{\partial^2 f}{\partial y \partial \beta} \beta'\right) = 0,$$

as a relation including the possible forms of α and β. If such possible forms are definitely determinable and can be associated with the equation $f = 0$, the new equation is such that, in consequence of $P = 0$, the value of p is the same as at first, and, in consequence of $Q = 0$, the value of q is the same as at first: so that the new curve thus obtained can be looked upon as an osculant of the original curve.

Now from the three equations

$$f = 0, \quad P = 0, \quad Q = 0,$$

we can eliminate x and y, and obtain a relation

$$F(\alpha, \beta, \beta') = 0:$$

an equation of the first order to determine the value of β in terms of α. Let its complete primitive be

$$G(\alpha, \beta, \theta) = 0,$$

where θ is the arbitrary constant of integration. The value of β' is given by

$$\frac{\partial G}{\partial \alpha} + \frac{\partial G}{\partial \beta} \beta' = 0,$$

and from the mode of derivation, this value agrees with the value given by $P = 0$: that is, we have

$$J(\alpha, \beta, \theta) = \frac{\partial f}{\partial \alpha} \frac{\partial G}{\partial \beta} - \frac{\partial f}{\partial \beta} \frac{\partial G}{\partial \alpha} = 0.$$

We now have three equations, viz.

$$f = 0, \quad G = 0, \quad J(f, G) = 0,$$

involving an arbitrary constant θ, the varied parameters α and β, and the two variables x and y; eliminating α and β between them, we have an equation

$$H(x, y, \theta) = 0,$$

which can be such that, from the mode of its derivation, it represents an osculant of the curves given by the equation

$$f(x, y, \alpha, \beta) = 0,$$

and that it satisfies the same differential equation of the second order as $f = 0$, though it is not included within the equation $f = 0$.

Again, the family of curves

$$H(x, y, \theta) = 0$$

may have an envelope; if they have, its equation is included in the eliminant of $H = 0$ and

$$\frac{\partial H}{\partial \theta} = 0.$$

At the point of contact of any of the curves with the envelope, the direction of the tangent is given by

$$\frac{\partial H}{\partial x} + p\frac{\partial H}{\partial y} = 0.$$

In general, the contact between a curve and its envelope is simple only, and the curvatures are different: thus, in general, the eliminant of $H = 0$ and $\frac{\partial H}{\partial \theta} = 0$ will not provide the same value of q as is provided by H, though it provides the same value of p; the original differential equation will then not be satisfied. But if, in obtaining the envelope, the value of θ be such as to give for the envelope at the point of contact the same curvature as for the curve of the family, then the original differential equation will be satisfied. From the original equation $H = 0$ we have, when θ is no longer merely parametric,

$$\frac{\partial H}{\partial x} + p\frac{\partial H}{\partial y} = 0$$

for the curve and the envelope, provided $\frac{\partial H}{\partial \theta} = 0$: and also

$$\frac{\partial^2 H}{\partial x^2} + 2p\frac{\partial^2 H}{\partial x \partial y} + p^2\frac{\partial^2 H}{\partial y^2} + q\frac{\partial H}{\partial y} = 0$$

for the curve and the envelope, provided

$$\left(\frac{\partial^2 H}{\partial x \partial \theta} + p \frac{\partial^2 H}{\partial y \partial \theta}\right) \frac{d\theta}{dx} = 0,$$

that is, (θ being variable), provided

$$\frac{\partial^2 H}{\partial x \partial \theta} + p \frac{\partial^2 H}{\partial y \partial \theta} = 0.$$

As the value of p is the same for the curve and the envelope, the last condition may be replaced by

$$K = \frac{\partial^2 H}{\partial x \partial \theta} \frac{\partial H}{\partial y} - \frac{\partial^2 H}{\partial y \partial \theta} \frac{\partial H}{\partial x} = 0.$$

If therefore the envelope of $H = 0$ is also an osculant of $H = 0$, the three equations

$$H = 0, \qquad \frac{\partial H}{\partial \theta} = 0, \qquad K = 0,$$

must coexist in virtue of a single relation between x and y. This relation includes the equation of the osculating envelope, if such an envelope exists.

The original equation $f = 0$ may represent a family of curves that possess no osculant, though perhaps they possess a non-osculating family of envelopes. Similarly the equation $H = 0$ may represent a family of curves that possess no envelope at all: the analytical equations are, however, of the same form, and they then correspond to other geometrical properties. Accordingly, in favourable cases, the analytical equations may contain, in addition to osculants and osculating envelopes for $f = 0$ and $H = 0$, other geometrical properties and loci. The whole subject merits further investigation.

Ex. 1.   Consider the equation

$$(y - a)^2 + (x - \beta)^2 = 2ax,$$

the complete primitive of the differential equation in Ex. 1, § 239.

The equation $P = 0$, on division by 2, is

$$y - a + (x - \beta)\beta' = 0.$$

The equation $Q = 0$, on division by 4, is

$$x - \beta - (y - a)\beta' = a,$$

so that

$$\frac{y - a}{-\beta'} = \frac{x - \beta}{1} = \frac{a}{1 + \beta'^2};$$

on substituting these values in the primitive, the equation to determine β in terms of a is found to be

$$\beta'^2 = -\frac{2\beta + a}{2\beta}.$$

This, when integrated, gives

$$\left.\begin{array}{l} -\beta = \tfrac{1}{4} a \left(1 + \cos 2\phi\right) \\ a - \theta = \tfrac{1}{4} a \left(2\phi + \sin 2\phi\right) \end{array}\right\},$$

from which ϕ is to be eliminated. For the osculant, we have

$$X = \beta + \frac{a}{1 + \beta'^2} = -\beta$$

$$= \tfrac{1}{4} a \left(1 + \cos 2\phi\right),$$

$$Y = a - \frac{a\beta'}{1 + \beta'^2};$$

so that

$$Y - \theta = \tfrac{1}{4} a \left(2\phi - \sin 2\phi\right).$$

The values of X and Y give the first singular integral of the differential equation as before.

There is no second singular integral.

Ex. 2. The equation[*]

$$(qy - \tfrac{1}{2}p^2)^3 = 4q \, (p - xq)^3$$

has

$$a\beta^3 y - (a^2 x + 1)^2 - 2\beta^2 = 0$$

for its complete primitive.

We take $f = 0$ in the form

$$f = y - \frac{(a^2 x + 1)^2}{a\beta^3} - \frac{2}{a\beta} = 0.$$

We have

$$\frac{\partial f}{\partial y} = 1, \qquad \frac{\partial f}{\partial x} = -\frac{2}{\beta^3} (a^3 x + a);$$

the equation $Q = 0$ thus is

$$-\frac{2}{\beta^3} (3a^2 x + 1) + \frac{6}{\beta^4} (a^3 x + a) \beta' = 0,$$

and therefore

$$3a^2 x = \frac{\beta - 3a\beta'}{a\beta' - \beta}.$$

The equation $P = 0$ is

$$-\frac{1}{\beta^3} (3a^2 x^2 + 2x) + \frac{1}{a^2 \beta^3} + \frac{2}{a^2 \beta} + \left\{ 3 \frac{(a^2 x + 1)^2}{a\beta^4} + \frac{2}{a\beta^2} \right\} \beta' = 0,$$

which, on substitution for x and reduction, leads to

$$a^2 \beta'^2 - \beta^2 - \tfrac{2}{3} = 0.$$

This can be integrated at once, with the result

$$\beta = \theta a - \frac{1}{6\theta a},$$

---

[*] This illustration is given by Dixon, *Phil. Trans.* (1895, A), p. 565: some slight changes in constants have been made.

where θ is the arbitrary constant of integration. To find the equation of the osculant, we eliminate a, β between this equation, $f=0$, and

$$3a^2 x = \frac{\beta - 3a\beta'}{a\beta' - \beta},$$

the value of β' being

$$\theta + \frac{1}{6\theta a^2}.$$

The result of the elimination is

$$H = xy + 12\theta x + 6\theta^2 y + 8\theta^3 = 0,$$

which accordingly is the osculant : and $H=0$ provides a first singular integral of the original differential equation.

The three equations which must coexist, if $H=0$ is to possess an osculating envelope and if the original differential equation is to possess a singular integral of the second kind, are

$$\left. \begin{aligned} 0 &= H \\ 0 &= \frac{\partial H}{\partial \theta} = 12\,(x + \theta y + 2\theta^2) \\ 0 &= K \;\; = 12\,(x - \theta y - 6\theta^2) \end{aligned} \right\}.$$

These are satisfied simultaneously by

$$x = 2\theta^2, \quad y = -4\theta :$$

that is, a singular integral of the second kind is possessed by the original differential equation, and it is given by

$$y^2 = 8x.$$

Ex. 3. Shew that the equation

$$(qy - 2p^2)^6 + 54p^3\,(qx + 2p)\,(qy - 2p^2)^3 + 243p^6\,(qy - 2p^2)^2 + 729p^6 q^2 = 0$$

has the equation

$$(y - \beta)\,(x - a) = 4\gamma$$

for its primitive, where

$$\beta^6 - 54a\beta^3\gamma + 243\beta^2\gamma^2 + 729\gamma^2 = 0.$$

Shew that an osculant of the curve, being a singular integral of the first kind of the differential equation, is given by

$$y^2 = 2\theta x + \theta^2 + 1.$$

Is the envelope of this singular integral a singular integral of the second kind of the differential equation ?

CHAPTER XVI.

Equations of the Second Order with Sub-uniform Integrals; with some General Considerations*.

246. It has already appeared (Chap. ix) that the determination of descriptive properties of the integral of a differential equation of the first order is somewhat laborious and inconclusive, as compared with the determination of analytic properties of the integral function. This divergence is still more marked for equations of higher order, partly owing to the mere added complexity of the analysis, but chiefly owing to the fundamental difference, to which reference already (§ 91) has been made, between equations of the first order and those of higher order. By Painlevé's theorem, it is known that a parametric value of the variable may be a pole or may be an algebraical critical point for the integral of an equation of the first order: it cannot be an essential singularity of the integral. But this property does not extend to equations of higher orders: as was seen from a simple

---

* As regards the earlier part (§§ 246—258) of this chapter, the following authors may be consulted:

> Picard, *Liouville*, 4me Sér., t. v (1889), pp. 135—319, *Acta Math.*, t. xvii (1893), pp. 297—300; the subject is developed more fully in the *Théorie des fonctions algébriques de deux variables indépendantes* (t. i, 1897, the later volume to contain the applications) by Picard and Simart:
>
> Mittag-Leffler, *Acta Math.*, t. xviii (1894), pp. 233—245:
>
> Fransén, *Stockh. Öfv.*, t. lii (1895), pp. 223—241:
>
> Wallenberg, *Crelle*, t. cxix (1898), pp. 87—113, *ib.* t. cxx (1899), pp. 113—131.

No attempt has been made to give an account of investigations that relate to descriptive properties of equations of the second order, such as are given in Painlevé's *Stockholm Lectures* (quoted at the beginning of Chap. xiv). The principal reason for this omission is the same as in the case of a single equation of the first order, stated in Chap. x.

example (§ 91), the integral of an equation of the second order can have a parametric essential singularity.

A limited class of equations of the first order was considered in Chapter IX, the assigned limitation being that the equations should have no parametric critical points. In the case of these equations, a parametric value of z is either a pole or an ordinary point of the integral, for it cannot be an essential singularity of the integral. Also, equations still more limited were considered, viz. those which have neither fixed nor parametric critical points; their integrals are uniform functions of the independent variable.

It is natural to enquire whether, taking account of the fundamental difference between equations of the first order and those of (say) the second order, it is possible to delimit corresponding classes of the latter. The simplest case to investigate obviously is that, in which the integrals either are uniform, or satisfy as many of the indirect tests of being uniform as can be framed. Thus it is possible to assign the conditions that no parametric value of the independent variable shall provide an algebraic critical point of the integral. We can also determine whether an integral of an equation of the second order is uniform in the vicinity of any point, when it is determined by the assignment of arbitrary values to itself and its first derivative at the point. But even if all the necessary conditions are satisfied, it does not follow that the integral is a uniform function: for the possibility that the integral may possess a parametric essential singularity, with either definite or indefinite branching, is not touched by either of the sets of conditions. It manifestly is not touched by the first, which deals with algebraical branching only; it manifestly is not touched by the second because, if the point be an essential singularity, the assigned values will be included among the unlimited number which can be acquired at an essential singularity.

For instance, consider the equation

$$-27ww'^5 = 2\,(w'^2 - ww'')^3.$$

If a point a could be an algebraical branch-point (with either finite or infinite values of the function there), the governing term would have a form

$$\frac{\kappa}{(z-a)^n},$$

where κ is a constant, and n is some (non-integer) positive or negative rational quantity. On the left-hand side, the index of the governing term is $-(6n + 5)$; on the right-hand side, it is $-(6n + 6)$. These two cannot be equal: and therefore no conditions have to be assigned to exclude any parametric point from being a branch-point. Also, taking the equation in the form

$$w'' = \frac{w'^2}{w} + 3w'\left(\frac{w'^2}{2w^2}\right)^{\frac{1}{3}},$$

and assigning initial arbitrary values to w' and w—the value zero may not be assigned to w, because for no number n can there be a leading term $\kappa(z-a)^{-n}$—then the equation determines three separate functions, each of them uniform: that is, it would be satisfied by three uniform functions of the variable. As a matter of fact, the general integral is not uniform: the complete primitive is

$$w = \alpha e^{(z-c)^{-\frac{1}{2}}},$$

where α and c are arbitrary: the parametric value c is an essential singularity.

As another example—it is propounded by Picard—consider the equation, which has

$$w = \wp\left\{\alpha + \log(z + \beta)\right\}$$

for its primitive, α and β being arbitrary. The point

$$-\beta + e^{-\alpha}$$

is a (parametric) pole of the integral, and the (parametric) point $-\beta$ is an essential singularity: there are no algebraical branch-points. The integral is uniform, only if $2\pi i$ is a period of the elliptic functions: a condition which would require a transcendental relation between the invariants g_2 and g_3.

Still, the equations, which satisfy the conditions, that no parametric point is an algebraic critical point, and that the assignment of initial arbitrary values determines an integral function (or a limited number of integral functions) uniform in the vicinity of the initial point, constitute a definite class. They are called by Picard* equations having their general integral *à apparence uniforme*: here, they will be called equations with *sub-uniform* integrals. The further tests, necessary that the sub-uniform

---

* "Mémoire sur la théorie des fonctions algébriques de deux variables indépendantes," *Liouville*, 4ᵐᵉ Sér., t. v (1889), p. 278; *Acta Mathematica*, t. xvii (1893), p. 298.

integral should really be a uniform integral, are as yet un-formulated for the general equation.

Further, among equations of the second order, there are two classes which admit of immediate formal reduction to the first order. One class is constituted by those, from which the dependent variable is explicitly absent, so that they have the form

$$\phi(z, w', w'') = 0:$$

by taking $w' = w_1$, the equation becomes

$$\phi(z, w_1, w_1') = 0,$$

and the general integral of the latter leads, after a quadrature, to the general integral of the former. The other is constituted by those, from which the independent variable is explicitly absent, so that they have the form

$$\phi(w, w', w'') = 0.$$

By taking $w' = W$, so that

$$w'' = W \frac{dW}{dw},$$

the equation becomes

$$\phi\left(w, W, W \frac{dW}{dw}\right) = 0,$$

which is of the first order and has w for the new independent variable. The disadvantage of this form for discussion of the functional properties of the integral lies in the fact, that the independent variable is not definite, when z acquires a value which is an essential singularity of w.

The latter class is the more important of the two. We shall deal with the case when it is of the first degree in the derivative of the second order.

PICARD'S EQUATIONS OF SECOND ORDER AND FIRST DEGREE, WITH SUB-UNIFORM INTEGRALS.

247. We proceed therefore to the consideration of such equations

$$\frac{d^2w}{dz^2} = R\left(w, \frac{dw}{dz}\right)$$

as have their integrals sub-uniform: R being rational in w and $\frac{dw}{dz}$, and being explicitly independent of z.

In the first place, R must be an integral function of $\dfrac{dw}{dz}$. It has been seen that, when R has the form

$$\frac{f(z, w, w')}{g(z, w, w')},$$

initial values of w and w', which make g vanish but not f, lead to a multiform function for the integral (§ 215): a result excluded by the hypothesis of the problem. We thus have

$$w'' = A_0 + A_1 w' + \ldots + A_n w'^n,$$

where A_0, A_1, ..., A_n are rational functions of w. Let m_r be the excess of the degree, in w, of the numerator in A_r over its denominator (for $r = 1, \ldots, n$).

Next, let a denote a value of z, which is a pole of w; as the integral is supposed to be uniform in the immediate vicinity of a, it can there be expanded in the form

$$w = \frac{\kappa}{(z - a)^\theta} + \ldots,$$

where κ is a constant, θ is a positive integer, and all the indices of the unexpressed terms are greater than $-\theta$. Then the expansion of A_r in ascending powers of $z - a$ begins with a term in $(z - a)^{-\theta m_r}$, and the expansion of $A_r w'^r$ begins with a term in $(z - a)^{-\theta m_r - r(\theta + 1)}$; and therefore the greatest negative indices in the successive terms on the right-hand side are

$$\theta m_0, \quad \theta m_1 + \theta + 1, \quad \theta m_2 + 2(\theta + 1), \quad \ldots.$$

The negative index of the single term on the left is

$$\theta + 2,$$

which accordingly must be equal to one or more than one (being the greatest) of the indices on the right. The most general case is that in which all the indices on the right are equal to one another, so that we can have

$$\theta m_0 = \theta m_r + r(\theta + 1),$$

and therefore

$$m_0 - m_r = r\left(1 + \frac{1}{\theta}\right),$$

for various values of r. Now θ must be unity at least: as all the indices m_r are integers, θ cannot be greater than unity, if the above relation holds for each of the numbers m; hence we have

$$m_0 = m_r + 2r.$$

If, however, the indices on the right-hand side are not equal to one another, then a modification must be made. For instance, let

$$\theta m_0 > \theta m_1 + \theta + 1,$$
$$\theta m_0 = \theta m_2 + 2(\theta + 1);$$

then

$$m_1 < m_0 - 1 - \frac{1}{\theta},$$

$$m_2 = m_0 - 2 - \frac{2}{\theta},$$

which would be satisfied if $\theta = 2$, by taking

$$m_1 < m_0 - \tfrac{3}{2} < m_0 - 2,$$
$$m_2 = m_0 - 3 :$$

and so on. It is easy to see that, in general, we have

$$m_0 \geqslant m_r + 2r,$$

if the net degree of A_0 in powers of $(z - a)^{-1}$ is not less than the net degree of any other term.

Also the highest negative index on the right-hand side must be equal to $\theta + 2$, so that

$$\theta + 2 = m_0 \theta,$$

and so

$$m_0 = 1 + \frac{2}{\theta}.$$

For the most general case, the value of θ is 1, so that

$$m_0 = 3 ;$$

and therefore $m_1 = 1$, $m_2 = -1$, and so on. The net degrees of the coefficients of the derivatives of w are thus $3, 1, -1, \ldots$.

Again, the integral is to be a uniform function in the vicinity of any point. Let z_0 be any point; let the integral function have a value w_0 assigned to it there; then in the immediate vicinity of z_0, we have

$$w - w_0 = \lambda (z - z_0)^\kappa + \ldots,$$

the unexpressed terms having indices greater than κ. As the integral is to be uniform, no such expansion can be possible in which κ shall have a fractional value; in particular, κ must not have a value $\frac{q}{p}$, where the integers q and p are prime to one another, and $q < p$. For such an expansion, the lowest term in $z - z_0$ on the left-hand side would have an index $\frac{q}{p} - 2$: and as $\frac{q}{p} - 1$ is negative, the lowest term in $z - z_0$ on the right-hand side would have an index $\left(\dfrac{q}{p} - 1\right) n$. The equation could be satisfied (with other conditions among coefficients) if

that is, if

$$\frac{q}{p} - 2 = \left(\frac{q}{p} - 1\right) n,$$

$$\frac{q}{p} = \frac{n-2}{n-1} = 1 - \frac{1}{n-1}.$$

For values of n greater than 2, such a fractional value would thus be possible. On the other hand, on the hypothesis adopted, such an expansion is to be impossible; and therefore n may not be greater than 2, that is, the right-hand side of the equation cannot have more than three terms. The equations therefore have the form

$$w'' = A_0 + A_1 w' + A_2 w'^2,$$

where A_0, A_1, A_2 are rational functions of w, of net degrees 3, 1, -1, respectively.

248. Transforming this equation momentarily to the form

$$v \frac{dv}{dw} = A_0 + A_1 v + A_2 v^2,$$

where v denotes w', we have an equation of the first order in v as a function of w. By the adopted hypothesis, w is to be regular in the vicinity of any value of z, and w' also will be regular there; when therefore a relation between w and w' is sought in that vicinity, it is a definite equation though the functions expressing the relation may be multiform. Thus, if

$$w - w_0 = (z - c)^n P(z - c),$$

where P is a regular function that does not vanish with its argument, then

$$w' = (z - c)^{n-1} P_1 (z - c),$$

and therefore

$$w' = (w - w_0)^{1 - \frac{1}{n}} Q\{(w - w_0)^{\frac{1}{n}}\},$$

that is, the value of w can be an ordinary point, or a definite branch-point, but it cannot be a point of indeterminateness, of w'. If $n = 1$, the point is an ordinary point, and w' is not zero there; if $n > 1$, the point is a branch-point, and w' is zero there.

To determine necessary conditions (if any), let w_0 be any point in the vicinity of which A_0, A_1, A_2 are regular. Let

$$w - w_0 = u,$$

so that the equation is

$$v\frac{dv}{du} = A_0 + vA_1 + v^2A_2.$$

When $n = 1$, the condition is that v is not zero, when $u = 0$; which can be satisfied by taking an arbitrary initial value for v: and then we have

$$v = R(u),$$

where R is a regular function of u, which does not vanish with u. We then have

$$z - c = \int\frac{du}{v} = \int\frac{du}{R(u)},$$

the inversion of which gives a general integral, uniform in the vicinity of $u = 0$. Hence, in this case, no condition need be imposed upon the original equation.

When $n > 1$, a condition is that v is zero, when $u = 0$. We have

$$\frac{du}{dv} = \frac{v}{A_0 + A_1v + A_2v^2},$$

so that, if w_0 is not a root of A_0 (say any arbitrary quantity), we have

$$2A_0(w_0)u = v^2 + \dots,$$

that is,

$$v = u^{\frac{1}{2}}\{2A_0(w_0)\}^{\frac{1}{2}} + \dots,$$

so that $n = 2$. Then

$$z - c = \int\frac{du}{v} = \int\frac{du}{u^{\frac{1}{2}}}R(u^{\frac{1}{2}}),$$

where R is a regular function of its argument, which does not vanish when $u = 0$. Effecting the quadrature, and inverting the relation, we have u (and therefore $w - w_0$) as a uniform function of u. If however w_0 is a root of $A_0(w)$, the equation still being

$$\frac{du}{dv} = \frac{v}{A_0 + A_1v + A_2v^2},$$

we have

$$u = v^{\frac{n}{n-1}}Q_1(v) = vT,$$

where Q_1 is a regular function of v, and T vanishes with v. Now taking the first form of u, we have

$$\frac{du}{dv} = v^{\frac{1}{n-1}} Q_2(v),$$

where Q_2 is regular, that is, $\frac{du}{dv} = 0$ when $v = 0$. As regards the fraction equal to $\frac{du}{dv}$, it is

$$= \frac{v}{A_1 v + A_2 v^2 + A_0' vT + \text{powers of } v}$$

$$= \frac{1}{A_1 + A_2 v + A_0' T + \text{powers of } v},$$

that is, it is not zero when $v = 0$. These results are contradictory: and therefore w_0 cannot be a zero of A_0, when $n > 1$. Hence when an initial value w_0 is assigned to w, this value being an ordinary point of A_0, A_1, A_2, the general integral is regular in the vicinity: either if an arbitrary value be assigned to w' for $w = w_0$; or if a zero value be assigned to w' for $w = w_0$, provided w_0 is not a root of $A_0 = 0$. This is the sole condition (and it is one of inequality) to be imposed upon the original equation, when w_0 is an ordinary point of each of the quantities A_0, A_1, A_2.

Next, let w_0 be a pole of one of the coefficients A_0, A_1, A_2, and let it be of multiplicity s; then the equation takes the form

$$(w - w_0)^s \frac{dv}{dw} = \phi(v, w),$$

where $\phi(v, w)$ is a uniform function of v and a regular function of w, of the form

$$B_2 v + B_1 + \frac{B_0}{v},$$

B_0, B_1, B_2 being regular functions of w, not all of which vanish when $w = w_0$. As the lowest power of $w - w_0$ in v has $1 - \frac{1}{n}$ for its index, the lowest index of powers of $w - w_0$ on the left-hand side is

$$s - \frac{1}{n}.$$

On the right-hand side, the lowest index in the first term is $1 - \frac{1}{n}$, if B_2 does not vanish; in the second term it is 0, if B_1 does not vanish; in the third term it is $-1 + \frac{1}{n}$, if B_0 does not vanish; and the lowest of these must be equal to $s - \frac{1}{n}$. If

$$s - \frac{1}{n} = -1 + \frac{1}{n},$$

so that

$$s + 1 = \frac{2}{n},$$

the only possible integer values are $n = 1$, $s = 1$. If

$$s - \frac{1}{n} = 0,$$

the only possible integer values are $s = 1$, $n = 1$. If

$$s - \frac{1}{n} = 1 - \frac{1}{n},$$

then $s = 1$.

Hence in every case $s = 1$; thus no one of the coefficients A_0, A_1, A_2 can have a multiple pole, if the integral of the equation is to have the specified character. Let $\alpha, \beta, \ldots, \lambda$ denote the aggregate of the poles of A_0, A_1, A_2, each of them being simple, and there being m in all; then the equation has the form

$$\frac{d^2 w}{dz^2} = \frac{P(w) + Q(w)\dfrac{dw}{dz} + R(w)\left(\dfrac{dw}{dz}\right)^2}{(w - \alpha)(w - \beta) \ldots (w - \lambda)},$$

where P, Q, R are polynomials in w, respectively of degrees $m + 3$, $m + 1$, $m - 1$ at most.

In particular, if $m = 0$, there is no function $R(w)$, and the term in $\left(\dfrac{dw}{dz}\right)^2$ disappears from the right-hand side.

249.   The conditions that have been obtained are necessary: but they merely settle the general type of the equation, which can have sub-uniform integrals; and they are not sufficient to secure this property.

The simple equation

$$\frac{d^2 w}{dz^2} = \frac{c}{w - \alpha}\left(\frac{dw}{dz}\right)^2,$$

which is an instance of the general type, has

$$w - a = A (z - a)^{\frac{1}{1-c}}$$

for its primitive, A and a being arbitrary. This functional expression for w is not sub-uniform (nor uniform), unless

$$c = 1 - \frac{1}{n},$$

where n is an integer (positive or negative). Thus one condition upon the constant c is necessary.

On the other hand, the equation

$$\frac{d^2w}{dz^2} = aw^2 + 2bw + c,$$

where a, b, c are constants, has

$$aw + b = 6\wp (z - \kappa, g_2, g_3)$$

for its primitive, the invariant g_2 being $\frac{1}{3}(b^2 - ac)$, and the arbitrary quantities being κ and the invariant g_3. The primitive manifestly gives w as a uniform function : and no condition needs to be imposed on the equation.

We proceed to obtain the further conditions that must be satisfied for the various forms.

The two cases, $m > 0$, $m = 0$, will be discussed separately.

Case I: $m > 0$.

250.　In the former case, let

$$D (w) = (w - \alpha) (w - \beta) \dots (w - \lambda),$$

so that the equation is

$$\frac{d^2w}{dz^2} = \frac{P (w)}{D (w)} + \frac{Q (w)}{D (w)} \frac{dw}{dz} + \frac{R (w)}{D (w)} \left(\frac{dw}{dz}\right)^2 ;$$

or writing W for $\dfrac{dw}{dz}$, we may take it more briefly in the form

$$w'' = W \frac{dW}{dw} = \frac{P + QW + RW^2}{D}.$$

Now if w is a sub-uniform function over the plane of the variable, so that it is uniform in the vicinity of any point, w, w', w'' are finite together and are infinite together; a finite value for any one of them requires a finite value for each of the other two, and likewise for an infinite value. Moreover, it is the general integral which is being discussed, so that there will be two arbitrary constants in its expression. One of these arises in connection with the independent variable, which manifestly will occur in a form $z - c$, where c is arbitrary; so that the integral must contain one other arbitrary constant, besides involving the combination $z - c$.

First, consider finite values of w. Let any value w_0 be assigned as an initial value of w, and any value W_0 be assigned as an initial value of W: so that W_0 is finite. If w_0 is not a root of D, then, (subject to a single condition of inequality, as has been seen), the equation is satisfied, in connection with these values, by an integral, which is a uniform function of z in the vicinity of the (arbitrarily assigned) initial value of z. Hence in connection with such values, no further condition needs to be imposed upon the equation.

If however w_0 is a root of D, say α, then the assignment of an arbitrary initial value to W will make the initial value of w'' infinite: and this is to be excluded. Consequently, the initial value of W, when α is the initial value of w, must satisfy the equation

$$P(\alpha) + Q(\alpha) \xi + R(\alpha) \xi^2 = 0.$$

Manifestly there are in general two values: say p_1 and p_2. As p_1 is definite, and as the integral is to be a general integral, it follows that an integral, which is determined by α and p_1 as initial values, must, in addition to c, contain an arbitrary constant entering otherwise than through the initial values; that is, there must be a (simple) infinitude of such integrals. Similarly, if p_2 is different from p_1, there must be a (simple) infinitude of integrals, determined by α and p_2 as initial values. The necessary conditions can be obtained as follows.

In the vicinity of these initial values, let

$$z - c = x.$$

Then assuming that $P(\alpha)$ is not zero, so that neither p_1 nor p_2 is zero and therefore w' does not vanish at $x = 0$, we have

$$w - \alpha = xI(x),$$

in the immediate vicinity, I being a regular function that does not vanish with x; and therefore

$$w' = J(x),$$

where J is a regular function that does not vanish with x. Hence, by reversing the first equation and substituting in the second, we have

$$w' = S(w - \alpha),$$

where S is a regular function that does not vanish when $w = \alpha$. This is evidently an integral equivalent of

$$W \frac{dW}{dw} = \frac{P + QW + RW^2}{\Theta};$$

so that the integral of this equation, which is determined by the initial values α and p_1 (or p_2), must be regular. By the preceding discussion, it must, contain an arbitrary constant; and therefore conditions must be assigned, in order to secure that there is a (simple) infinitude of regular integrals. Let

$$w - \alpha = u, \quad W - p = V,$$

where p stands for p_1 or p_2; therefore V must vanish with u, it is a regular function of u, and it contains one arbitrary constant. Now

$$\frac{dW}{dw} = \frac{dV}{du};$$

$$\Theta = u\Theta'(\alpha)\{1 + \Phi\},$$

where Φ vanishes with u; also

$$A + BW + CW^2 = u\{A'(\alpha) + pB'(\alpha) + p^2C'(\alpha)\}$$
$$+ V\{B(\alpha) + 2pC(\alpha)\} + ...,$$

the unexpressed terms being higher powers of u and V. Thus the equation is

$$u\frac{dV}{du} = \frac{A + BW + CW^2}{\Theta'(\alpha)\{1 + \Phi\}}$$
$$= \kappa u + \lambda V + ...,$$

where

$$\kappa = \frac{A'(\alpha) + pB'(\alpha) + p^2C'(\alpha)}{\Theta'(\alpha)},$$

$$\lambda = \frac{B(\alpha) + 2pC(\alpha)}{\Theta'(\alpha)},$$

and the other terms are higher powers of u and V. The equation thus obtained is the first typical reduced form, as discussed in

Chapter VI; and it appears that the equation can have a single infinitude of integrals expressing V as a regular function of u, which vanishes with u, only if

    (i) λ is a positive integer: and

    (ii) some relation holds among the constant coefficients on the right-hand side.

Thus if $\lambda = 1$, the relation is

$$\kappa = 0.$$

If $\lambda = 2$, and the equation is

$$u \frac{dV}{du} = \kappa u + 2V + (a_2, b_2, c_2 \backslash\!\!\!\backslash V, u)^2 + \ldots,$$

the condition is

$$(a_2, b_2, c_2 \backslash\!\!\!\backslash \kappa, -1)^2 = 0.$$

If $\lambda = 3$, and the equation is

$$u \frac{dV}{du} = \kappa u + 3V + (a_2, b_2, c_2 \backslash\!\!\!\backslash V, u)^2 + (a_3, b_3, c_3, d_3 \backslash\!\!\!\backslash V, u)^3 + \ldots,$$

the condition is

$$2 (a_2, b_2, c_2 \backslash\!\!\!\backslash -\tfrac{1}{2}\kappa, 1)^2 . (a_2, b_2 \backslash\!\!\!\backslash -\tfrac{1}{2}\kappa, 1) = (a_3, b_3, c_3, d_3 \backslash\!\!\!\backslash -\tfrac{1}{2}\kappa, 1)^3;$$

and so on (§ 73, Ex. and Corollary).

This holds for each of the two values p_1 and p_2. It also holds for each of the m roots of $\Theta = 0$. Thus in connection with these particular finite values, it is necessary to impose upon the equation

    (i) $2m$ descriptive conditions of the form, $\lambda = $ integer,

    (ii) $2m$ relations among coefficients,

in order to secure that the general integral is uniform in the plane: provided (as has been assumed) that $A(\alpha)$ is not zero, and that the roots p_1 and p_2 are unequal.

251. Consider the latter incidental conditions separately. If the roots p_1 and p_2 are equal, then

$$B(\alpha) + 2pC(\alpha) = 0,$$

so that, in the reduction for the common value, λ becomes zero. The equation is then of the second typical form in Chapter VI: and it appears (§§ 76—80) that the equation does not possess a (simple) infinitude of regular integrals vanishing with u. Accordingly, in this case, the general integral of the original equation cannot be sub-uniform; and therefore the condition, that *the quantity*

$$B^2(\xi) - 2A(\xi) C(\xi)$$

shall not vanish for any root ξ of $\Theta = 0$, must be imposed upon the equation, in order to secure that the general integral shall be sub-uniform.

The other incidental proviso is, that $A(\alpha)$ is not zero. To see whether this must be maintained, we consider the alternative. When α is a root of $A(w) = 0$, one of the roots of the quadratic in ξ is zero, and the other is different from zero. To the latter, the general investigation applies as above: and we therefore discuss the zero root. In the immediate vicinity of the initial values, we have

$$w - \alpha = x^n P(x),$$

where $P(x)$ does not vanish with x; the integer n is greater than 1, because W initially is zero. Thus

$$W = w' = x^{n-1} P_1(x),$$

$$w'' = x^{n-2} P_2(x);$$

and therefore, in the immediate vicinity of the initial values,

$$w'' = (w - \alpha)^{1 - \frac{2}{n}} R\{(w - \alpha)^{\frac{1}{n}}\},$$

$$w' = (w - \alpha)^{1 - \frac{1}{n}} Q\{(w - \alpha)^{\frac{1}{n}}\}.$$

Now the root α of Θ is to be a zero of A_0. This can happen in two ways: first, if A_0 is identically zero: and secondly, if α is a root (simple or multiple) of A_0.

When A_0 is identically zero, the equation is

$$w'' = \frac{A_1 w' + A_2 w'^2}{\Theta},$$

so that, taking $w' = v$, our equation becomes divisible by v in the form

$$\frac{dv}{dw} = \frac{A_1}{\Theta} + \frac{A_2}{\Theta} v.$$

This can be integrated by quadratures in the form

$$v e^{-\int \frac{A_2}{\Theta} dw} = K + \int \frac{A_1}{\Theta} e^{-\int \frac{A_2}{\Theta} dw} dw,$$

where K is arbitrary. Let

$$\frac{A_2(\alpha)}{\Theta'(\alpha)} = p;$$

then

$$v = (w - \alpha)^p \, \Psi \left[K + \int \frac{A_1 dw}{\Theta \, (w - \alpha)^p \, \Psi} \right],$$

where Ψ is a regular function of $w - \alpha$, which does not vanish with
its argument. The form, that should be obtained in this case, is

$$v = (w - \alpha)^{1 - \frac{1}{n}} \, Q \, \{(w - \alpha)^{\frac{1}{n}}\},$$

n being a positive integer $\geqslant 2$. Since Θ contains a factor $w - \alpha$,
the above expression is of the appropriate form, only if

$$A_1(\alpha) = 0,$$

$$\frac{A_2(\alpha)}{\Theta'(\alpha)} = p = 1 - \frac{1}{n} :$$

and when these conditions are satisfied, K being an arbitrary constant, the general integral is uniform in the vicinity considered.

When A_0 is not identically zero, take $w - \alpha = u$, so that

$$A_0 = uB,$$

and B is a polynomial in u. The equation now is

$$uv \, \frac{dv}{du} = \frac{uB + vA_1 + v^2 A_2}{\Theta'(\alpha) + \dots}.$$

Also

$$u = v^{\frac{n}{n-1}} \, Q \, (v^{\frac{1}{n-1}});$$

this must satisfy

$$\frac{du}{dv} = \frac{uv \, \{\Theta'(\alpha) + \dots\}}{uB + vA_1 + v^2 A_2}.$$

If $A_1(\alpha)$ is not zero, then $u = 0$ (a persistent zero) is the integral
of this equation. Manifestly, this is not a general integral.

If $A_1(\alpha)$ is zero, then A_1 has the form uB_1, where B_1 is of the
same nature as B. In this case, it is easy to see that the equation
cannot have the solution in the required form, unless $n = 2$: and
then the solution is

$$u = v^2 T,$$

where T is a regular function of v, which does not vanish with
v. Substituting this value, the equation becomes, on reduction,

$$v \, \frac{dT}{dv} + 2T = T \, \frac{\Theta'(\alpha) + \dots}{A_2 + BT + vB_1}.$$

In order that this may be satisfied by a function T, which is regular and does not vanish with v, the initial value κ is given by

$$2 = \frac{\Theta'(\alpha)}{A_2(\alpha) + \kappa B(\alpha)},$$

which determines κ unless $B(\alpha) = 0$. Every subsequent coefficient in T is then determinate; so that the integral does not contain an arbitrary constant, and it therefore ceases to lead to a general integral. If however $B(\alpha) = 0$, so that $w = \alpha$ is a repeated zero of A_0, then κ is arbitrary, provided

$$2 = \frac{\Theta'(\alpha)}{A_2(\alpha)};$$

the integral contains an arbitrary constant, thus leading to a general integral.

252. Summing up the results, we find that the general integral of the equation remains sub-uniform for finite initial values of w, in the following cases:—

(i)   When A_0 and Θ have no common factor:

(ii) When a factor $w - \alpha$ of Θ occurs in at least the second degree in A_0, occurs in at least the first degree in A_1, and when

$$\Theta'(\alpha) = 2A_2(\alpha):$$

(iii) When A_0 is identically zero, a factor $w - \alpha$ of Θ occurs in at least the first degree in A_1, and

$$1 - \frac{A_2(\alpha)}{\Theta'(\alpha)}$$

is the reciprocal of a positive integer $\geqslant 2$:

the conditions obtained earlier with (i) being satisfied.

Lastly, w' cannot be infinite when w is finite, if the general integral is sub-uniform; so that it must not happen that the equation

$$-\frac{dV}{dw} = \frac{PV^3 + QV^2 + RV}{\Theta},$$

obtained by substituting $w'V = 1$, should have a variable integral, which vanishes for a finite value of w.

If w_0 be the finite value and it be not a root of Θ, then since $e^{\int \frac{R}{\Theta} dw}$ is a regular function of w in the vicinity of w_0, the only solution of the equation is $V = 0$. This is not a variable integral: that is, no condition is thus imposed.

If α, any root of Θ, be the finite value, and if $\dfrac{R(\alpha)}{\Theta'(\alpha)}$ be negative or have its real part negative, there is such an integral, which has

$$(w - \alpha)^{-\frac{R(\alpha)}{\Theta'(\alpha)}}$$

as its leading term. This is to be excluded: and therefore the real part of $\dfrac{R(\alpha)}{\Theta'(\alpha)}$ must be positive.

253. Next, consider infinite values of w. We take

$$w = \frac{1}{u};$$

because w is sub-uniform, u also is sub-uniform for the same range of the independent variable: but we now need consider only values of u in the vicinity of a zero. We have

$$w' = -\frac{u'}{u^2}, \qquad w'' = -\frac{u''}{u^2} + 2\frac{u'^2}{u^3}:$$

so that, writing

$$\frac{P(1/u)}{\Theta(1/u)} = \frac{1}{u^3} P_1(u),$$

$$\frac{Q(1/u)}{\Theta(1/u)} = \frac{1}{u} Q_1(u),$$

$$\frac{R(1/u)}{\Theta(1/u)} = u R_1(u),$$

where P_1, Q_1, R_1 are rational functions of u, which are regular in the vicinity of $u = 0$, the equation acquires the form

$$u'' = \frac{1}{u} \left[-P_1(u) + u'Q_1(u) + u'^2 \{2 - R_1(u)\} \right].$$

The discussion is similar to that for values of w in the vicinity of a root of $\Theta(w)$ in the former case: and the result is that, in order to have the general integral sub-uniform, two descriptive conditions must be satisfied, and two relations among coefficients must hold.

Case II : $m = 0$.

254.   We now pass to the case when $m = 0$; the equation is

$$w'' = aw^3 + bw^2 + cw + f + (kw + h) w'.$$

As before, we first take finite values of w into consideration.

If an arbitrary initial value w_0 and an arbitrary initial value w_0' be assigned to w and w' respectively, the general integral so determined is uniform in the vicinity of the initial values. No condition to be satisfied by the original equation thus arises.

If, still with an initial arbitrary value w_0 for w, a zero value be assigned to w', the equation can be taken in the form

$$v \frac{dv}{dw} = aw^3 + bw^2 + cw + f + (kw + h) v,$$

where $v = w'$; then we have

$$v^2 = 2l (w - w_0) I (w - w_0).$$

where l is a constant equal to $aw_0^3 + bw_0^2 + cw_0 + f$, and I is a regular function, which is equal to 1 when $w = w_0$.   Then

$$z - c = \int \frac{dw}{v}$$

$$= \left(\frac{2}{l} \right)^{\frac{1}{2}} (w - w_0) \, J (w - w_0)^{\frac{1}{2}},$$

where J is a regular function, which is equal to 1 when $w = w_0$.   The reversion of this gives an integral, that is uniform and contains two arbitrary constants : it is a general integral.   The one condition necessary is that l should not vanish : and this is satisfied when w_0 is arbitrarily assigned.

Now suppose that w_0 is not arbitrarily assigned but is a root of $l = 0$: this can effectively be secured by taking $f = 0$.   There then must be a (simple) infinitude of integrals of the equation

$$v \frac{dv}{dw} = aw^3 + bw^2 + cw + (kw + h) v,$$

which satisfy the condition that $v = 0$, when $w = 0$. Let U be a new variable, such that

$$cw + hv = vU\ ;$$

then

$$v\frac{dU}{dv} + U - h = c\frac{dw}{dv}$$

$$= \frac{cv}{aw^3 + bw^2 + cw + kwv + hv}$$

$$= \frac{c}{U + vQ\,(v,\ U)},$$

where Q is an integral polynomial in v and U. The initial conditions leave U merely as a regular function in the vicinity of $v = 0$. If its value for $v = 0$ be λ, then

$$\lambda - h = \frac{c}{\lambda},$$

say

$$\lambda^2 - h\lambda - c = 0.$$

Let $U = \lambda + T$, where T must vanish with v and be a regular function of v: we have

$$v\frac{dT}{dv} = -\left(1 + \frac{c}{\lambda^2}\right)T + \gamma v + \ldots,$$

the unexpressed terms being higher powers. Quite independently of the integrals of this equation—that is, whether there is a simple infinitude or not—we have

$$cw = v\,(U - h)$$

$$= v\,(\lambda - h + T),$$

and therefore

$$dz = \frac{dw}{v} = \frac{\lambda - h + T}{cw}\,dw,$$

so that z cannot approach an assigned (finite) value when $w = 0$. The suggested initial values do not, in fact, permit the integral to be general; and therefore no condition is imposed upon the original equation on this score.

The equation must, of itself, exclude the possibility of an infinite value of w' for a finite value of w, arising as part of a general integral. Taking

$$w' = \frac{1}{V},$$

so that

$$w'' = w' \frac{dw'}{dw} = -\frac{1}{V^3}\frac{dV}{dw},$$

we have

$$-\frac{dV}{dw} = (aw^3 + bw^2 + cw + d)V^3 + (kw + h)V^2,$$

with the condition that $V = 0$ when w is a finite quantity. The only solution of the equation satisfying this condition is a persistent zero value of V, independent of the value of w. No general integral is thus determined: and no condition is imposed upon the original equation on this score.

255.   Passing next to the consideration of infinite values of w, we write

$$w = \frac{1}{y}, \quad w' = -\frac{y'}{y^2}, \quad w'' = -\frac{y''}{y^2} + 2\frac{y'^2}{y^3},$$

in the equation: and we discuss values of y in the vicinity of its zero, y being a sub-uniform function, and y', y'' being therefore limited to finite values for such a range of variation of y. The equation is

$$y'' = \frac{-(a + by + cy^2 + dy^3) + (k + hy)y' + 2y'^2}{y}.$$

The value of y'' is not to be infinite when $y = 0$: this limitation will not be satisfied, unless y' has an initial value θ such that

$$2\theta^2 + k\theta - a = 0.$$

This gives to y' a definite initial value: and therefore, in order that the integral under consideration may be general, there must be an arbitrary constant in the integral of the equation

$$p\frac{dp}{dy} = \frac{-(a + by + cy^2 + dy^3) + (k + hy)p + 2p^2}{y},$$

(where $p = y'$), determined by the condition that $p = \theta$ when $y = 0$.

256.   First, suppose that the roots of the quadratic in θ are distinct, and that neither of them is zero. Denoting either of them by θ, and remembering that θ is the initial value of p, we write

$$p = \theta + P,$$

so that P vanishes with y. The equation for P, after reduction, is found to be

$$y \frac{dP}{dy} = \frac{(k + 4\theta) P + (h\theta - b) y + 2P^2 + hyP - cy^2 - dy^3}{\theta + P}$$

$$= \left(4 + \frac{k}{\theta}\right) P + \left(h - \frac{b}{\theta}\right) y + \dots,$$

the unexpressed terms being of higher order in y and P. Now as regards the relation between y and P, let $z = c$ be the value, for which w is infinite, and y therefore zero: then, as p is θ when $y = 0$, we have

$$y = \theta (z - c) I (z - c),$$

where I is a function regular in the vicinity of $z = c$, and is equal to 1 at the point c. Hence

$$p = \theta J (z - c),$$

where J is of the same character as I; and therefore

$$p = \theta R (y),$$

where R is a regular function of y, which is equal to 1 when y vanishes. Thus we have

$$P = - \theta + \theta R (y),$$

that is, the integral of the equation between P and y must give P as a regular function of y, which vanishes with y and involves an arbitrary constant.

In order that this may be the character of the integral, two conditions are necessary in connection with a particular root θ, viz.

(i)   the quantity $4 + \dfrac{k}{\theta}$ must be a positive integer:

(ii)   a relation among the coefficients, affected by the magnitude of this integer, must be satisfied:

(Ex. and Cor. §§ 72, 73, 75).

As the condition of being sub-uniform is to apply to every general integral of the equation, there are two sets of conditions to be imposed; one set in association with each root of the quadratic. Let θ_1 and θ_2 denote the two roots; write

$$\frac{k}{\theta_1} = m_1, \quad \frac{k}{\theta_2} = m_2,$$

so that m_1 and m_2 are integers, each $\geqslant -3$.   Then

$$-\frac{a}{2} = \theta_1\theta_2 \quad = \frac{k^2}{m_1 m_2},$$

$$-\frac{k}{2} = \theta_1 + \theta_2 = \frac{k}{m_1} + \frac{k}{m_2}.$$

The latter relation can be satisfied by the (limited integer) values of m_1 and m_2, only if

$$m_1,\ m_2 = -1,\ 2\ ;$$

the values of the coefficients $4 + \dfrac{k}{\theta_1}$, $4 + \dfrac{k}{\theta_2}$, are 3 and 6 respectively; and then the first relation gives

$$k^2 = a.$$

In addition to this relation, there are two other conditions associated with the coefficients 3 and 6 for the respective conditions : when these are satisfied, the general integral of

$$w'' = aw^3 + bw^2 + cw + d + (kw + h)\,w'$$

is sub-uniform.

257.   These results have been obtained on the supposition that the roots are unequal, and different from zero.

If the roots could be equal, and the common root be different from zero, let it be denoted by ϑ.   In connection with this root, there is a corresponding reduction : and in the equation

$$y\frac{dP}{dy} = \left(4 + \frac{k}{\vartheta}\right)P + \left(h - \frac{k}{\vartheta}\right)y + \ldots$$

the coefficient $4 + \dfrac{k}{\vartheta}$ is zero.   No integral of this equation, vanishing with y and containing an arbitrary constant, exists in the form of a regular function : consequently the integral of the original equation ceases to be uniform.   *The roots of the quadratic must therefore be unequal :* which is a condition of inequality.

Next, consider the possibility of a zero root of the quadratic : that this may be possible, we must have

$$a = 0.$$

In this case, let

$$y = (z - c)^n\, Y_1,$$

$$p = (z - c)^{n-1}\, Y_2,$$

where the positive integer n is greater than unity: thus the relation between p and y must be of the form

$$p = y^{1-\frac{1}{n}} I(y^{\frac{1}{n}}),$$

where I is a regular function of its argument, which does not vanish when $y = 0$. The equation between p and y is

$$p\frac{dp}{dy} = \frac{-(by + cy^2 + dy^3) + (k + hy)p + 2p^2}{y},$$

and therefore

$$y\frac{dp}{dy} = -\frac{by + cy^2 + dy^3}{p} + k + hy + 2p.$$

The left-hand side vanishes when $y = 0$; in order that the right-hand side also may vanish, we must have

$$k = 0.$$

Instead of proceeding further with this equation, we introduce the two conditions $a = 0$, $k = 0$, so that the original equation is

$$w'' = bw^2 + cw + d + hw'.$$

We assume that b is not zero; otherwise we have a linear equation with constant coefficients, the general integral of which is known to be uniform. Let a new variable u be determined by the relation

$$w + \tfrac{1}{2}\frac{c}{b} = \frac{6}{b} u,$$

so that the properties of u are similar to those of w; the equation for u is

$$u'' = 6u^2 + hu' + f,$$

where

$$f = \tfrac{1}{24}(4bd - c^2).$$

What is required is the condition (if any) between f and h, which shall make the general integral of this equation sub-uniform. The preceding investigation shews that it is only in connection with variations of the variables, in the vicinity of $u = \infty$, that the condition arises if at all.

Denoting by c, which can be an arbitrary quantity, the value of z for which u is infinite, let the leading term in the expression for u in the vicinity of c be

$$\frac{\beta}{(z - c)^n},$$

where β is not zero, and n is a positive integer. Then in u'', the leading term is

$$\frac{n(n+1)\beta}{(z-c)^{n+2}};$$

in $6u^2$, it is

$$\frac{6\beta^2}{(z-c)^{2n}},$$

and in hu', it is

$$\frac{-nh\beta}{(z-c)^{n+1}}.$$

In order that the equation may be satisfied, we must have

$$n + 2 = 2n,$$
$$n(n+1)\beta = 6\beta^2;$$

the former gives $n = 2$, and the latter then gives $\beta = 1$, for β cannot be zero*. Taking now $z - c = x$, writing

$$u = \frac{1}{x^2} + \frac{\gamma}{x} + c_0 + c_1 x + c_2 x^2 + c_3 x^3 + c_4 x^4 + \dots,$$

and choosing the coefficients so that the equation is satisfied, the relation

$$\frac{6}{x^4} + \frac{2\gamma}{x^3} + 2c_2 + 6c_3 x + 12c_4 x^2 + 20c_5 x^3 + \dots$$

$$+ \frac{2h}{x^3} + \frac{h\gamma}{x^2} - hc_1 - 2hc_2 x - 3hc_3 x^2 - 4hc_4 x^3 - \dots - f$$

$$= \frac{6}{x^4}(1 + \gamma x + c_0 x^2 + c_1 x^3 + c_2 x^4 + c_3 x^5 + c_4 x^6 + \dots)^2$$

must be an identity. It is easy to see that, equating the coefficients of successive powers of x, we determine the coefficients in the expression for u in succession—with one exception: in the coefficient of x^2 on the two sides, the terms $12c_4$ cancel. The coefficient c_4 remains arbitrary; instead of a determining relation, we have a relation of condition which, in fact, is the condition sought. We have

$$\gamma = \tfrac{1}{5}h, \quad c_0 = -\tfrac{1}{12}\gamma^2, \quad c_1 = \tfrac{1}{12}\gamma^3, \quad c_2 = -\tfrac{1}{10}f - \tfrac{7}{48}\gamma^4,$$
$$c_3 = \tfrac{11}{30}f\gamma + \tfrac{73}{144}\gamma^5;$$

the relation of condition, arising from the coefficient of x^2, becomes

$$10f\gamma^2 + 15\gamma^6 = 0,$$

---

* It was in order to have unity as the coefficient of the leading term, that the particular transformation from w to u was adopted.

on reduction. Thus either $h = 0$, leading to the equation connected with elliptic functions as already (§ 249) shewn : or

$$f = -\tfrac{3}{2}\gamma^4,$$

so that the equation is

$$u'' = 6u^2 + 5\gamma u' - \tfrac{3}{2}\gamma^4,$$

and this equation certainly has its integral of a sub-uniform character.

The method just adopted is due to Mittag-Leffler[*]; and is applied by him also to the general equations of § 254.

258. As already pointed out (§ 246), the process does not exclude the possibility of transcendental branch-points: so that we cannot make a further inference that the integral is really uniform, and not merely sub-uniform, over the plane. One method of testing the result would be, when possible, to obtain the actual integral of the deduced equation.

Picard[†] has devised a curious and ingenious method of integration, which is effective for such an equation, if its integral be uniform and not merely sub-uniform.

A rational integral function of u and u' is formed with disposable constants; these constants are determined so as to diminish as far as possible the multiplicity of the pole of the function thus formed; the order of the function in u and u' is chosen, so that the point can be a simple pole and may be an ordinary point. If an ordinary point, then an expression has been obtained which has no poles for any finite value. If a pole, then such a constant multiple of the function can be taken, that the residue[‡] of the pole is unity; the function in this form being denoted by F, then

$$G = e^{\int F dz}$$

is a function, which has no poles in the plane. Thus if the integral of the original equation really be uniform, the function F in the first case, and the function G in the second case, is a function of z, which is regular in the finite part of the plane.

Now let

$$\theta = F;$$

we have

$$\theta' = F_1,$$

[*] *Acta Math.*, t. xviii (1894), pp. 233—245.

[†] See the first memoir quoted at the beginning of this chapter; in particular, (*l.c.*), pp. 283—287.

[‡] *Th. Fns.*, p. 42.

where initially F_1 involves u, u', u''; on substitution for u'' from the original differential equation, it becomes a function of u, u'. Again,

$$\theta'' = F_2,$$

where F_2 can be made a function of u and u' in the same manner as F_1. Eliminating u and u' between these three equations, we have a relation between θ, θ', θ'', that is, a differential equation of the second order, which is known to be satisfied by a regular function of z. Deducing this function and substituting in $\theta = F$, the new form is an equation of the first order, which is, in fact, a first integral of the differential equation.

Mittag-Leffler, following Picard, has applied (*l.c.*) this process to the equation

$$u'' = 6u^2 + 5\gamma u' - \tfrac{3}{2}\gamma^4.$$

It is to be noted that the order of the parametric pole for u is 2, and for u' is 3; accordingly the degree of the highest power in the expression for u' is to be even. In the expression

$$u'^2 + u'(a_1 + a_2 u) + a_3 u + a_4 u^2 + a_5 u^3,$$

the order of the pole is 6, when the coefficients a_1, ..., a_5 are arbitrary; but the coefficients can be chosen so as to reduce the order to 1, and they are such as to give

$$F = u'^2 - (2\gamma^3 + 4\gamma u) u' + \gamma^4 u - 2\gamma^2 u^2 - 4u^3.$$

Moreover taking account of the limit of definite development of u, the coefficient c_4 being the first that is undetermined, we find the residue of the simple pole of F to be definite; the coefficient of $\dfrac{1}{z-c}$ is, in fact,

$$-(6f\gamma + 9\gamma^5),$$

where f is $-\tfrac{3}{2}\gamma^4$: in other words, the residue is zero, and therefore F has no pole for any finite value of z. Now F is known to satisfy a differential equation of the second order, having as its solution a function of z, which is regular over the finite part of the plane. Forming F', and eliminating u'' from it by the original equation, we have

$$F' = 6\gamma u'^2 - 12\gamma^4 u' - 24\gamma^2 uu' + 6\gamma^5 u - 12\gamma^3 u^2 - 24\gamma u^3 + 3\gamma^7$$
$$= 6\gamma F + 3\gamma^7,$$

so that, in this case, we do not require to proceed so far as the equation of the second order, which is satisfied by F. This gives

$$F = Ae^{6\gamma z} - \tfrac{1}{2}\gamma^6,$$

where A is arbitrary: thus verifying the regular character of F in the finite part of the plane. A first integral of the equation is

$$u'^2 - 4u'\gamma (u + \tfrac{1}{2}\gamma^2) + \tfrac{1}{2}\gamma^6 + \gamma^4 u - 2\gamma^2 u^2 - 4u^3 = A e^{6\gamma z}.$$

This equation is of genus unity, as between u and u' for its variables, and the integral is known to have no parametric branch-points: it is therefore (§ 124) integrable by transformation and quadrature. Let

$$v = u + \tfrac{1}{2}\gamma^2:$$

the equation is

$$(v' - 2\gamma v)^2 - 4v^3 = A e^{6\gamma z},$$

or since

$$v' - 2\gamma v = e^{2\gamma z} \frac{d}{dz} (v e^{-2\gamma z}),$$

we take

$$v e^{-2\gamma z} = V,$$

and then we have

$$V'^2 = (4V^3 + A) e^{2\gamma z}.$$

This, when integrated by quadrature, gives

$$V = \wp \left(\frac{e^{\gamma z} - B}{\gamma}, \ 0, \ -A \right),$$

where B is another arbitrary constant, and $0, -A$ are the invariants of the elliptic functions. Thus the general integral of the equation

$$u'' = 6u^2 + 5\gamma u' - \tfrac{3}{2}\gamma^4$$

is

$$u + \tfrac{1}{2}\gamma^2 = \wp \left(\frac{e^{\gamma z} - B}{\gamma}, \ 0, \ -A \right):$$

it is a uniform function of z and not merely sub-uniform.

Ex. 1.   Prove that, if the general integral of the equation

$$w'' = aw^3 + bw^2 + cw + f + (kw + h) w'$$

is sub-uniform, then the equation can be changed, by a transformation of the dependent variable that does not affect the character of the integral, into one of the forms

(i) $w'' + 3ww' + w^3 = b(w' + w^2) + cw + f: \cdot$

(ii) $w'' + 2ww' = b(w' + w^2) + f:$

(iii) $w'' = 2w^3 - 2g^2 w + 3gw':$

(iv) $w'' = w^3 - ww' + cw:$

(v) $w'' = w^3 - ww' + b(w' + w^2) - \tfrac{4}{25}b^2 w - \tfrac{4}{25}b^3:$

(vi) $w'' = w^3 - ww' + b(w' + w^2) - \tfrac{1}{25}b^2 w - \tfrac{1}{25}b^3.$

Integrate the first of these completely, and the second as far as possible, using the transformation

$$w = \frac{y'}{y}.$$

Shew that, for the third, the function F is given by

$$F = (w' - gw)^2 - w^4,$$

and that it satisfies the equation

$$F'' = 4gF;$$

hence integrate the third equation.

Shew that the other three can be transformed to equations, which are discussed in the text, by taking

$$w' + w^2 = 6y + a,$$

and choosing the constant a appropriately. Hence integrate the equations.

(Mittag-Leffler.)

Ex. 2. If the coefficients A_0, A_1, A_2 in

$$w'' = A_0 + A_1 w' + A_2 w'^2$$

have only a single pole, it may be taken at $w = 0$: so that the equation is of the form

$$w'' = \frac{u_4 + u_2 w' + cw'^2}{w},$$

where c is a constant, and u_4, u_2 are a quartic and a quadratic in w respectively.

Develop the conditions that must be satisfied, so as to secure that the general integral of this equation may be sub-uniform.

Ex. 3. If the coefficients A_0, A_1, A_2 have two poles, at a and b say, then the transformation

$$w - a = (b - a) W$$

does not affect the character of the equation, and the new poles are at 0, 1. Thus the equation takes the form

$$w'' = \frac{u_5 + u_3 w' + u_1 w'^2}{w(w-1)},$$

where u_1, u_3, u_5 are algebraical polynomials of degrees 1, 3, 5 respectively.

Develop the conditions that must be satisfied, so as to secure that the general integral of this equation may be sub-uniform.

Ex. 4. Prove that the complete primitive of the equation

$$F(w'', w) = 0,$$

where F is polynomial in its arguments, cannot be a uniform function of the independent variable, unless F is linear in w''.

(Wallenberg.)

Ex. 5. A uniform function, having no essential singularity in the finite part of the plane, satisfies a differential equation of the second order involving only the function and its second derivative; prove that the function is either rational, simply-periodic, or doubly-periodic.

(Picard; Wallenberg.)

Ex. 6. The equation $F(w'', w) = 0$ is of degree greater than unity, and $w = a$ is a root of the discriminant of F with regard to w'', which makes at least two of the roots w'' of F equal; find the conditions necessary and sufficient to secure that an integral of the equation, equal to a for a parametric value of the variable, shall be a sub-uniform function of the independent variable in the finite part of the plane.

Illustrate the conditions in connection with the equations

$$w''^3 = aw^4 (w - b),$$

$$w''^3 = c (w - a) \{w - \tfrac{1}{3} (2a + b)\}^3 (w - b)^5.$$

(Wallenberg.)

Ex. 7. Let $F(w'', w', w)$ denote a homogeneous polynomial in its arguments; shew that the conditions, necessary and sufficient to secure that the primitive of the equation

$$F(w'', w', w) = 0$$

shall be a sub-uniform function for all finite values of the variable, are

(i) the equation in v, which results from the transformation

$$v = \frac{w'}{w},$$

must satisfy the Briot and Bouquet conditions of § 136;

(ii) the roots u' of the equation, which results from the transformation

$$u = \frac{w}{w'},$$

must be either of at least the first degree in u, or they must be the reciprocals of integers when u vanishes.

Illustrate these results in connection with the equations

$$ww''^2 - 2w'^2 w'' + (1 + c)\, ww'^2 - cw^3 = 0,$$

(for $c = 0$, 1 and $0 < c < 1$); and

$$ww''^2 - 2w'^2 w'' - w^3 = 0.$$

(Wallenberg.)

Ex. 8. Let $F(w'', w', w, z)$ denote a homogeneous polynomial in the arguments w'', w', w, and let the coefficients be analytical functions of z: shew that the conditions, necessary and sufficient to secure that the primitive of the equation

$$F(w'', w', w, z) = 0$$

shall, so far as regards parametric singularities, be sub-uniform, are

(i) the equation in v, which results from the transformation

$$v = \frac{w'}{w},$$

must satisfy the Fuchs conditions of § 121;

(ii) the roots u' of the equation, which results from the transformation

$$u = \frac{w}{w'},$$

must, when $u = 0$, be independent of z, and be either the reciprocals of integers or, vanishing with u, of at least the first degree in u.

Illustrate these results in connection with the equations:—

$$z(ww'' - w'^2) + \kappa ww' = 0, \quad \text{for} \quad \kappa = 1, 2;$$

$$ww'' - w'^2 - Pw^2 = 0, \quad \text{for} \quad P = -\frac{1}{2z^2}, \ \frac{1}{z}, \ \frac{2}{z^3}.$$

(Wallenberg.)

General Considerations on Integrals of $F = 0$.

259. Consider now more generally the relation, which is borne to the equation

$$F(w'', w', w, z) = 0,$$

assumed rational in w'', w', and w, by any other equation

$$G(\dots, w''', w'', w', w, z) = 0,$$

where G involves derivatives of any order, and is not deducible from $F = 0$ by direct processes of differentiation and elimination. By means of $F = 0$, the value of w''', and the value of every higher derivative of w, can be obtained explicitly and uniquely in terms of w'', w', w, z; and therefore w''' and every higher derivative of w can be removed from G: let the new form be denoted by $\bar{G}$.

If when this change is made, $\bar{G}$ involves only w and z (it cannot be an identity because it is not deducible from $F = 0$), then $\bar{G} = 0$ is a primitive: the range of generality of the primitive depends upon the number of arbitrary constants it involves.

If when the change is made, $\bar{G}$ involves w', w and z but not w'', then $\bar{G} = 0$ is an intermediary integral: it is a general or a special intermediary integral, according as it does or does not involve an arbitrary parameter.

If when the change is made, $\bar{G}$ involves w'', w', w, z, it is an equation compatible with $F = 0$. By elimination, w'' can be regarded as removable: and the eliminant would be an intermediary integral. Even if $\bar{G}$ were algebraical, its degree would be changed by the process of elimination: and therefore it is simpler in the first instance to consider $\bar{G}$ as involving w'', w', w, z.

Denoting by

$$r\,(w'',\,w',\,w,\,z) = 0$$

any equation that is compatible with $F = 0$, we may regard the two equations as determining w'' and w' in terms of w and z. When these determinate values are substituted in $r = 0$ and $F = 0$, each of the equations becomes an identity: and therefore, on differentiating, we have

$$\frac{\partial F}{\partial w''}\,w''' + \frac{\partial F}{\partial w'}\,w'' + \frac{\partial F}{\partial w}\,w' + \frac{\partial F}{\partial z} = 0,$$

$$\frac{\partial r}{\partial w''}\,w''' + \frac{\partial r}{\partial w'}\,w'' + \frac{\partial r}{\partial w}\,w' + \frac{\partial r}{\partial z} = 0\,;$$

so that, eliminating w''', we have

$$-\frac{\partial r}{\partial w''}\left(w''\frac{\partial F}{\partial w'} + w'\frac{\partial F}{\partial w} + \frac{\partial F}{\partial z}\right) + \frac{\partial r}{\partial w'}\,w''\frac{\partial F}{\partial w''} + \frac{\partial r}{\partial w}\,w'\frac{\partial F}{\partial w''} + \frac{\partial r}{\partial z}\frac{\partial F}{\partial w''} = 0,$$

say $(r,\,F) = 0$, an equation that is satisfied simultaneously with $F = 0$, $r = 0$.

260. This new equation is manifestly characteristic of any equation $r = 0$ compatible with $F = 0$. Moreover, if $F = 0$ is irreducible, as manifestly may be assumed, it is clear that the new equation is not satisfied solely in virtue of $F = 0$: and therefore the partial differential equation is an equation, to which the form of any compatible equation is subject. Also, as the singular solutions (if any) of the original equation have already been taken into account, we shall assume that $\dfrac{\partial F}{\partial w''}$ is not zero.

The subsidiary system of the partial differential equation is

$$\frac{dw''}{w''\dfrac{\partial F}{\partial w'} + w'\dfrac{\partial F}{\partial w} + \dfrac{\partial F}{\partial z}} = \frac{-\,dw'}{w''\dfrac{\partial F}{\partial w''}} = \frac{-\,dw}{w'\dfrac{\partial F}{\partial w''}} = \frac{-\,dz}{\dfrac{\partial F}{\partial w''}}.$$

It is easy to see that $F = 0$ is satisfied in virtue of these equations. Any other integral of the system will give a solution of the differential equation; denoting such an one by $r\,(w'',\,w',\,w,\,z)$, then

$$r\,(w'',\,w',\,w,\,z) = a,$$

where a is an arbitrary constant, is an equation compatible with $F = 0$. The elimination of w'' between this new equation and

$F = 0$ leads to a general intermediary integral of the original differential equation.

Let another equation compatible with $F = 0$ be given by

$$s(w'', w', w, z) = 0.$$

Then since this is compatible also with $r = 0$, we have

$$(s, F) = 0, \quad (s, r) = 0;$$

and in order that these may coexist, we must have

$$(r, F) = 0,$$

a condition that is satisfied by the mode of derivation of r. Let a common integral of the two equations $(s, F) = 0$, $(s, r) = 0$, derived in accordance with the regular Jacobian theory of simultaneous partial differential equations of the first order, be denoted by $s(w'', w', w, z)$: then

$$s(w'', w', w, z) = b,$$

where b is an arbitrary constant, is another equation compatible with $F = 0$. It is compatible also with $r = a$; it is independent of the latter, in the sense that a functional relation of the form $\psi(r, s) = 0$ does not exist. The elimination of w'' between $F = 0$ and $s = b$ leads to a general intermediary integral of the original equation, which is functionally distinct from the general intermediary integral given by the elimination of w'' between $F = 0$ and $r = a$. The elimination of w'' and w' between

$$F = 0, \quad r = a, \quad s = b,$$

leads to a relation between w and z, which accordingly is a primitive of the original equation; as it involves two independent arbitrary parameters, it is the complete primitive.

(It might happen that an integral r of the subsidiary system could be obtained not involving w''; then $r = a$ of itself, and without any process of elimination, would be a general intermediary integral.)

261. Let another equation, conceived as similarly obtained to be compatible with $F = 0$, be denoted by

$$t(w'', w', w, z) = c,$$

where c is an arbitrary constant. Then the relation

$$(\theta, F) = 0$$

is satisfied when $\theta = F$, $\theta = r$, $\theta = s$, $\theta = t$: remembering that $\dfrac{\partial F}{\partial w''}$ is not zero, and eliminating determinantally from the four equations, we have

$$ J \left(\frac{F, \; r, \; s, \; t}{w'', \; w', \; w, \; z} \right) = 0. $$

Now $J = 0$ is manifestly not satisfied in virtue of $r = a$, $s = b$, $t = c$; for the arbitrary constants occur only on the right-hand sides of the equations, and they do not occur in J. Further, the analysis throughout would be unaltered, if the equation $F = 0$ were replaced by $F = \kappa$, where κ is an arbitrary constant; it is clear that $J = 0$ is not satisfied in virtue of $F = \kappa$, and therefore generally not in virtue of $F = 0$. Hence the relation $J = 0$ must be satisfied identically; and therefore between the four quantities F, r, s, t, regarded as involving the four arguments w'', w', w, z, there exists a functional relation

$$ \Theta \, (F, \; r, \; s, \; t) = 0, $$

the coefficients of which are free from w'', w', w, z, and involve only the permanent constants in F, r, s, t. For our present purposes, $F = 0$ is a permanent equation, with which $r = a$, $s = b$, $t = c$ are simultaneously compatible: so that

$$ \Theta \, (0, \; a, \; b, \; c) = 0, $$

say

$$ \theta \, (a, \; b, \; c) = 0, $$

a relation satisfied in association with the equation $F = 0$. It might happen that the functional relation $\Theta = 0$ does not involve F, and then $\theta = 0$ would be satisfied independently of $F = 0$; most generally $\Theta = 0$ does involve F, and then $\theta = 0$ would be satisfied in virtue of $F = 0$.

One or two statements may be made; their proof is omitted as being simple and immediate. In the first place, if three equations $r = a$, $s = b$, $t = c$ are compatible with $F = 0$, so that a functional relation

$$ \theta \, (a, \; b, \; c) = 0 $$

is satisfied in association with $F = 0$, then any arbitrary functional combination

$$ u = \phi \, (r, \; s, \; t), $$

(which need not involve all the three quantities r, s, t), leads to another compatible equation

$$u = k,$$

where k is an arbitrary constant; and the relation between the arbitrary constants is

$$k = \phi\,(a,\,b,\,c).$$

In the second place, when two compatible equations $r = a$, $s = b$ have been obtained, then an unlimited number of other compatible equations can be deduced from the forms

$$\left.\begin{array}{l} \Theta\,(F,\,r,\,s,\,t) = 0 \\ u = \phi\,(r,\,s,\,t) \end{array}\right\};$$

and, as in the corresponding investigation associated with equations of the first order (§ 139), a question arises as to the simplest forms that may be chosen as forms of reference.

CHAPTER XVII.

GENERAL THEOREMS ON ALGEBRAIC INTEGRALS: BRUNS'S THEOREM*.

262. WE have indicated, in the preceding chapter, some forms of the equation $F(w'', w', w, z) = 0$, the integral of which is a sub-uniform function of the variable. A wider question arises when there is a demand for the conditions, which must be imposed upon that equation, supposed rational in w'', w', and w, in order to secure that its complete primitive shall be an algebraical relation between w and z. When we have to deal with an equation of order n, or with a system of n equations of the first order, the corresponding investigation of the conditions, under which algebraical primitives are possible, is more complicated, from the nature of the case. We do not propose even to initiate such a discussion here.

---

* References are given on p. 312. A similar but more extended result, in connection with a more limited problem, has been established by Poincaré, *Méc. Cél.*, t. I (1892), pp. 253, 254.

To a large extent I have followed the interesting and valuable memoir, in which Bruns first expounded his theorem. At the same time, substantial modifications in the argument have been made in several places: and some deficiencies in Bruns's work, which are required to justify the earlier theory of algebraic integrals possessed by equations of a less special form, have been supplied. The chief among these changes are: the reduction of quantities, which (p. 321) satisfy the differential equations and contain arbitrary constants, to quantities, which also satisfy those equations and are free from arbitrary constants: the construction of Poincaré's result to shew that, in integrals free from arbitrary constants, the aggregate of terms of highest order in the first derivatives is rational in the variables themselves: and the mode of dealing with the two sets of partial differential equations, which occur later in the investigation.

The argument is strictly limited to the establishment of the general theorems, and of Bruns's theorem in particular. No attempt is made to indicate some further developments, which are given by Bruns in the later part of his memoir.

There is, however, one investigation of a cognate character, which we proceed to reproduce with certain modifications. It does not belong to the class of discussion which is the matter of preceding chapters; it will consequently appear somewhat isolated. Sufficient reasons for its inclusion are to be found in the importance of the result and in the completeness of the investigation. It leads up to a theorem, due to Bruns[*], which relates to the algebraical integrals of the differential equations of the motion of n mutually attracting bodies. Certain integrals are known, all of an algebraical character. These are the energy-equation, the three equations expressing the constancy of the moments of momentum round any three rectangular axes, the three equations expressing the constancy of the momentum of the centre of gravity of the system, and the three equations (derived by time-quadrature from the last three) expressing the rectilinear path of the centre of gravity. As with Painlevé, these integrals will be called the *classical* integrals of the problem.

All attempts to obtain additional algebraical integrals of a general type[†] have failed: and consequently have led to the surmise that the classical set constitutes the aggregate of algebraical

---

[*] The memoir is entitled "Ueber die Integrale des Vielkörperproblems," *Acta Math.*, t. xi (1887); the part of special importance for the immediate purpose is pp. 25—59. One defect in the investigation has been remedied by Poincaré, *Comptes Rendus*, t. cxxiii (1896), pp. 1224—1228.

There is also a memoir by Painlevé, *Bulletin Astronomique*, t. xv (1898), pp. 81—113, dealing partly with the same question as Bruns, partly with a wider question.

[†] That is to say, involving no particular relations among the variables. Algebraical integrals not of a general type are known. Thus there is the case of three bodies in the same plane throughout the motion, represented by the additional integrals

$$z_1 = 0, \quad z_2 = 0, \quad z_3 = 0.$$

There is the case of Lagrange's three particles in a fixed plane, represented by the additional integrals

$$z_1 = 0, \quad z_2 = 0, \quad z_3 = 0, \quad r^2_{12} = r^2_{23} = r^2_{31}.$$

There is also the case of three particles in one straight line. There are, moreover, other cases with possible algebraical integrals, when special relations among the masses are satisfied.

All such particular cases, where there is any deviation from full generality, will be omitted (§ 273). The aim of the investigation is to shew that all algebraical integrals of a quite general type, which exist, can be compounded of the classical integrals.

integrals, independent of one another. This surmise has been established as a definite result by Bruns.

It is convenient to select for discussion only those integrals, which do not explicitly involve the time. From the three classical integrals, which express the position of the centre of gravity at any moment, the time can be eliminated: and then two integrals survive from them, which shew that the path of the centre of gravity is a straight line. Any one of the three integrals can then be used, to introduce the time into a complete system of integrals, and also to eliminate the time from every other integral in the system. Accordingly, our question is the determination of all the algebraical integrals, which do not involve the time and are independent of one another.

263. The investigation is somewhat long ; its course may be rendered clearer by the following brief statement of the principal results.

In the first instance, differential equations of the same type as, but somewhat more general than, those of the astronomical problem are considered : but only for the first two of the results as stated. It is proved that all algebraical integrals, which do not involve the time-variable, can be compounded algebraically from integrals ϕ, which possess special properties. The chief among these properties are :—

(i) Certain simple transformations, homogeneous in character, leave the differential equations unaltered ; the corresponding invariantive property is used to simplify the integrals that need be considered. They are said to be of the homogeneous class (§ 272):

(ii) Every integral is a rational integral function of the velocities ; it is a rational function of the variables, and (possibly) a single irrational quantity (§§ 267—276):

(iii) If ϕ_0 denote the aggregate of terms, which are of the highest order in the velocities, the coefficients of the various combinations of the velocities in ϕ_0 are polynomials in the coordinates, not having any common factor : this result being definitely established only for the astronomical problem (§§ 277—285):

(iv)   The quantity ϕ_0 involves the coordinates only in polynomial combinations of the form

$$\frac{dx_1}{dt}\theta - \frac{d\theta}{dt}x_1,$$

where x_1 and θ are any two of the coordinates: and, with the possible exception of a negative power of $\dfrac{dx_1}{dt}$, the coefficients of such combinations are rational integral coefficients of the velocities (§ 286):

(v)   The combinations in the preceding result occur only through the three moments of momentum (say A, B, C), and the integrals defining the linear path of the centre of gravity of the system (say A', B', C'): and ϕ_0 is a rational integral function of A, B, C, A', B', C', and the velocities (§§ 287—290):

(vi)   In ϕ_0, the velocities (besides occurring in the integrals A, B, C, A', B', C') occur in the forms of the three components of linear momentum, and of the kinetic energy, and they occur in these alone (§§ 291, 292).

It is thence inferred that, outside the classical integrals, no algebraical integrals of the equations exist (§ 293).

Various explanations and distinctions as to integrals and (what I propose to call*) sub-integrals will be made from time to time. The fifth and sixth stages of the proof will shew how the precise number of integrals (equal to the number of classical integrals) arises in the two sets respectively: and will also shew why the results do not necessarily (as it is known that they do not) apply to the problem of two bodies.

THE CLASSICAL INTEGRALS OF THE DIFFERENTIAL EQUATIONS.

264.   Let m_1, ..., m_n denote the masses of the n bodies, supposed concentrated in points; let x_s, y_s, z_s be the coordinates of m_s at any time, and let X_s, Y_s, Z_s be the components of its velocity, referred to any three rectangular axes; also let r_{ij}

---

* Bruns calls them *Integralgleichungen.*

denote the distance between the masses m_i and m_j at that time. The equations of motion are

$$\frac{dx_s}{dt} = X_s, \quad \frac{dy_s}{dt} = Y_s, \quad \frac{dz_s}{dt} = Z_s,$$

$$\left.\begin{array}{l} \dfrac{dX_s}{dt} = \sum\limits_{a=1}^{n} m_a \dfrac{x_a - x_s}{r^3_{as}} = A_s \\[2ex] \dfrac{dY_s}{dt} = \sum\limits_{a=1}^{n} m_a \dfrac{y_a - y_s}{r^3_{as}} = B_s \\[2ex] \dfrac{dZ_s}{dt} = \sum\limits_{a=1}^{n} m_a \dfrac{z_a - z_s}{r^3_{as}} = C_s \end{array}\right\}, \qquad (s = 1, \ldots, n),$$

the term corresponding to $\alpha = s$ being absent from each summation.

The classical integrals are as follows. First, we have the integrals, which give the three components of momentum, say

$$L' = \sum_{s=1}^{n} m_s X_s, \quad M' = \sum_{s=1}^{n} m_s Y_s, \quad N' = \sum_{s=1}^{n} m_s Z_s;$$

and the three further integrals defining the path of the centre of gravity, say

$$\left.\begin{array}{l} L'' = \sum\limits_{s=1}^{n} m_s x_s - L't = L - L't \\[2ex] M'' = \sum\limits_{s=1}^{n} m_s y_s - M't = M - M't \\[2ex] N'' = \sum\limits_{s=1}^{n} m_s z_s - N't = N - N't \end{array}\right\}.$$

As integrals independent of the time are desired, we take

$$\left.\begin{array}{l} A' = M''N' - M'N'' = MN' - M'N \\ B' = N''L' - N'L'' = NL' - N'L \\ C' = L''M' - L'M'' = LM' - L'M \end{array}\right\};$$

we keep the three forms A', B', C', and note that there is the relation

$$L'A' + M'B' + N'C' = 0.$$

There are, further, the three moments of momentum, viz.

$$\left.\begin{array}{l} A = \sum\limits_{s=1}^{n} m_s (y_s Z_s - z_s Y_s) \\[2ex] B = \sum\limits_{s=1}^{n} m_s (z_s X_s - x_s Z_s) \\[2ex] C = \sum\limits_{s=1}^{n} m_s (x_s Y_s - y_s X_s) \end{array}\right\};$$

and there is the energy-equation, viz.

$$E = \tfrac{1}{2} \sum_{s=1}^{n} m_s (X_s^2 + Y_s^2 + Z_s^2) - \sum_{i=1}^{n} \sum_{j=1}^{n} \frac{m_i m_j}{r_{ij}} = T - U,$$

all simultaneous combinations of (unequal) values $i, j = 1, \ldots, n$ occurring in the double summation. The symbols used in these integrals will be used throughout the investigation in their respective significations.

265. It is convenient, partly for purposes of illustration of the general argument, partly because the case is really the sole general exception to Bruns's theorem, to deal with the equations of two bodies: and we shall take only the simplest form, viz. when both of them move in the plane of x, y. Denoting the distance between them by r, and the sum of their masses by μ, the equations are

$$\left.\begin{aligned} \frac{dx_1}{dt} &= X_1 \\[4pt] \frac{dy_1}{dt} &= Y_1 \\[4pt] \frac{dX_1}{dt} &= \frac{m_2 (x_2 - x_1)}{r^3} \\[4pt] \frac{dY_1}{dt} &= \frac{m_2 (y_2 - y_1)}{r^3} \end{aligned}\right\} , \qquad \left.\begin{aligned} \frac{dx_2}{dt} &= X_2 \\[4pt] \frac{dy_2}{dt} &= Y_2 \\[4pt] \frac{dX_2}{dt} &= -\frac{m_1 (x_2 - x_1)}{r^3} \\[4pt] \frac{dY_2}{dt} &= -\frac{m_1 (y_2 - y_1)}{r^3} \end{aligned}\right\} .$$

The path of each of the particles, relative to the centre of gravity, is an ellipse, of which that point is a focus; hence

$$\frac{m_1}{\mu} r = a \frac{m_1}{\mu} (x_2 - x_1) + b \frac{m_1}{\mu} (y_2 - y_1) + c',$$

that is,

$$r = a (x_2 - x_1) + b (y_2 - y_1) + c,$$

is an integral relation compatible with the differential equations. Differentiating with regard to the independent variable, we have

$$\frac{1}{r} \{(x_2 - x_1)(X_2 - X_1) + (y_2 - y_1)(Y_2 - Y_1)\} = a (X_2 - X_1) + b (Y_2 - Y_1),$$

which must be satisfied. It is not satisfied in virtue of the integral relation from which it is derived, because it does not involve c; nor is it satisfied as a mere algebraical combination of the differential equations; it therefore is effectively another integral, which may be taken in the form

$$(X_2 - X_1)(x_2 - x_1 - ar) + (Y_2 - Y_1)(y_2 - y_1 - br) = 0 \ldots \text{(i)}.$$

Differentiating again with regard to the independent variable, and using the differential equations, we have

$$(X_2 - X_1)^2 + (Y_2 - Y_1)^2$$

$$-\frac{1}{r}\{(x_2 - x_1)(X_2 - X_1) + (y_2 - y_1)(Y_2 - Y_1)\}\{a(X_2 - X_1) + b(Y_2 - Y_1)\}$$

$$= \frac{\mu}{r} - \frac{\mu}{r^2}\{a(x_2 - x_1) + b(y_2 - y_1)\}$$

$$= \frac{\mu}{r} - \frac{\mu}{r^2}(r - c) = \frac{\mu c}{r^2} \quad\ldots\ldots\ldots\ldots\ldots\ldots\ldots\ldots\ldots\ldots\ldots\text{(ii)}.$$

From (i), we have

$$\frac{X_2 - X_1}{-(y_2 - y_1 - br)} = \frac{Y_2 - Y_1}{x_2 - x_1 - ar} = \theta \quad\ldots\ldots\ldots\text{(iii)},$$

say; so that

$$(X_2 - X_1)^2 + (Y_2 - Y_1)^2 = \theta^2 [r^2 - 2r\{a(x_2 - x_1) + b(y_2 - y_1)\} + r^2(a^2 + b^2)]$$

$$= \theta^2 \{r^2(a^2 + b^2 - 1) + 2cr\},$$

and

$$\{(x_2 - x_1)(X_2 - X_1) + (y_2 - y_1)(Y_2 - Y_1)\}\{a(X_2 - X_1) + b(Y_2 - Y_1)\}$$

$$= \theta^2 r\{b(x_2 - x_1) - a(y_2 - y_1)\}^2$$

$$= \theta^2 r[(a^2 + b^2)\{(x_2 - x_1)^2 + (y_2 - y_1)^2\} - \{a(x_2 - x_1) + b(y_2 - y_1)\}^2]$$

$$= \theta^2 r\{(a^2 + b^2 - 1)r^2 + 2cr - c^2\}.$$

Substituting these in (ii), we have

$$\theta^2 = \frac{\mu}{cr^2} \quad\ldots\ldots\ldots\ldots\ldots\ldots\ldots\ldots\ldots\text{(iv)}.$$

Now one of the classical integrals in the present case is

$$C = m_1(x_1 Y_1 - y_1 X_1) + m_2(x_2 Y_2 - y_2 X_2);$$

and another is

$$C' = (m_1 x_1 + m_2 x_2)(m_1 Y_1 + m_2 Y_2) - (m_1 y_1 + m_2 y_2)(m_1 X_1 + m_2 X_2).$$

Also

$$\mu C - C' = m_1 m_2\{(Y_2 - Y_1)(x_2 - x_1) - (X_2 - X_1)(y_2 - y_1)\}$$

$$= m_1 m_2 \Phi,$$

say, so that Φ is an integral; and then, by (iii) and (iv),

$$\Phi = \theta\{(x_2 - x_1)(x_2 - x_1 - ar) - (y_2 - y_1)(y_2 - y_1 - br)\}$$

$$= \theta\{r^2 - r(r - c)\}$$

$$= \theta rc = \sqrt{\mu c}.$$

Lastly,

$$Y_2 - Y_1 = \theta(x_2 - x_1 - ar) = \sqrt{\frac{\mu}{c}}\left(\frac{x_2 - x_1}{r} - a\right),$$

$$X_2 - X_1 = -\theta(y_2 - y_1 - br) = -\sqrt{\frac{\mu}{c}}\left(\frac{y_2 - y_1}{r} - b\right);$$

so that

$$(Y_2 - Y_1)\Phi = \mu\left(\frac{x_2 - x_1}{r} - a\right),$$

$$(X_2 - X_1)\Phi = -\mu\left(\frac{y_2 - y_1}{r} - b\right);$$

and therefore three integrals are

$$\left.\begin{array}{c} (X_2 - X_1)\Phi + \mu\,\dfrac{y_2 - y_1}{r} = \mu b \\[2mm] -(Y_2 - Y_1)\Phi + \mu\,\dfrac{x_2 - x_1}{r} = \mu a \\[2mm] \Phi = \sqrt{\mu c} \end{array}\right\},$$

which give the significance of a, b, c in the integral relation

$$r - a(x_2 - x_1) - b(y_2 - y_1) - c = 0.$$

These results will be used later.

As regards Bruns's theorem in the present case, the quantity, denoted in the statement of § 263 by ϕ_0, is

$$(X_2 - X_1)\Phi, \quad -(Y_2 - Y_1)\Phi,$$

in the respective cases; and if the theorem holds in the present case, both of these quantities should be expressible rationally and integrally in terms of C, C', T, L', M', where

$$2T = m_1(X_1^2 + Y_1^2) + m_2(X_2^2 + Y_2^2),$$

$$L' = m_1 X_1 + m_2 X_2, \quad M' = m_1 Y_1 + m_2 Y_2.$$

Now $\Phi = \dfrac{1}{m_1 m_2}(\mu C - C')$; so that $X_2 - X_1$ and $Y_2 - Y_1$ would be expressible rationally in terms of C, C', T, L', M', the coefficients in the expression being mere constants. This manifestly is impossible; and the theorem therefore does not hold for two bodies.

266. When a relation among the variables x, y, z and their first derivatives X, Y, Z, say in a form

$$\phi = 0,$$

(ϕ not explicitly involving the time), is compatible with the differential equations, then the equation

$$\frac{d\phi}{dt} = \sum_{s=1}^{n} \left(X_s \frac{\partial\phi}{\partial x_s} + Y_s \frac{\partial\phi}{\partial y_s} + Z_s \frac{\partial\phi}{\partial z_s} + A_s \frac{\partial\phi}{\partial X_s} + B_s \frac{\partial\phi}{\partial Y_s} + C_s \frac{\partial\phi}{\partial Z_s} \right) = 0$$

must be satisfied: and this may occur in one of three ways.

The equation may be satisfied identically: thus in the case of two bodies, we have

$$\frac{d}{dt} \left\{ (X_2 - X_1)\,\Phi + \mu\,\frac{y_2 - y_1}{r} \right\} = 0.$$

We then call ϕ an *integral* of the system of differential equations, and the equation $\phi = 0$ a *solution*.

It may happen that the equation $\dfrac{d\phi}{dt} = 0$ cannot be satisfied without using the equation $\phi = 0$; thus

$$\frac{d}{dt} \left\{ r\,(X_2 - X_1)\,\Phi + \mu\,(y_2 - y_1) \right\} = 0$$

is satisfied only in connection with the relation

$$r\,(X_2 - X_1)\,\Phi + \mu\,(y_2 - y_1) = 0.$$

It will appear that such relations are of two classes: one of the classes contains relations that can be rejected, the other contains relations out of which integrals (in the preceding sense of the word) can be constructed. A function ϕ belonging to the latter class, so that an integral can be constructed from it, will be called a *sub-integral* of the system of differential equations; and $\phi = 0$ will still be called a *solution*.

Lastly, it may happen that the equation $\dfrac{d\phi}{dt} = 0$ is satisfied, though it is not satisfied either identically or in virtue of $\phi = 0$. It therefore can be satisfied only in virtue of other equations of the same kind; it will be proved that, in all cases which need be retained, $\phi = 0$ is then a combination of solutions.

Thus, in the case of two bodies, the equation

$$\frac{d}{dt} \left\{ r - a\,(x_2 - x_1) - b\,(y_2 - y_1) - c \right\} = 0$$

is satisfied; it is not satisfied either identically or in virtue of

$$r - a\,(x_2 - x_1) - b\,(y_2 - y_1) - c = 0;$$

but it is satisfied in virtue of

$$(X_2 - X_1)\, \Phi + \mu\, \frac{y_2 - y_1}{r} = \mu b, \quad -(Y_2 - Y_1)\, \Phi + \mu\, \frac{x_2 - x_1}{r} = \mu a.$$

Also

$$\mu\phi = \quad (y_2 - y_1)\left\{ \quad (X_2 - X_1)\,\Phi + \mu\, \frac{y_2 - y_1}{r} - \mu b \right\}$$

$$+ (x_2 - x_1)\left\{ -(Y_2 - Y_1)\,\Phi + \mu\, \frac{x_2 - x_1}{r} - \mu a \right\}$$

$$+ (\Phi + \sqrt{\mu c})\, \{\Phi - \sqrt{\mu c}\},$$

that is, $\phi = 0$ is a combination of other solutions. When it is necessary to refer to such a case, $\phi = 0$ will be called a *composite solution*.

Again, consider the problem of three bodies, when they remain in one plane throughout the motion. Taking it to be the plane of x, y, we should have

$$z_3 = 0,$$

as an algebraical equation compatible with the general differential equations, and therefore the equation

$$\frac{dz_3}{dt} = 0$$

must be satisfied in association with the differential equations. It is not satisfied identically, nor in virtue of $z_3 = 0$ alone; it is satisfied in virtue of

$$z_1 = 0, \quad z_2 = 0, \quad z_3 = 0,$$

because, in general,

$$\frac{dz_3}{dt} = \frac{m_1}{r^2_{13}} z_1 + \frac{m_2}{r^2_{23}} z_2 - \left(\frac{m_1}{r^2_{13}} + \frac{m_2}{r^2_{23}}\right) z_3.$$

Similarly, for $z_1 = 0$ and $\frac{dz_1}{dt} = 0$; also for $z_2 = 0$ and $\frac{dz_2}{dt} = 0$.

Clearly $z_3 = 0$ is not a combination of other solutions, as was $\phi = 0$ in the preceding example. When it is necessary to refer to such a case, the solution will be called *particular*.

Consider once more the special case of the three bodies in one plane, known as Lagrange's three particles. We have

$$r_{12} = r_{23} = r_{31},$$

in that instance; a side of the equilateral triangle formed by the particles can be of any magnitude, so that the equation

$$r^2_{12} = \alpha^2,$$

where α is arbitrary, is compatible with the general differential equations. Then

$$r_{12}\frac{dr_{12}}{dt} = (x_1 - x_2)(X_1 - X_2) + (y_1 - y_2)(Y_1 - Y_2) = 0$$

must be satisfied. It is not satisfied identically in association with the differential equations, nor in virtue of $r^2_{12} = \alpha^2$: but it is satisfied in virtue of

$$X_1 - X_2 = \theta(y_1 - y_2), \quad Y_1 - Y_2 = -\theta(x_1 - x_2),$$

where

$$\theta^2 = \frac{m_1 + m_2 + m_3}{r^3_{12}} - m_3\{(x_1 - x_2)U + (y_1 - y_2)V\},$$

and

$$U = x_1\left(\frac{1}{r^3_{12}} - \frac{1}{r^3_{13}}\right) + x_2\left(\frac{1}{r^3_{23}} - \frac{1}{r^3_{12}}\right) + x_3\left(\frac{1}{r^3_{13}} - \frac{1}{r^3_{23}}\right),$$

$$V = y_1\left(\frac{1}{r^3_{12}} - \frac{1}{r^3_{13}}\right) + y_2\left(\frac{1}{r^3_{23}} - \frac{1}{r^3_{12}}\right) + y_3\left(\frac{1}{r^3_{13}} - \frac{1}{r^3_{23}}\right),$$

with the relations

$$r^2_{12} - r^2_{13} = 0, \quad r^2_{23} - r^2_{12} = 0, \quad r^2_{13} - r^2_{23} = 0.$$

And

$$X_1 - X_2 = \theta(y_1 - y_2), \quad Y_1 - Y_2 = -\theta(x_1 - x_2),$$

are equations compatible with the differential **equations**: their derivatives with respect to t vanish, in virtue of

$$r^2_{12} = \alpha^2,$$

and in virtue of

$$U = 0, \quad V = 0,$$

the latter holding because of the relations

$$r^2_{12} = r^2_{23} = r^2_{31}.$$

Clearly $r^2_{12} = \alpha^2$ is not a combination of other **solutions, as was** $\phi = 0$ in the first example. In such a case, where **parameters** can occur in compatible equations, but where **also compatible** equations arise which involve variables only **and not parameters,** the solution will still be called *particular,* (on **account of the latter** class of compatible equations).

In every case, where $\frac{d\phi}{dt} = 0$ is compatible with the differential equations in some one of the foregoing modes, we shall say that ϕ *satisfies* the differential equations.

Note. A relation $\phi = 0$, which involves only the variables x, y, z and not the variables X, Y, Z, is a special case of the more general form, which involves all of them. As we are dealing with algebraic integrals, we are bound to consider the possibility of an algebraic relation among the variables x, y, z alone; thus it is not inconceivable that each of these variables might be a transcendental function of t, and yet that (as in the comparison of transcendents, established by Abel's Theorem) algebraic relations among the variables could exist, not explicitly involving the time. It will be seen that such a relation is not an integral: it is a composite solution.

SOME PROPERTIES OF MORE GENERAL EQUATIONS.

267. Certain of the fundamental properties connected with algebraic integrals belong to equations of the same type as those in the astronomical problem, and do not depend upon the special form within the type. We therefore consider the set of equations

$$\frac{dx_r}{dt} = y_r, \quad \frac{dy_r}{dt} = A_r(x_1, \ldots, x_m), \qquad (r = 1, \ldots, m),$$

where A_r is a homogeneous algebraical (but not necessarily rational) function of its arguments. The order of A_r is taken to be even, equal to $2N$: thus $N = -1$ for the astronomical problem; and the coefficients are real. Moreover, it is assumed that A_r does not explicitly involve t.

When A_r is irrational in the variables x, it can be made rational by the introduction of a single new variable. Let this variable be s, which is irrational in the variables x; and suppose that it is defined by an equation

$$F(s, x_1, \ldots, x_m) = s^n + S_1 s^{n-1} + \ldots + S_{n-1}s + S_n = 0,$$

where the coefficients S are rational integral functions* of the m variables x, and involve only such constant coefficients as occur in the functions A.

Thus, in the problem of three bodies, we take

$$s = r_{12} + r_{23} + r_{31}.$$

---

* It will be convenient to denote a rational function by R, and a rational integral function by G, when account is taken merely of generic character.

Each of the quantities r_{12}, r_{23}, r_{31} is then expressible as a rational function of s and the coordinates of the bodies. For instance, we have

$$s^2 + r^2_{12} + r^2_{23} - r^2_{31} = 2s(r_{12} + r_{23}) - 2r_{12}r_{23},$$

$$(s^2 + r^2_{12} + r^2_{23} - r^2_{31} + 2r_{12}r_{23})^2 = 4s^2(r^2_{12} + r^2_{23}) + 8s^2 r_{12}r_{23};$$

the second of these expresses $r_{12}r_{23}$ as a rational function, and thence the first expresses $r_{12} + r_{23}$ as a rational function, say

$$r_{12} + r_{23} = \sigma.$$

Then

$$r_{12} = \frac{1}{2\sigma}(\sigma^2 + r^2_{12} - r^2_{23}):$$

and so for r_{23}, r_{31}. Also, s itself is given as the root of an equation of degree 8, of the same form as F above.

In the problem of n bodies, we take

$$s = \sum_{i=1}^{n} \sum_{j=1}^{n} r_{ij}.$$

Each of the quantities r_{ij} is similarly expressible as a rational function of s and the coordinates of the bodies; and the degree in s of the equation $F = 0$ is $2^{\frac{1}{2}n(n-1)}$.

REDUCTION OF ALGEBRAIC INTEGRALS.

268. The most general manner, in which we conceive an algebraical magnitude to be defined, gives it as a root of an algebraical equation. Accordingly, beginning with the most general form of our quantity ϕ, (whether it be an integral or a sub-integral, or lead to a composite solution, or to a particular solution), we take it as a root of an equation

$$\phi^p + B_1\phi^{p-1} + \dots + B_{p-1}\phi + B_p = 0,$$

where the coefficients B are rational functions of the variables x, y, and may be denoted by the generic symbol $R(x, y)$. Without loss of generality, the equation may be regarded as irreducible, that is, the left-hand side cannot be resolved into factors which are of its own type. In association with ϕ, we are required to have $d\phi/dt = 0$, since ϕ must satisfy the differential equations under one of the modes in § 266; and therefore, differentiating the above equation,

$$\frac{dB_1}{dt}\phi^{p-1} + \dots + \frac{dB_{p-1}}{dt}\phi + \frac{dB_p}{dt} = 0.$$

If the coefficient of every power of ϕ in this equation vanishes identically, then B_1, B_2, ..., B_p satisfy the equations. They are rational functions of x, y; and then ϕ is an algebraical combination of rational quantities of the character $R(x, y)$. It therefore would be sufficient to discuss quantities of the latter class.

If not all the coefficients of powers of ϕ in the derived equation vanish identically, then each such non-vanishing coefficient acquires a form $R(x, y, s)$, on substituting for dx/dt, dy/dt, from the differential equations. Thus the original equation of degree p in ϕ, and the derived equation of degree $\lessgtr p - 1$, coexist, they are satisfied by the same value (or by several the same values) of ϕ.

If they are satisfied without reference to any other equation, it follows that the first equation, which is irreducible solely by the variables x, y, is reducible by the variables x, y, s; and the roots, which it possesses in common with the derived equation, can be represented as the roots of a new equation

$$\phi^q + C_1\phi^{q-1} + \dots + C_{q-1}\phi + C_q = 0,$$

where the coefficients C are of the form $R(x, y, s)$. If this equation is reducible, we take its several irreducible component equations, each of which is of the same form as itself: thus no generality is lost, if we regard the preceding equation as one of the irreducible components. The equation still defines the quantity ϕ, so that we must have $d\phi/dt = 0$: and so

$$\frac{dC_1}{dt}\phi^{q-1} + \dots + \frac{dC_q}{dt} = 0,$$

compatible with the former equation. But no equation is compatible with the former equation if it is of lower degree, on account of the irreducibility; and therefore the new equation must be evanescent, that is, each of the coefficients dC/dt is zero, and each of the coefficients C satisfies the differential equations. These coefficients C are rational functions of x, y, s; and then ϕ is an algebraical combination of quantities, which are of the character $R(x, y, s)$. It therefore would be sufficient to consider quantities of the latter class.

Since functions $R(x, y)$, that arise in the former case, are only special forms of functions $R(x, y, s)$, that arise in the latter case, it will be sufficient for both alternatives, that we discuss quantities of the type $R(x, y, s)$, which satisfy the differential equations in some one of the modes of § 266.

If the equation, which defines ϕ, and its derived equation are satisfied only in virtue of (one or more) other equations, the derived equation can be used to replace one of the latter. It then can be regarded as a new initial equation, which has the form

$$\phi^{p-r} + C_1 \phi^{p-r-1} + \ldots = 0, \qquad (r \geqslant 1),$$

being of lower degree than the original equation; and it possesses a derived equation which coexists with it. If the coexistence be independent of other equations, then, as above, ϕ is an algebraical combination of quantities of the form $R(x, y)$ or $R(x, y, s)$. If the coexistence is not independent of other equations, one of the latter can be replaced by the derived equation. We proceed as before. Because p is a finite integer, the number of occurrences of the second alternative is limited. At each of the stages, there is the possibility that ϕ can then be inferred to be an algebraical combination of quantities of the form $R(x, y)$ or $R(x, y, s)$; and in the least favourable case of all, the last equation would be of degree one in ϕ, that is, ϕ would then be explicitly given in the form $R(x, y, s)$.

These quantities may or may not contain arbitrary constants, i.e. constants which do not occur in the original differential equations.

If they do not, the quantity $R(x, y, s)$ is a function of the variables and of any (fixed) constants that occur in the differential equations.

If arbitrary constants occur in $R(x, y, s)$, the occurrence may be rational or irrational. When it is irrational, let it be, e.g., by a constant c, defined as a root of an irreducible equation, say

$$c^\kappa + c_1 c^{\kappa-1} + c_2 c^{\kappa-2} + \ldots + c_\kappa = 0,$$

where the coefficients $c_1, c_2, \ldots, c_\kappa$ do not involve c. Since this equation is irreducible, our quantity $R(x, y, s)$, which is supposed to involve c, can (by the customary methods) be expressed in the form

$$R = F_0 + F_1 c + \ldots + F_{\kappa-1} c^{\kappa-1},$$

where $F_0, F_1, \ldots, F_{\kappa-1}$ do not involve c.

By substituting the κ roots c in turn, we obtain κ quantities R, each of which satisfies the differential equations and is expressible in the foregoing form in terms of $F_0, F_1, \ldots, F_{\kappa-1}$. By means of these κ expressions, $F_0, F_1, \ldots, F_{\kappa-1}$ can each be expressed as a linear combination of the quantities R with constant coefficients;

that is, each of the quantities F_0, F_1, ..., $F_{\kappa-1}$ satisfies the differential equations. Each of them is rational in x, y, s, and is independent of c; so that R, when irrational in an arbitrary constant c, can be algebraically composed from quantities that do not involve c.

Similarly for any other irrational arbitrary constant, if it occur in $R(x, y, s)$.

It therefore is sufficient to consider quantities $R(x, y, s)$, which satisfy the differential equations, and either do not involve any arbitrary constants, or else involve them only rationally. We now proceed to shew that, when $R(x, y, s)$ does contain a number of arbitrary constants (which, by what has just been proved, enter rationally into its expression), *it can be algebraically composed of quantities u, which satisfy the differential equations, which are rational in x, y, s, and which involve no arbitrary constants.*

269. Accordingly, let ϕ denote a quantity, which satisfies the differential equations, and which involves an arbitrary constant b rationally. Then ϕ is expressible in one of the forms

$$G(b), \quad G_1(b) \div G_2(b),$$

where G, G_1, G_2, are polynomials in b, the coefficients of the powers of which are rational in x, y, s. Each such coefficient, though it does not involve b, may involve other arbitrary constants rationally.

First, let ϕ be of the form $G(b)$; as it satisfies the differential equations, we have

$$\frac{d}{dt}\{G(b)\} = 0.$$

Now ϕ may have arisen in such a way that

$$G(b) = 0,$$

that is, $\phi = 0$ is a solution of the differential equations. Consequently, the equation

$$\frac{d}{dt}\{G(b)\} = 0$$

may be satisfied independently of $G(b) = 0$, or in virtue of $G(b) = 0$.

(i)　When it is satisfied independently of $G(b) = 0$, so that it is

$$\frac{dI_0}{dt} + b\frac{dI_1}{dt} + b^2\frac{dI_2}{dt} + \ldots + b^\kappa\frac{dI_\kappa}{dt} = 0,$$

then this may be satisfied identically, without regard to any equation involving the constants b. Since b is an arbitrary constant, we must, in this case, have

$$\frac{dI_0}{dt} = 0, \quad \frac{dI_1}{dt} = 0, \quad ..., \quad \frac{dI_\kappa}{dt} = 0,$$

that is, the quantity $G(b)$ is then a linear combination of this set of quantities, no one of which involves the constant b, and each of which satisfies the differential equations.

(ii) When the equation $\dfrac{d}{dt}\{G(b)\} = 0$ is satisfied in virtue of $G(b) = 0$, then the equations

$$\left.\begin{array}{l} G(b) = I_0 + bI_1 + ... + b^\kappa I_\kappa = 0 \\[2mm] \dfrac{dI_0}{dt} + b\dfrac{dI_1}{dt} + ... + b^\kappa \dfrac{dI_\kappa}{dt} = 0 \end{array}\right\}$$

coexist with one another and with the differential equations. No generality is lost by assuming that $G(b)$ is irresoluble into factors of its own type: otherwise each such factor would be discussed in turn; and we may therefore regard the equation as irreducible. Hence the eliminant of the two equations vanishes, say $\Delta = 0$. In order to form this eliminant, it is sufficient to use the dialytic process. We construct the equations

$$b^s G(b) = 0, \qquad b^s \frac{d}{dt} G(b) = 0, \qquad (s = 0, 1, ..., \kappa - 1);$$

the determinant of these equations is Δ, and when $2\kappa - 1$ of the equations are used, they give

$$b^r = B_r,$$

where the functions B_r are rational in the coefficients of $G(b)$, that is, are of the form $R(x, y, s)$, the rational function involving other parameters that occur in G. Now $b = B_1$ makes G vanish, provided $\Delta = 0$; accordingly we divide $G(b)$ by $b - B_1$ and obtain a remainder, which does not involve b and must vanish when $\Delta = 0$, that is, must be divisible by Δ. Thus

$$G(b) = (b - B_1)\Theta_1 + \Delta\Phi_1;$$

and similarly

$$\frac{d}{dt}\{G(b)\} = (b - B_1)\Theta_2 + \Delta\Phi_2.$$

Conversely,

$$b - B_1 = \theta_1 G(b) + \phi_1 \frac{d}{dt}\{G(b)\},$$

$$\Delta = \theta_2 G(b) + \phi_2 \frac{d}{dt}\{G(b)\}.$$

Now since $\frac{d}{dt}\{G(b)\} = 0$ is an equation, which is consistent with $G(b) = 0$ and the differential equations, and since it is not an identity, we have

$$\frac{d^2}{dt^2}\{G(b)\} = 0,$$

which is consistent with the solution $G(b) = 0$ and the derived equation $\frac{d}{dt}\{G(b)\} = 0$. Hence

$$\frac{d}{dt}(b - B_1) = 0, \qquad \frac{d}{dt}\Delta = 0;$$

that is, $b - B_1 = 0$, $\Delta = 0$ are solutions: and therefore B_1 and Δ are quantities that satisfy the differential equations. But

$$G(b) = (b - B_1)\Theta_1 + \Delta\Phi_1;$$

whence $G(b)$ is algebraically composed of quantities B_1 and Δ, which satisfy the differential equations and do not contain the arbitrary constant b.

(iii) If the equation

$$\frac{d\phi}{dt} = \frac{d}{dt}\{G(b)\} = 0$$

is satisfied, neither identically nor in virtue of $G(b) = 0$, but in virtue of other equations, which are consistent with the differential equations, and therefore are solutions, it must (by the ordinary theory of elimination) be expressible in some such form as

$$\frac{d\phi}{dt} = P_1\phi_1 + P_2\phi_2 + \ldots + P_\kappa\phi_\kappa = 0,$$

where $\phi_1 = 0$, ..., $\phi_\kappa = 0$ are the other solutions indicated, being such that

$$\frac{d\phi_1}{dt} = 0, \quad \ldots, \quad \frac{d\phi_\kappa}{dt} = 0.$$

Now as $\frac{d\phi}{dt}$ is a complete differential, so also must $P_1, \ldots, P_\kappa$ be perfect differentials in the present case: suppose $P_i = dp_i/dt$, for $i = 1, ..., \kappa$. Then

$$\phi = p_1\phi_1 + p_2\phi_2 + ... + p_\kappa\phi_\kappa + \phi_{\kappa+1},$$

where $\phi_{\kappa+1}$ is another quantity which satisfies the differential equations.

If each of the quantities ϕ_1, ϕ_2, ..., $\phi_{\kappa+1}$, is independent of b, so that the parameter enters only through the coefficients p, then ϕ is expressed as an algebraical combination of quantities, which are free from the parameter b.

If a solution, say $\phi_1 = 0$, involves b, then, when $\phi_1 = 0$ and $\dfrac{d\phi_1}{dt} = 0$ are treated simultaneously in the same manner, as were $G(b) = 0$ and $\dfrac{d}{dt}\{G(b)\} = 0$ in the preceding case, it appears that ϕ_1 is expressible in a form

$$\phi_1 = (b - B_2)u_1 + \Delta_2 v_1,$$

where B_2 and Δ_2 are quantities, which satisfy the differential equations and do not involve b. Similarly for any other solution $\phi_2 = 0$ that involves b. Accordingly when we substitute for each of such expressions ϕ_1, ϕ_2, ..., its equivalent, the full form of ϕ becomes

$$\phi = \Sigma\, b\,\{u(b - B) + v\Delta\},$$

that is, it is an algebraical combination of quantities B and Δ, which satisfy the differential equations and do not involve b.

270.   Secondly, let ϕ be of the form $G_1(b) \div G_2(b)$, where $G_1(b)$ and $G_2(b)$ may be assumed to have no common factor. The equation $\dfrac{d\phi}{dt} = 0$ is to be satisfied.

(a)   The equation $\dfrac{d\phi}{dt} = 0$ may be satisfied, solely in virtue of $\phi = 0$, or in virtue of $\phi = 0$ as well as of other equations. In either case, because $\phi = 0$, we must have $G_1(b) = 0$; which accordingly is the equation to be discussed, being the effective form of $\phi = 0$. Moreover, substituting $G_1(b) = 0$ in $\dfrac{d\phi}{dt} = 0$, we have

$$\frac{dG_1(b)}{dt} = 0,$$

concurrently with $G_1(b) = 0$. Now $G_1(b)$ is a polynomial in b, and therefore the argument of § 269 applies to it; we therefore infer that the effective part of ϕ, in the present case, is an algebraical combination of quantities, which satisfy the differential equations and do not involve b.

(β)　The equation $\dfrac{d\phi}{dt} = 0$ may be satisfied, neither identically nor in virtue of $\phi = 0$, but solely in virtue of other equations.

The third part of the argument, viz. (iii), in § 269 applies to this case. The only modification, that may be needed, arises when, at the last stage, a quantity ϕ_1 is fractional, being such that $\phi_1 = 0$ and $\dfrac{d\phi_1}{dt} = 0$ simultaneously; we then apply to ϕ_1 the argument (α) above. The inference is the same as before.

(γ)　The equation $\dfrac{d\phi}{dt} = 0$ may be satisfied identically. Let

$$G_1(b) = g_0 b^p + g_1 b^{p-1} + \ldots + g_p \Big\} \; ;$$
$$G_2(b) = h_0 b^q + h_1 b^{q-1} + \ldots + h_q \Big\}$$

then we have

$$b^{p+q}\left(h_0 \frac{dg_0}{dt} - g_0 \frac{dh_0}{dt}\right) + \ldots = 0.$$

We proceed exactly as in § 272, and we infer that each of the quantities g_r/h_0, h_s/h_0, for $r = 0, 1, \ldots, p$, $s = 1, \ldots, q$, satisfies the differential equations. In other words, ϕ is an algebraical combination of quantities, which satisfy the differential equations and do not involve the arbitrary constant.

Hence, in every case, ϕ is expressible in the specified form.

271.　It therefore follows that a quantity ϕ, which satisfies the differential equations, and involves a number of arbitrary constants b, c, ..., each of them rationally, can be algebraically composed from quantities ψ, which satisfy the differential equations and do not involve the arbitrary constant b, though they may involve the remaining arbitrary constants $c, \ldots$ rationally. All the quantities ϕ and ψ are rational in x, y, s.

To each of these quantities ψ, the similar argument can be applied, with reference to the arbitrary constant c. It leads to the result that each of the quantities ψ can be algebraically composed from other quantities χ, which satisfy the differential equations and do not involve the arbitrary constant c (nor b), though they may involve the arbitrary constants, other than b and c, which occur in ϕ. Each such quantity χ is rational in the variables x, y, s.

And so on, for each of the arbitrary constants in succession : until, at the last stage, the quantities, which are subsidiary to the algebraical composition of the preceding set, are free from all arbitrary constants. Proceeding backwards through the successive sets, we conclude that the initial quantity ϕ, which satisfies the differential equations and involves any number of arbitrary constants, can be algebraically composed of quantities, which satisfy the differential equations, involve no arbitrary constants, and are rational functions of the variables x, y, s.

Note 1.   The number of independent arbitrary constants in such a quantity ϕ can only be finite. If it were infinite, then whether ϕ has the form $G(b)$ or the form $G_1(b) \div G_2(b)$, the number of combinations of the variables would be unlimited, that is, ϕ would be a transcendental function of the variables.

Note 2.   Further, it has been assumed tacitly in the course of § 268, that ϕ is either rational or only algebraically irrational in each of its constants, and that it is not transcendental in those constants. What has just been remarked about the number of combinations of the variables applies also to each one of the constants ; so that, if ϕ be transcendental in a constant a, it must be because one or more transcendental functions of a occur in its expression. For example, let such a function be denoted by A, so that ϕ is rational in A, a : then

$$\frac{d}{dt}\phi(x, y, s, a, A) = 0.$$

An argument similar to that of § 269 would shew that the further treatment of this equation would lead to equations

$$a = \psi(x, y, s), \quad A = \chi(x, y, s),$$

where ψ and χ are rational, which are consistent with the differential equations and with one another. These forms shew that A cannot be a transcendental function of a : in other words, the functions ϕ involve arbitrary constants only in algebraic forms.

Ex. Consider, as an example, the equations in the problem of two bodies. It has been seen that solutions of the differential equations occur in the forms

$$u_3 = \Phi - \sqrt{\mu c} = (Y_2 - Y_1)(x_2 - x_1) - (X_2 - X_1)(y_2 - y_1) - \sqrt{\mu c} = 0$$

$$u_1 = (X_2 - X_1)\Phi + \mu \frac{y_2 - y_1}{r} - \mu b = 0$$

$$u_2 = -(Y_2 - Y_1)\Phi + \mu \frac{x_2 - x_1}{r} - \mu a = 0$$

One solution is known in the form

$$\phi = r - a(x_2 - x_1) - b(y_2 - y_1) - c = 0 ;$$

we have

$$\mu\phi = (y_2 - y_1)u_1 + (x_2 - x_1)u_2 + (\Phi + \sqrt{\mu c})u_3 = 0.$$

Another solution is known in the form

$$\psi = (X_2 - X_1)(x_2 - x_1 - ar) + (Y_2 - Y_1)(y_2 - y_1 - br) = 0 ;$$

we have

$$\mu\psi = r\{(X_2 - X_1)u_2 + (Y_2 - Y_1)u_1\} = 0.$$

Both $\dfrac{d\phi}{dt} = 0$, $\dfrac{d\psi}{dt} = 0$, are satisfied, in consequence of $u_1 = 0$, $u_2 = 0$. Each of the quantities ϕ and ψ, which satisfy the differential equations because $\dfrac{d\phi}{dt} = 0$ and $\dfrac{d\psi}{dt} = 0$, is algebraically composed of u_1, u_2, u_3. But $u_1 = U_1 - \mu a$, $u_2 = U_2 - \mu b$, $u_3 = U_3 - \sqrt{\mu c}$; and U_1, U_2, U_3 are quantities, which satisfy the differential equations and involve no arbitrary constants.

The general theorem is therefore verified for these particular cases.

272. The quantities, which satisfy the differential equations, can be still further resolved, by utilising a property of homogeneity possessed by the differential equations. Let κ denote any constant, and replace x, s, y, t by $x\kappa^a$, $s\kappa^a$, $y\kappa^\gamma$, $t\kappa^\beta$ respectively, choosing α, β, γ so that the differential equations conserve their form unaltered. In order that this may be the case for an equation $\dfrac{dx_r}{dt} = y_r$, we must have

$$\alpha - \beta = \gamma ;$$

and that it may be the case for an equation $\dfrac{dy_r}{dt} = A_r$, we must have

$$\gamma - \beta = 2N\alpha.$$

Hence

$$\frac{\alpha}{2} = \frac{\beta}{1 - 2N} = \frac{\gamma}{1 + 2N},$$

and we may therefore take

$$\alpha = 2, \quad \beta = 1 - 2N, \quad \gamma = 1 + 2N.$$

Hence when x, s, y, t are replaced by $x\kappa^2$, $s\kappa^2$, $y\kappa^{1+2N}$, $t\kappa^{1-2N}$ respectively, the arbitrary constant κ disappears from the differential equations. Consequently, when these substitutions are made in any quantity, which satisfies the differential equations, the new form of the quantity must still satisfy them, whatever be the arbitrary constant κ.

To infer the significance of this result, consider a quantity u, which satisfies the differential equations, and is a rational function of x, y, s, devoid of all parameters. The general expression of such a function is

$$u = \frac{G_1(x, y, s)}{G_2(x, y, s)},$$

where G_1 and G_2 are polynomials, that may be assumed to have no common non-homogeneous factor. When the indicated transformations are effected upon u, we have

$$u = \frac{L_0\kappa^p + L_1\kappa^{p-1} + \dots + L_p}{M_0\kappa^q + M_1\kappa^{q-1} + \dots + M_q} = \frac{L}{M},$$

where, in the numerator and the denominator, all terms involving any (the same) power of κ are gathered together; so that L_0 may be regarded as homogeneous of dimension p, L_1 as homogeneous of dimension $p-1$, and so on. From what has been said, the modified form of u still satisfies the differential equation; and therefore

$$\frac{d}{dt}\left(\frac{L}{M}\right) = 0,$$

that is,

$$M\frac{dL}{dt} - L\frac{dM}{dt} = 0,$$

and therefore

$$\kappa^{q+p}\left(M_0\frac{dL_0}{dt} - L_0\frac{dM_0}{dt}\right)$$
$$+ \kappa^{q+p-1}\left(M_0\frac{dL_1}{dt} + M_1\frac{dL_0}{dt} - L_1\frac{dM_0}{dt} - L_0\frac{dM_1}{dt}\right) + \dots = 0.$$

The quantity κ is arbitrary, and the coefficients of powers of κ are devoid of all arbitrary constants; accordingly, the equation can be satisfied only if

$$M_0\frac{dL_0}{dt} - L_0\frac{dM_0}{dt} = 0,$$

$$M_0\frac{dL_1}{dt} + M_1\frac{dL_0}{dt} - L_1\frac{dM_0}{dt} - L_0\frac{dM_1}{dt} = 0,$$

....

which are $q + p + 1$ equations, homogeneous and linear in the derivatives of the $q + p + 2$ quantities $L_0, \ldots, L_p, M_0, \ldots, M_q$. These equations are equivalent to the set of $q + p + 1$ equations given by

$$\frac{1}{L_0}\frac{dL_0}{dt} = \frac{1}{L_1}\frac{dL_1}{dt} = \frac{1}{L_2}\frac{dL_2}{dt} = \ldots = \frac{1}{L_p}\frac{dL_p}{dt}$$

$$= \frac{1}{M_0}\frac{dM_0}{dt} = \frac{1}{M_1}\frac{dM_1}{dt} = \ldots = \frac{1}{M_q}\frac{dM_q}{dt},$$

unless the eliminant of L and M vanishes, that is, unless L and M have some common factor, say $P\kappa + Q$. But taking $\kappa = 1$, we have

$$G_1 = L_0 + L_1 + \ldots + L_p,$$
$$G_2 = M_0 + M_1 + \ldots + M_q:$$

if L and M have the common factor $P\kappa + Q$, then G_1 and G_2 would have the common (non-homogeneous) factor $P + Q$, contrary to the hypothesis that they have no common factor.

The actual reduction to the above forms for such a case as $p=2$, $q=2$, is as follows. Writing $\dot\theta$ for $d\theta/dt$ in the case of every quantity concerned, we have

$$\dot L_0 = \theta_1 L_0, \quad \dot M_0 = \theta_1 M_0,$$

from the first equation. Substituting for $\dot L_0$ and $\dot M_0$ in the second, we have

$$M_0(\dot L_1 - \theta_1 L_1) - L_0(\dot M_1 - \theta_1 M_1) = 0,$$

and therefore

$$\dot L_1 = \theta_1 L_1 + \theta_2 L_0, \quad \dot M_1 = \theta_1 M_1 + \theta_2 M_0.$$

Substituting for $\dot L_0$, $\dot M_0$, $\dot L_1$, $\dot M_1$, in the third, we have

$$M_0(\dot L_2 - \theta_1 L_2 - \theta_2 L_1) - L_0(\dot M_2 - \theta_1 M_2 - \theta_2 M_1) = 0,$$

and therefore

$$\dot L_2 = \theta_1 L_2 + \theta_2 L_1 + \theta_3 L_0, \quad \dot M_2 = \theta_1 M_2 + \theta_2 M_1 + \theta_3 M_0.$$

The fourth equation becomes

$$\theta_3(L_0 M_1 - M_0 L_1) + \theta_2(L_0 M_2 - M_2 L_0) = 0,$$

and the fifth becomes

$$\theta_3(L_0 M_2 - M_0 L_2) + \theta_2(L_1 M_2 - M_1 L_2) = 0.$$

Now the determinant

$$\begin{vmatrix} L_0 M_1 - M_0 L_1, & L_0 M_2 - M_2 L_0 \\ L_0 M_2 - M_2 L_0, & L_1 M_2 - M_1 L_2 \end{vmatrix}$$

does not vanish. It is the eliminant of L and M, in the Bezout-Cayley form: and it cannot vanish, because L and M have no common factor. Hence $\theta_2 = 0$, $\theta_3 = 0$; that is,

$$\frac{\dot L_j}{L_j} = \frac{\dot M_k}{M_k} = \frac{\dot L_0}{L_0},$$

for all the values of j and k.

The mode of reduction in the general case is a mere generalisation of the preceding process.

We at once have

$$\frac{d}{dt}\left(\frac{M_k}{L_j}\right) = 0,$$

for all values of k and j; in other words, u is composed algebraically of quantities of the form

$$\frac{M_k(x, y, s)}{L_j(x, y, s)}.$$

The numerator and the denominator of every such quantity v are homogeneous, so that v is homogeneous; each such quantity satisfies the differential equations, so that $\frac{dv}{dt} = 0$ is satisfied in some one of the modes of § 266.

Accordingly, it is sufficient to consider quantities u, which satisfy the differential equations, which are rational functions of x, y, s, and are homogeneous in those variables for the transformations

$$x' = x\kappa^2, \quad s' = s\kappa^2, \quad y' = y\kappa^{1+2\nu},$$

and which involve no arbitrary constants, that is, constants which do not occur in the differential equations.

Quantities u of this type are called *homogeneous*; and when u is an integral (§ 266) of the differential equations, it is called a homogeneous integral.

DISCRIMINATION AMONG THE VARIOUS FUNCTIONS.

273. At this stage, it is desirable to discriminate among homogeneous quantities u, which are rational in the variables and satisfy the differential equations. The discrimination is made according to the mode in which the differential equations are satisfied (§ 266).

We have seen that, for our purpose, it is sufficient to retain only quantities u, which are devoid of arbitrary constants.

(i) When the equation

$$\frac{du}{dt} = 0$$

is satisfied identically in connection with the differential equations, then u is an *integral* of the system. The integral equation manifestly is

$$u = \alpha,$$

where α is an arbitrary constant; and $u - \alpha = 0$ is a solution.

(ii)   When the equation

$$\frac{du}{dt} = 0$$

is satisfied, not identically but in virtue of the equation $u = 0$ and without reference to any other quantity that satisfies the differential equations, that is to say, solely in virtue of the equation $u = 0$, we call u a *sub-integral*, and we call $u = 0$ a solution.

Further investigation will be required for quantities u of this type; and we shall prove that either they lead (after appropriate analysis) to integrals or they may be rejected.

(iii)   When the equation

$$\frac{du}{dt} = 0$$

is satisfied, not identically, nor solely in virtue of $u = 0$, but in virtue of the equations $v = 0$, $w = 0$, ..., (with or without the use of $u = 0$), where v, w, ... themselves satisfy the differential equations, then we must have some relation of the form

$$\frac{du}{dt} = Pu + Qv + Rw + \dots.$$

Thus the equations

$$v = 0, \quad w = 0, \dots$$

must persist, concurrently with the differential equations. They involve the variables, or some of them; and they must be satisfied for initial values of the variables. Moreover, no one of them contains an arbitrary constant; consequently, in any such case, there must be relations among the initial values of the variables.

The results of the present discussion are to be applied to the investigation of the independent algebraic integrals of the astronomical problem of n bodies. The investigation will be limited to the case, when the solution of the problem (whatever it may be) deals with a completely general case, in which the initial positions of the n bodies are quite arbitrary and their initial velocities are also quite arbitrary, as regards both magnitude and direction. No relations, whether algebraic or otherwise, can subsist between the initial values of the variables; and consequently, equations such as $v = 0$, $w = 0$, ... cannot be satisfied for initial values. We accordingly exclude all cases where there is any deviation from full generality; that is, we no longer retain cases, when the equation

$$\frac{du}{dt} = 0$$

is satisfied in the mode just considered.

In passing, two remarks call for notice. One of them is that the result of the investigation can apply to only the most general case, and does not necessarily apply to any other : e.g. it would not necessarily apply to the case of three bodies in any plane describing relative periodic orbits. The other of them is an explanation of the retention of the second class of cases, in spite of the indicated rejection of the third class; for $u = 0$ in itself implies a particular relation, and therefore a limitation upon the generality of the problem. The reason is that, save for some exceptional forms, an integral can be constructed from u, and the case can therefore be transferred to the first class: and it will be seen that the exceptional forms are limited in number and, being partly imaginary in expression, can be rejected (§ 276). The same use cannot be made of the integrals $u, v, w, \ldots$ in the third class of cases considered.

We thus retain functions u, such that u is either an integral or a sub-integral : that is, such that the equation

$$\frac{du}{dt} = 0$$

is satisfied either identically or in virtue of $u = 0$. The functions u, so retained, are homogeneous and rational in the variables x, y, s.

Moreover, *every such function u must involve some of the variables y.* For since $u - a = 0$ or $u = 0$ is a solution of the differential equations, we have

$$\frac{du}{dt} = 0,$$

that is,

$$\sum_{r=1}^{m} y_r \frac{\partial u}{\partial x_r} + \sum_{r=1}^{m} A_r \frac{\partial u}{\partial y_r} + \frac{\partial u}{\partial s} \frac{ds}{dt} = 0.$$

Now s is given by

$$F = s^n + S_1 s^{n-1} + \ldots + S_n = 0,$$

so that

$$\frac{ds}{dt} \frac{\partial F}{\partial s} + \sum_{p=1}^{n} \sum_{r=1}^{m} s^{n-p} y_r \frac{\partial S_p}{\partial x_r} = 0;$$

hence the equation satisfied by u is

$$\sum_{r=1}^{m} y_r \frac{\partial u}{\partial x_r} + \sum_{r=1}^{m} A_r \frac{\partial u}{\partial y_r} - \frac{\frac{\partial u}{\partial s}}{\frac{\partial F}{\partial s}} \sum_{p=1}^{n} \sum_{r=1}^{m} s^{n-p} y_r \frac{\partial S_p}{\partial x_r} = 0,$$

and it must be satisfied either identically, or in consequence of $u = 0$. If therefore u is independent of all the variables y, we must have

$$\frac{\partial F}{\partial s} \frac{\partial u}{\partial x_r} = \left(\sum_{p=1}^{n} s^{n-p} \frac{\partial S_p}{\partial x_r} \right) \frac{\partial u}{\partial s} = \frac{\partial F}{\partial x_r} \frac{\partial u}{\partial s},$$

satisfied for all the m values of r. Now as u is supposed a variable quantity, we have

$$du = \frac{\partial u}{\partial s} ds + \sum_{r=1}^{m} \frac{\partial u}{\partial x_r} dx_r$$

$$= \frac{\frac{\partial u}{\partial s}}{\frac{\partial F}{\partial s}} \left(\frac{\partial F}{\partial s} ds + \sum_{r=1}^{m} \frac{\partial F}{\partial x_r} dx_r \right)$$

$$= \frac{\frac{\partial u}{\partial s}}{\frac{\partial F}{\partial s}} dF,$$

that is, du is zero when dF is zero, or u is constant when F is constant. Eliminating any one of the variables s, x_1, ..., x_n from u by means of $F = 0$, all the others will disappear with that one variable: and we have

$$u = f(F),$$

where f denotes some functional form. But F is persistently zero: hence u is an actual constant, so that u is independent of the variables x, that is, if it is independent of the variables y; and therefore it ceases to provide a solution. Accordingly, the homogeneous functions u must involve some of the variables y.

274. Our homogeneous function u is of the form

$$\frac{G_1(x, y. s)}{G(x, y, s)},$$

where the numerator or the denominator or both of them certainly involve some of the variables y and, involving them, are polynomials in those variables. Now it may be the case that G_1 is resoluble into a product of polynomials in y, the coefficients of the powers and combinations being rational functions of x and s; let ψ_1, ψ_2, ... denote such factors, which can now be regarded as irresoluble, and let the degrees of occurrence be λ_1, λ_2, ... respectively. Thus

$$G_1 = Q_1 \psi_1{}^{\lambda_1} \psi_2{}^{\lambda_2} \dots,$$

where Q_1 is a rational function of x and s. Similarly, if G is resoluble, we shall have

$$G = Q_2 \theta_1{}^{\mu_1} \theta_2{}^{\mu_2} \dots,$$

where μ_1, μ_2, ... are positive integers, Q_2 is a rational function of x and s, and θ_1, θ_2, ... are polynomials, irresoluble when regarded as functions of the variables y. Thus

$$u = Q \psi_1{}^{\lambda_1} \psi_2{}^{\lambda_2} \dots \theta_1{}^{-\mu_1} \theta_2{}^{-\mu_2} \dots,$$

where $Q_1 = Q_1/Q_2$, is a rational function of x and s.

When the solution is $u = 0$, it can arise solely through the factors ψ; and $\psi = 0$ is then consistent with the differential equations, so that

$$\frac{d\psi}{dt} = 0$$

is satisfied. This may be satisfied identically, or it may be satisfied only in virtue of $\psi = 0$. In either case, we have to deal with an irreducible polynomial ψ, such that the equations

$$\psi = 0, \quad \frac{d\psi}{dt} = 0,$$

coexist with one another and with the differential equations; in the polynomial, the coefficients of the various powers of y are rational functions of x and s.

When the solution is $u - a = 0$, so that

$$\frac{du}{dt} = 0$$

identically, we have

$$\frac{1}{u}\frac{du}{dt} = 0;$$

in this case, we shall take into account every one of the factors ψ and every one of the factors θ. We have

$$\frac{1}{Q}\frac{dQ}{dt} + \Sigma \frac{\lambda_r}{\psi_r}\frac{d\psi_r}{dt} - \Sigma \frac{\mu_r}{\theta_r}\frac{d\theta_r}{dt} = 0,$$

satisfied identically. Hence multiplying up by any of the factors ψ, say ψ_1, we have

$$\lambda_1 \frac{d\psi_1}{dt} + \frac{\psi_1}{Q}\frac{dQ}{dt} + \Sigma\lambda_r \frac{\psi_1}{\psi_r}\frac{d\psi_r}{dt} - \Sigma\mu_r \frac{\psi_1}{\theta_r}\frac{d\theta_r}{dt} = 0.$$

Now consider the equation $\psi_1 = 0$. Since ψ_1, ψ_2, ... are irreducible polynomials, the roots of $\psi_1 = 0$, regarded as an equation in (say) y_1, are distinct from the roots of $\psi_r = 0$; for otherwise, ψ_1 and ψ_r would have a common factor $\Pi(y_1 - \eta)$, the product extending over those common roots, which are not rational in x, s, and the variables $y_2, ..., y_r$. Hence when $\psi_1 = 0$, ψ_r is not zero; and therefore each one of the quantities $\dfrac{\psi_1}{\psi_r}$ is zero with ψ_1. Similarly, each one of the quantities $\dfrac{\psi_1}{\theta_r}$ is zero with ψ_1. Therefore we have

$$\frac{d\psi_1}{dt} = 0,$$

at the same time as $\psi_1 = 0$; and these two equations coexist with the differential equations.

Similarly, we have

$$\frac{d\theta_1}{dt} = 0,$$

at the same time as $\theta_1 = 0$; and these two equations also coexist with the differential equations. Likewise for every other quantity ψ and every other quantity θ.

Further, the equation $\dfrac{d\psi}{dt} = 0$ exists at the same time as $\psi = 0$; but $\dfrac{d\psi}{dt}$ may be an identical zero, so that $\dfrac{d\psi}{dt} = 0$ is then satisfied identically and not in virtue of $\psi = 0$. Similarly for $\dfrac{d\theta}{dt} = 0$; it may be satisfied identically and not in virtue of $\theta = 0$.

It therefore appears that, for a solution either of the form $u - a = 0$ or of the form $u = 0$, the homogeneous function u in each case can be algebraically composed of polynomials in the variables y. If ψ be such a polynomial, the equation

$$\frac{d\psi}{dt} = 0$$

is satisfied; but it may be satisfied either identically or in virtue of $\psi = 0$. The coefficients in the polynomials are rational functions of x and s; and the polynomials themselves are homogeneous in the variables *.

Such irreducible polynomials in y, rational in x and s, homogeneous in all the variables, have arisen through the function u in a solution $u - a = 0$ or $u = 0$. As u can be algebraically composed from them, it is sufficient for our purpose now to neglect their origin: we may limit our discussion strictly to the inferences from their characteristic property that, in connection with the differential equations, the relation

$$\frac{d\psi}{dt} = 0$$

is satisfied, either identically, or in virtue of the equation $\psi = 0$.

FUNCTIONS, WHICH ARE INTEGRAL IN THE VARIABLES.

275. When the equation

$$\frac{d\psi}{dt} = 0$$

is satisfied identically, then

$$\psi = \text{arbitrary constant}$$

is an equivalent of that equation; and therefore the solution is of the form

$$\psi - \alpha = 0,$$

that is, with our definition of § 266, ψ is an integral of the differential equations.

When the equation

$$\frac{d\psi}{dt} = 0$$

is satisfied, not identically, but only in virtue of $\psi = 0$, then ψ cannot be claimed as an integral. Now the polynomial $\dfrac{d\psi}{dt}$, which does not vanish identically, is one degree higher than ψ in the

---

* This is easily seen to follow from the fact that each of the polynomials under consideration is a factor of either the numerator or the denominator of u; and both the numerator and the denominator are homogeneous for the transformations in § 272. A non-homogeneous polynomial would, after those transformations, become

$$\kappa^\lambda p_0 + \kappa^{\lambda+1} p_1 + \kappa^{\lambda+2} p_2 + \dots,$$

which could not be a factor of $\kappa^p L$ or of $\kappa^q M$, where L and M do not involve κ.

variables y; also ψ and $\dfrac{d\psi}{dt}$ vanish together. This concurrent evanescence can take place only in virtue of some common factor; and ψ is irreducible; hence the common factor must be ψ itself. The remaining factor of $\dfrac{d\psi}{dt}$, say ω, can only be of the first degree in the variables y: let it be

$$\omega = \omega_0 + \omega_1 y_1 + \ldots + \omega_m y_m,$$

where $\omega_0, \omega_1, \ldots, \omega_m$ are homogeneous rational functions of x and s. Then

$$\frac{d\psi}{dt} = \omega\psi.$$

As our functions are rational and homogeneous, this equation will persist when we make the transformations

$$x' = x\kappa^2, \quad s' = s\kappa^2, \quad y' = y\kappa^{1-2N}, \quad t' = t\kappa^{1+2N}.$$

When these are effected upon $\dfrac{1}{\psi}\dfrac{d\psi}{dt}$, there is a factor κ^{-1-2N} to be associated with it; this factor must be the factor to be associated with ω, that is, with

$$\omega_0 + \omega_1 y_1 + \ldots + \omega_m y_m.$$

As the functions $\omega_0, \omega_1, \ldots, \omega_m$ are rational and homogeneous in x and s, they acquire an even power of κ as a factor after transformation, while $y_1, \ldots, y_m$ acquire an odd power. It therefore appears that the first term would acquire an even power of κ, while it ought to acquire an odd power κ^{-1-2N}; hence $\omega_0 = 0$, and then

$$\frac{d\psi}{dt} = (\omega_1 y_1 + \ldots + \omega_m y_m)\,\psi = \omega\psi.$$

276. One further limitation may be imposed upon the quantities ψ about to be considered. There clearly is no loss of generality in assuming, from the beginning of the whole investigation, that, when the differential equations contain only real quantities, all the integrals are real; for if

$$\frac{d}{dt}(P + iQ) = 0,$$

identically, then

$$\frac{dP}{dt} = 0, \quad \frac{dQ}{dt} = 0,$$

both identically, that is, the real quantities P and Q are integrals. All the operations (reductions, transformations, and the like), which have been effected until the last stage, have introduced no imaginary element: and therefore the polynomial u at the beginning of that last stage is such, that all the coefficients are real rational functions of x and s. In resolving the polynomial into factors, which are rational in x and s, it is possible that the imaginary $\sqrt{-1}$ will be introduced: an example indeed will be given (§ 279). But as the polynomial is real, complex factors of this kind will enter in conjugate pairs; the product of a conjugate pair is a real polynomial, such that

$$\frac{1}{\psi\psi'}\frac{d}{dt}(\psi\psi') = \frac{1}{\psi}\frac{d\psi}{dt} + \frac{1}{\psi'}\frac{d\psi'}{dt} = \omega + \omega' = \Omega,$$

an equation of the same form. Accordingly, we shall assume that, in resolving the quantity u into component polynomials, we take only such a resolution as gives real factors ψ; and the quantity ω also then is real. Account will afterwards be taken of the limitation (if any), which this assumption implies in the case of the astronomical problem.

Another property of the homogeneous polynomial ψ may be noted. Let two terms be chosen from ψ, which are of dimensions q and q' in the variables y, and are of dimensions r and r' in the variables x and s. When the homogeneous transformation is effected upon ψ, the former of the terms acquires the factor

$$\kappa^{q(1+2N)+2r},$$

and the latter of the terms acquires the factor

$$\kappa^{q'(1+2N)+2r'}.$$

On account of the homogeneity, we have

$$q(1 + 2N) + 2r = q'(1 + 2N) + 2r',$$

that is, $q - q'$ is an even integer. It therefore follows that, when ψ is arranged in aggregates of terms, each aggregate containing all the terms that are of the same dimensions in y, and when the successive aggregates are arranged in descending order of dimensions in y, the descent in order is by even differences[*]. If the highest order be p, the other orders are $p - 2$, $p - 4$, ..., so that we can take

$$\psi = \psi_0 + \psi_2 + \psi_4 + \dots,$$

where ψ_0 is the aggregate of terms of order p, ψ_2 that of terms of order $p - 2$, and so on: ψ_0, ψ_2, ψ_4, ..., being polynomials, which have real rational functions of x and s for their coefficients.

* This remark was first made by Poincaré, *Comptes Rendus*, t. cxxiii (1896), p. 224.

INTEGRALS: SUB-INTEGRALS.

277. The case, when the equation $\frac{d\psi}{dt} = 0$ is satisfied identically, has already been mentioned: it needs no further discussion at this stage, ψ being an integral. We proceed to the case when that equation is satisfied only in virtue of $\psi = 0$: and shall prove that, *on multiplying ψ by an appropriate factor, which depends on x and s alone, say $R(x, s)$, so that $\psi R(x, s) = \Psi$, then Ψ is, in general, an integral of the differential equations.* On this account, ψ is called a *sub-integral*, as already indicated (§ 266).

It is easy to see how this result is suggested. We have

$$\frac{1}{\psi}\frac{d\psi}{dt} = \omega_1 y_1 + \omega_2 y_2 + \ldots + \omega_m y_m$$
$$= \omega_1 \frac{dx_1}{dt} + \omega_2 \frac{dx_2}{dt} + \ldots + \omega_m \frac{dx_m}{dt},$$

from the differential equations. The quantities $\omega_1, \ldots, \omega_m$ are rational functions of $x_1, \ldots, x_m, s$. If therefore the right-hand side is a perfect differential, say

$$-\frac{1}{R(x, s)}\frac{d}{dt}\{R(x, s)\},$$

we should have

$$\frac{1}{\psi}\frac{d\psi}{dt} + \frac{1}{R(x, s)}\frac{d}{dt}\{R(x, s)\} = 0$$

identically, that is,

$$\frac{d\Psi}{dt} = \frac{d}{dt}\{\psi R(x, s)\} = 0$$

identically; or Ψ would be an integral.

For example, in the case (§ 265) of the problem of two bodies, if

$$\psi = r(X_2 - X_1)\Phi + \mu(y_2 - y_1),$$

we have

$$\frac{d\psi}{dt} = \frac{dr}{dt}(X_2 - X_1)\Phi - r\Phi\frac{\mu}{r^3}(x_2 - x_1) + \mu(y_2 - y_1)$$
$$= \frac{dr}{dt}(X_2 - X_1)\Phi + \mu\frac{y_2 - y_1}{r^2}r\frac{dr}{dt},$$

after reduction; that is,

$$\frac{d\psi}{dt} = \frac{dr}{dt}\frac{\psi}{r},$$

and therefore

$$\frac{d}{dt}\left(\frac{\psi}{r}\right)=0.$$

Thus

$$\frac{\psi}{r}, \ =(X_2-X_1)\,\Phi+\mu\frac{y_2-y_1}{r},$$

is an integral.

To establish the general result, it will be sufficient to prove that

$$\omega_1 dx_1 + \dots + \omega_m dx_m$$

is a perfect differential. This will be shewn to be the case when ψ_0, the aggregate of terms of highest order in y contained in ψ, does not explicitly involve the irrational variable s; it is not necessarily the case (as will be seen by a simple example) if ψ_0 does explicitly involve s. It will appear that all the possible instances of the latter form can be rejected from the solution of the astronomical problem.

Let p denote the order of the terms in ψ that contain the variables y in highest dimension: then ψ can be expressed in the form

$$\psi = \psi_0 + \psi_2 + \psi_4 + \dots,$$

where ψ_0 is the aggregate of terms of dimension p in the variables y, ψ_2 the aggregate of those of dimension $p-2$, and so on. Now the equation

$$\frac{d\psi}{dt} = \omega\psi$$

is satisfied identically; hence the terms of highest order in y on the two sides must be the same. In $\dfrac{d\psi}{dt}$, the part that contains the variables y to order $p+1$ is

$$\sum_{r=1}^{m} y_r \frac{\partial \psi_0}{\partial x_r},$$

while in $\omega\psi$ it is $\omega\psi_0$; accordingly, we must have

$$\sum_{r=1}^{m} y_r \frac{\partial \psi_0}{\partial x_r} = \omega\psi_0,$$

and therefore

$$\omega = \sum_{r=1}^{m} y_r \frac{\partial \log \psi_0}{\partial x_r}.$$

This equation also must be identically satisfied.

278. First, suppose that ψ_0 does not explicitly involve the irrational variable s. We proceed to shew that the quantities $\omega_1, \ldots, \omega_m$ in ω satisfy the conditions of integrability of

$$\omega_1 dx_1 + \ldots + \omega_m dx_m :$$

and we deduce the integral which can be constructed from the function ψ.

It is clear that, as derivatives of ψ_0 with regard to x alone are required in the equation

$$\omega\psi_0 = \sum_{r=1}^{m} y_r \frac{\partial \psi_0}{\partial x_r},$$

any factor of ψ_0, which is a function of the variables y alone, can be removed, without affecting the value of ω: in particular, any product of powers. Let all such factors be gathered together, say

$$\psi_0 = f \cdot \chi,$$

where f depends upon the variables y only; thus

$$\omega\chi = \sum_{r=1}^{m} y_r \frac{\partial \chi}{\partial x_r},$$

where χ is of course homogeneous in those variables, say of order Q.

If χ does not contain the variables y, we have $\omega_r \chi = \dfrac{\partial \chi}{\partial x_r}$ for $r = 1, \ldots, m$; that is, the conditions of integrability are satisfied.

When χ does contain the variables y, select all its terms which are free from some one of those variables, say from y_m: so that we may write

$$\chi = y_m A + \chi' f_1,$$

where χ' is a non-evanescent quantity, having no factor that involves variables y alone, and f_1 (if different from unity) involves no variables except y. Substituting, we find

$$\chi' \sum_{r=1}^{m-1} \omega_r y_r = \sum_{r=1}^{m-1} y_r \frac{\partial \chi'}{\partial x_r},$$

among other equations.

If χ' does not contain the variables y, the conditions of integrability among $\omega_1, \ldots, \omega_{m-1}$ are satisfied. When it does contain those variables, it is resolved, with reference to some variable y_{m-1}, as χ was resolved with reference to y_m; and corresponding alternatives are possible, in succession. It therefore appears that the least favourable case will, in the last resort, occur when the function $\chi^{(m-2)}$, say, involves two of the variables, say y_1 and y_2; so that, denoting their aggregate by θ, we have

$$y_1 \frac{\partial \theta}{\partial x_1} + y_2 \frac{\partial \theta}{\partial x_2} = (\omega_1 y_1 + \omega_2 y_2)\,\theta.$$

Here θ has no factor which involves variables y alone; it therefore may be taken in the form

$$\theta = c_1 y_1{}^q + c_2 y_2{}^q + \dots,$$

where neither c_1 nor c_2 vanishes. Proceeding as before, we find

$$\omega_1 = \frac{\partial \log c_1}{\partial x_1}, \quad \omega_2 = \frac{\partial \log c_2}{\partial x_2};$$

and we have to prove that the condition of integrability is satisfied.

Take $\theta = c_1 \vartheta$, so that ϑ is

$$y_1{}^q + \frac{c_2}{c_1} y_2{}^q + \dots;$$

the coefficients of powers of y have no common factor, since the coefficient of $y_1{}^q$ is unity. These coefficients are rational functions of the variables x; let D denote the least common multiple of their denominators, so that $\Theta, = D\vartheta$, is then a polynomial in x as well as in y_1 and y_2, the coefficients of the various powers of y having no common factor. Then we easily find

$$y_1 \frac{\partial \Theta}{\partial x_1} + y_2 \frac{\partial \Theta}{\partial x_2} = \left[y_1 \frac{1}{D} \frac{\partial D}{\partial x_1} + y_2 \frac{\partial}{\partial x_2} \left\{ \log \left(D\frac{c_2}{c_1} \right) \right\} \right] \Theta = \Omega\Theta,$$

say: the left-hand side is a polynomial in x and y, and the right-hand side must be also of that form. Now D is the coefficient of $y_1{}^q$ in Θ, and $D\dfrac{c_2}{c_1}$ is the coefficient of $y_2{}^q$ in Θ; denote the latter of them—each is a polynomial in the variables x—by C.

If X_1 be an irreducible factor of D involving x_1 (it might be D itself), then $y_1 \dfrac{1}{X_1} \dfrac{\partial X_1}{\partial x_1}$ is meromorphic, and no other part of Ω can combine with it so as to give a function that is integral in the variables x. Hence, if there is such a term in Ω, X_1 must be a factor of Θ in order that $\Omega\Theta$ may be polynomial in x: a condition contrary to the property that the coefficients of powers of y in Θ have no common factor. Thus there is no such term in Ω.

In the same way, we infer that there is no term $y_2 \dfrac{1}{C} \dfrac{\partial C}{\partial x_2}$ in Ω : so that $\Omega = 0$, and therefore

$$y_1 \frac{\partial \Theta}{\partial x_1} + y_2 \frac{\partial \Theta}{\partial x_2} = 0,$$

which is satisfied identically. This partial differential equation shews that Θ is restricted to be any function of the three arguments y_1, y_2, $y_1 x_2 - y_2 x_1$. Now Θ is a polynomial in x and y, which certainly has a term involving y_1^q and a term involving y_2^q; and therefore Θ is of the form

$$\Theta = \underset{\lambda \; \mu \; \rho}{\Sigma \Sigma \Sigma} \, C_{\lambda \mu \rho} \, y_1^\lambda y_2^\mu \, (y_1 x_2 - y_2 x_1)^\rho,$$

where $\lambda + \mu + \rho = q$. But Θ is homogeneous for the transformations of § 272; so that

$$\lambda (1 + 2N) + \mu (1 + 2N) + \rho \{(1 + 2N) + 2\}$$

must be the same for every term, that is, ρ must be the same for every term. We have seen that factors of ψ_0 (and therefore also of χ, θ, Θ in turn), which involve the variables y alone, may be omitted, without affecting ω; hence

$$\Theta = (y_1 x_2 - y_2 x_1)^q.$$

Thus

$$D = x_2^q, \quad D \frac{C_2}{C_1} = C_1 = (-x_1)^q \, ;$$

and therefore

$$\frac{\partial \omega_2}{\partial x_1} - \frac{\partial \omega_1}{\partial x_2} = \frac{\partial^2}{\partial x_1 \partial x_2} (\log c_2 - \log c_1) = 0,$$

that is, the condition of integrability is satisfied, so far as concerns ω_1 and ω_2. Moreover

$$\frac{\partial}{\partial x_2} \left(\log \frac{c_2}{c_1} \right) = - \frac{q}{x_2},$$

where the integer q is the order of the aggregate of the terms in χ that depend only upon y_1 and y_2, when all factors (if any) involving y_1 and y_2 (but not the variables x) are removed from that aggregate.

A corresponding investigation with each pair, framed by combining each of the variables y with y_1, leads to similar results. In particular,

$$\frac{\partial}{\partial x_r} \left(\log \frac{c_r}{c_1} \right) = - \frac{q_r}{x_r},$$

where the aggregate of the terms in χ that depend only upon y_1 and y_r, after all factors independent of the variables x have been removed, is

$$c_1 y_1{}^{q_r} + \ldots + c_r y_r{}^{q_r} + \ldots ;$$

also

$$\omega_r = \frac{\partial \log c_r}{\partial x_r}.$$

Returning now to our original function ψ, we had

$$\frac{1}{\psi} \frac{d\psi}{dt} = \omega_1 y_1 + \omega_2 y_2 + \ldots + \omega_m y_m$$

$$= y_1 \frac{\partial \log c_1}{\partial x_1} + y_2 \frac{\partial \log c_2}{\partial x_2} + \ldots + y_m \frac{\partial \log c_m}{\partial x_m}.$$

But

$$\frac{1}{c_1} \frac{dc_1}{dt} = y_1 \frac{\partial \log c_1}{\partial x_1} + y_2 \frac{\partial \log c_1}{\partial x_2} + \ldots + y_m \frac{\partial \log c_1}{\partial x_m} ;$$

and therefore

$$\frac{1}{\psi} \frac{d\psi}{dt} - \frac{1}{c_1} \frac{dc_1}{dt} = y_2 \frac{\partial}{\partial x_2} \left(\log \frac{c_2}{c_1} \right) + \ldots + y_m \frac{\partial}{\partial x_m} \left(\log \frac{c_m}{c_1} \right)$$

$$= -\sum_{r=2}^{m} y_r \frac{q_r}{x_r}$$

$$= -\sum_{r=2}^{m} \frac{q_r}{x_r} \frac{dx_r}{dt} ;$$

consequently

$$\frac{d}{dt} \left\{ \log \left(\frac{\psi}{c_1} x_2{}^{q_2} x_3{}^{q_3} \ldots x_m{}^{q_m} \right) \right\} = 0,$$

an equation which is satisfied identically. Hence

$$\phi, = \frac{\psi}{c_1} x_2{}^{q_2} x_3{}^{q_3} \ldots x_m{}^{q_m},$$

is an integral of the differential equations. In the present instance c_1 is the coefficient of a term in ψ_0, supposed rational in the variables x; and therefore it appears, *when the aggregate of terms of highest order in the variables y in our homogeneous function ψ is rational in the variables x, that ψ can be changed* [*] *into an integral, on multiplication by an appropriate rational function of the variables x.* As ψ is a polynomial in y, so also is the integral ϕ: and as the coefficients in the terms of the highest order in the polynomial ψ are rational functions of x,

---

[*] It is on this account that ψ is called a *sub-integral*.

so also are those in the integral ϕ. Moreover, the appropriate factor, which is

$$\frac{1}{c_1} x_2{}^{q_2} x_3{}^{q_3} \dots x_m{}^{q_m},$$

is determinable from ψ_0 by inspection.

It is to be expected, from the course of the preceding proof, that the aggregate of terms of the highest order in y which ϕ contains will be a polynomial in quantities of the type $x_r y_1 - x_1 y_r$: this property can be established briefly as follows. As ϕ is a polynomial in y, differing only from ψ by the associated factor, it can be arranged in the form

$$\phi = \phi_0 + \phi_2 + \phi_4 + \dots,$$

where ϕ_0 is the aggregate of terms of highest order in y, say p, ϕ_2 is of order $p - 2$, and so on. Since the equation

$$\frac{d\phi}{dt} = \frac{d\phi_0}{dt} + \frac{d\phi_2}{dt} + \dots = 0$$

is to be satisfied identically, the terms of highest order must vanish by themselves. These terms arise from $\dfrac{d\phi_0}{dt}$ alone, and they give the equation

$$y_1 \frac{\partial \phi_0}{\partial x_1} + y_2 \frac{\partial \phi_0}{\partial x_2} + \dots + y_m \frac{\partial \phi_0}{\partial x_m} = 0.$$

Now ϕ_0 is a polynomial in y, and it is a rational function of the variables x. This partial differential equation shews that ϕ_0 is restricted to be a function of the $2m - 1$ quantities $y_1, \dots, y_m$, $x_2 y_1 - x_1 y_2, \dots, x_m y_1 - x_1 y_m$: the variables x occurring only through the last $m - 1$ combinations. It cannot be a polynomial in y if it involves the last $m - 1$ combinations in fractional forms; and therefore ϕ_0 is a rational integral function of

$$x_2 y_1 - x_1 y_2, \dots, x_m y_1 - x_1 y_m,$$

the coefficients of the various combinations of these quantities being rational functions of $y_1, \dots, y_m$. As a matter of fact, these coefficients are integral functions of the variables y, save as to a possible power of y_1 occurring as a denominator. For we have seen that ϕ_0 is an integral function of the variables y, and it is rational in the variables x. Denoting $x_r y_1 - x_1 y_r$ by p_r, we have

$$x_r = \frac{p_r}{y_1} + x_1 \frac{y_r}{y_1}, \qquad (r = 2, \dots, m);$$

let these expressions be substituted for $x_2, \ldots, x_m$ in ϕ_0. Now ϕ_0, so far as concerns the variables x, involves them only in the combinations $p_2, \ldots, p_m$, of which it is a rational integral function; hence when the substitutions for $x_2, \ldots, x_m$ are made, x_1 disappears. Regarding the two forms of ϕ_0, the first as integral in y and rational in x, and the second as integral in p and rational in y, it follows that ϕ_0 is integral in p and integral in the variables y, except possibly for a power of y_1 occurring as a denominator.

279. We now take the alternative of the hypothesis at the beginning of § 278; we assume that ψ_0 is not rational in the variables x alone. As ψ_0 is rational in s and the variables x, it follows that s will then occur explicitly in ψ_0. The equation

$$\frac{1}{\psi_0} \frac{d\psi_0}{dt} = y_1 \omega_1 + y_2 \omega_2 + \ldots + y_m \omega_m$$

is still satisfied identically; but it is no longer the fact that $\sum\limits_{r=1}^{m} \omega_r dx_r$ is necessarily a perfect differential*. As a simple instance, consider the problem of three bodies. Denoting their coordinates by $x_1, x_2, x_3;\ x_4, x_5, x_6;\ x_7, x_8, x_9$; merely for the present purpose, and writing

$$\xi_1 = x_1 - x_4, \quad \xi_2 = x_2 - x_5, \quad \xi_3 = x_3 - x_6,$$
$$\eta_1 = y_1 - y_4, \quad \eta_2 = y_2 - y_5, \quad \eta_3 = y_3 - y_6,$$

we have

$$\phi = (\eta_1 \xi_2 - \eta_2 \xi_1)^2 + (\eta_2 \xi_3 - \eta_3 \xi_2)^2 + (\eta_3 \xi_1 - \eta_1 \xi_3)^2$$

as an integral of the differential equations. A factor of ϕ is

$$\psi = (\eta_1 \xi_2 - \eta_2 \xi_1)\, \xi_2 + (\eta_1 \xi_3 - \eta_3 \xi_1)\, \xi_3 + i r_{12} (\eta_3 \xi_2 - \eta_2 \xi_3),$$

where r_{12}, the distance between the first and second bodies, is a rational function of s, and i denotes $\sqrt{-1}$. It is not difficult to verify that

$$\sum_{r=1}^{9} y_r \frac{\partial \psi}{\partial x_r} = \psi \Omega,$$

where

$$\Omega = \eta_2 \frac{\xi_2 - \dfrac{i}{r_{12}} \xi_1 \xi_3}{\xi_2^{\,2} + \xi_3^{\,2}} + \eta_3 \frac{\xi_3 + \dfrac{i}{r_{12}} \xi_1 \xi_2}{\xi_2^{\,2} + \xi_3^{\,2}}$$

$$= \eta_2 \Omega_2 + \eta_3 \Omega_3,$$

---

* This was first pointed out by Poincaré, *Comptes Rendus*, t. cxxiii (1896), p. 1228, where he enunciated (without proof) the rule, which eliminates these solutions from the astronomical problem.

say. Then

$$\sum_{r=1}^{9} \omega_r dx_r = 0 \cdot d\xi_1 + \Omega_2 d\xi_2 + \Omega_3 d\xi_3,$$

manifestly not a perfect differential *.

In this particular instance, the quantity ψ_0 is complex: and therefore, in accordance with § 276, the quantity ϕ would be left unresolved, for the product of ψ_0 by the conjugate complex is $(\xi_2^2 + \xi_3^2) \phi$: and the case would not occur. We proceed to shew that all the possible instances, which can occur in the astronomical problem, are of the same character: viz. every function ψ_0 of the type under consideration, which involves the irrational variable s explicitly, the variables x rationally, is a polynomial in the variables y, and satisfies the equation

$$\frac{d\psi_0}{dt} = \omega\psi_0,$$

is complex, and (by § 276) is therefore merged in other functions.

280. The quantity ψ_0 is a homogeneous polynomial in the variables y; it is rational in the variables x and s, where s is given by the irreducible equation

$$F = s^n + S_1 s^{n-1} + \ldots + S_n = 0 ;$$

and it satisfies the equation

$$\sum_{r=1}^{m} y_r \frac{\partial \psi_0}{\partial x_r} = \Omega\psi_0.$$

When the n values of s (they are explicitly known for the astronomical problem) are substituted in ψ_0 in turn, the latter will acquire a number of distinct values, say $\psi_1, \psi_2, \ldots, \psi_l$, where l either is n or is a factor of n. For each of these values, we have

$$\frac{1}{\psi_a} \sum_{r=1}^{m} y_r \frac{\partial \psi_a}{\partial x_r} = \Omega_a,$$

where Ω_a is linear in the variables y. (In general, Ω will involve s, and the l values of Ω will be derivable from any one of them: but this need not always be the case.) Taking

$$\phi = \psi_1 \psi_2 \ldots \psi_l,$$

---

* The explanation of the difference from the case of § 278 is the occurrence of the irrational variable s.

so that ϕ, obviously a homogeneous polynomial in y, is a symmetric function of the values of ψ_0 for all the values of s, and therefore is a rational function of the variables x alone, we have

$$\frac{1}{\phi} \sum_{r=1}^{m} y_r \frac{\partial \phi}{\partial x_r} = \sum_{a=1}^{l} \left\{ \frac{1}{\psi_a} \sum_{r=1}^{m} \left(y_r \frac{\partial \psi_a}{\partial x_r} \right) \right\} = \sum_{a=1}^{l} \Omega_a = \Omega,$$

where Ω is linear in the variables y and, because ϕ is rational in the variables x, so also is Ω. Thus

$$\sum_{r=1}^{m} y_r \frac{\partial \phi}{\partial x_r} = \Omega \phi.$$

Now ϕ is a homogeneous polynomial in the variables y and is rational in the variables x; hence it is subject to the theorem established in § 278, that is, ϕ is a polynomial in the quantities

$$x_2 y_1 - x_1 y_2, \quad x_3 y_1 - x_1 y_3, \quad \ldots, \quad x_m y_1 - x_1 y_m,$$

and it is only through these quantities that the variables x enter into the expression for ϕ.

Thus ϕ certainly exists when ψ_0 exists. Conversely, when ϕ has been found, the various irresoluble real factors of ϕ will be the various values ψ_1, ψ_2, ..., ψ_l. Hence if all the functions ϕ of this type be determined, all the possible functions ψ_0 of the character under consideration will be derivable.

Accordingly let Φ denote such a function, irresoluble into factors of its own form which are rational in x alone, but resoluble into such factors, which are rational in x and s, and are real: the factors being the quantities ψ. These quantities ψ are changed into one another by appropriate changes of the roots s of $F = 0$ into one another: any two of them become the same, when corresponding roots s are equal. When two roots s are equal, there is a relation among the variables x_1, ..., x_m: so that, when this relation is satisfied, Φ contains a repeated factor. By assigning all possible equalities among the roots s in turn, we secure all possible repetitions* of the factors of Φ; all its factors will thus be considered in turn. They are obtained as follows.

281. Since $\Phi = 0$ is homogeneous in the variables y, we may divide throughout by any one of them, say by y_1: and, owing to

---

* It should be noted that an equality among the roots s does not necessarily make two factors of Φ equal: the equality might leave each factor unaltered.

the property proved at the end of § 278, the resulting expression is rational and integral in the variables y_r/y_1. But

$$\frac{y_r}{y_1} = \frac{dx_r}{dt} \div \frac{dx_1}{dt} = \frac{dx_r}{dx_1} = q_r,$$

say, so that the equation $\Phi = 0$ becomes

$$G\left(x_2 - x_1 q_2, \ \ldots, \ x_m - x_1 q_m, \ q_2, \ \ldots, \ q_m\right) = 0.$$

This equation may be regarded as expressing q_2 in terms of $q_3, \ldots, q_m$: and it has been seen that, for a certain relation among the variables x, it contains a repeated factor, so that the value of q_2 is repeated, or a set of values of q_2 is repeated; accordingly, these satisfy

$$\frac{\partial G}{\partial q_2} = 0.$$

But our function Φ is resoluble into factors that are homogeneous and integral in $y_1, \ldots, y_m$, that is, G is resoluble into factors that are integral in $q_2, \ldots, q_m$. Consider the repeated factor: equated to zero, it gives a repeated value of q_2 in terms of the rest, and therefore it gives a repeated value of q_r in terms of $q_2, \ldots, q_{r-1}, q_{r+1}, \ldots, q_m$, for all values of r. Hence when $\Sigma = 0$, the repeated factor in $G = 0$ is such that

$$\frac{\partial G}{\partial q_r} = 0, \qquad\qquad (r = 2, \ldots, m).$$

Moreover, when we determine $q_2, \ldots, q_m$ from the equations

$$\frac{\partial G}{\partial q_2} = 0, \ \ldots, \ \frac{\partial G}{\partial q_m} = 0,$$

as functions of $x_1, \ldots, x_m$ and substitute in $G = 0$, the latter (if originally resoluble in the manner supposed) must contain $\Sigma = 0$ or be $\Sigma = 0$, where $\Sigma = 0$ is the relation among the variables equivalent to that equality of roots s, which gives rise to the repeated factor in G. Accordingly, we have

$$G = A_2 \frac{\partial G}{\partial q_2} + \ldots + A_m \frac{\partial G}{\partial q_m} + B\Sigma,$$

from the general theory of elimination.

Take the instance of § 279 as an example. Let

$$\phi = (x_2 - x_1 q_2)^2 + (x_3 q_2 - x_2 q_3)^2 + (x_3 - x_1 q_3)^2 = 0,$$

which is resoluble into two factors linear in q_2 and q_3; the two factors become the same when

$$\Sigma = x_1{}^2 + x_2{}^2 + x_3{}^2 = 0.$$

It is easy to verify that

$$x_1^2\phi = (x_3q_2 - x_2q_3)^2 \Sigma - \tfrac{1}{2}\{x_1(x_2 - x_1q_2) + x_3(x_3q_2 - x_2q_3)\}\frac{\partial\phi}{\partial q_2}$$

$$- \tfrac{1}{2}\{x_1(x_3 - x_1q_3) - x_2(x_3q_2 - x_2q_3)\}\frac{\partial\phi}{\partial q_3}.$$

Using now the vocabulary of geometry of m dimensions, we have the equation of a line in the form

$$\frac{\xi_1 - x_1}{y_1} = \frac{\xi_2 - x_2}{y_2} = \ldots = \frac{\xi_m - x_m}{y_m}.$$

Hence the quantities

$$x_2 - x_1q_2, \ldots, x_m - x_1q_m, q_2, \ldots, q_m$$

specify a line through the point $x_1, \ldots, x_m$; and the equation $G = 0$ therefore represents an aggregate of lines. The lines through a point on $\Sigma = 0$, which correspond to the repeated factor, are such that

$$\frac{\partial G}{\partial q_2} = 0, \ldots, \frac{\partial G}{\partial q_m} = 0.$$

Taking a neighbouring point $x_1 + dx_1, x_2 + dx_2, \ldots, x_m + dx_m$, we have

$$d(x_r - x_1q_r) = q_r dx_1 - q_r dx_1 - x_1 dq_r$$

$$= - x_1 dq_r;$$

so that

$$dG = - \sum_{r=2}^{m} \frac{\partial G}{\partial x_r} x_1 dq_r + \sum_{r=2}^{n} \frac{\delta G}{\delta q_r} dq_r,$$

where $\dfrac{\delta G}{\delta q_r}$ denotes the partial derivative of G with respect to q_r owing to the occurrence of q_r outside the combinations $x_r - x_1q_r$. But

$$\frac{\partial G}{\partial q_r} = \frac{\delta G}{\delta q_r} + \frac{\partial G}{\partial x_r}\frac{\partial(x_r - x_1q_r)}{\partial q_r}$$

$$= \frac{\delta G}{\delta q_r} - x_1\frac{\partial G}{\partial x_r};$$

and therefore

$$dG = \sum_{r=2}^{m} \frac{\partial G}{\partial q_r} dq_r.$$

For the selected lines in question, all the quantities $\dfrac{\partial G}{\partial q_r}$ vanish:

and therefore for their directions through the point, we have $dG = 0$; that is, we have, for the directions of these lines,

$$0 = \Sigma d\left(A_r \frac{\partial G}{\partial q_r}\right) + Bd\Sigma + \Sigma dB.$$

We have seen that $\dfrac{\partial G}{\partial q_r} = 0$: when the direction-quantities are substituted in these equations, each is identically satisfied, so that

that is,
$$d\left(\frac{\partial G}{\partial q_r}\right) = 0,$$

$$d\left(A_r \frac{\partial G}{\partial q_r}\right) = 0.$$

Also the point $x_1, \ldots, x_m$ is on $\Sigma = 0$; and therefore we have $Bd\Sigma = 0$, that is, the directions of the selected lines through the point satisfy $d\Sigma = 0$, or all of them lie in the tangent-plane to the surface $\Sigma = 0$ at the point.

Hence the two (or more) factors of $\Phi = 0$, which become equal to one another when $\Sigma = 0$, represent lines, which are included among the tangent-lines to the surface $\Sigma = 0$. Similarly for another surface $\Sigma' = 0$, such that other factors of $\Phi = 0$ become repeated; the corresponding lines in $G = 0$ are included among the tangent-lines to $\Sigma' = 0$. By taking all the surfaces such as $\Sigma = 0$, we secure the consideration of every factor in Φ, (it may be that a factor is considered more than once), and we have a corresponding result. We thus infer the rule, due to Poincaré:

Take every surface $\Sigma = 0$, which is a locus of points where s has equal roots: and construct the aggregate of tangent-lines, say $T = 0$, to the surface. Then if T be resoluble, it includes those factors of Φ, which become the same for values of the variables x satisfying $\Sigma = 0$; and the complete set of the equations $T = 0$ certainly includes all the equations $\Phi = 0$.

It must not be assumed that each equation $T = 0$ is resoluble into factors: what has been proved is that, if an equation $\Phi = 0$ of the specified type exists, it can be obtained in the way indicated in the rule. All the possible cases will be obtained, when all the possible surfaces $\Sigma = 0$ and their aggregates of tangent-lines have been considered.

282. The equation
$$F(s, x_1, \ldots, x_n) = 0,$$
irresoluble as an equation in s, will have equal roots when
$\dfrac{\partial F}{\partial s} = 0$, that is, when its discriminant with regard to s vanishes,
say when
$$\Delta(x_1, \ldots, x_n) = 0.$$
This discriminant may be capable of resolution into a number of functions, each rational in the variables x and irresoluble; let these be
$$\Delta_1 = 0, \quad \Delta_2 = 0, \quad \ldots,$$
manifestly finite in number. Each of these last equations gives a critical equation, represented by $\Sigma = 0$ in the preceding investigation.

To obtain the equation, interpreted as the aggregate of tangents to the surface, we take any line $x_1 + ty_1$, $x_2 + ty_2$, $\ldots$, $x_m + ty_m$; where it meets a surface $\Delta = 0$, we have
$$\Delta(x_1 + ty_1,\ x_2 + ty_2,\ \ldots,\ x_m + ty_m) = 0.$$
If it is a tangent-line to the surface, two points of intersection are coincident, that is, t has two equal values. We therefore form the discriminant of Δ; it is *
$$\Phi = 0,$$
where Φ has the general significance in § 281.

As was remarked at the end of § 281, the quantity Φ, rational in the variables x and y, and homogeneous in the variables $\dot{y}$, is not necessarily resoluble into factors that are homogeneous in the variables y, and rational in x, y, s. If it should be so resoluble, then any such factor is a function ψ_0.

In the absence of the specific form of the equation
$$F(s, x_1, \ldots, x_n) = 0,$$
we are not in a position to discuss the functions ψ_0 on a general basis. All that has been proved is that, when
$$\frac{1}{\psi_0} \frac{d\psi_0}{dt} = \omega_1 y_1 + \ldots + \omega_m y_m,$$

---

* From the fact that the discriminant is given by $\dfrac{d\Delta}{dt} = 0$, that is,
$$y_1 \frac{\partial \Delta}{\partial x_1} + \ldots + y_n \frac{\partial \Delta}{\partial x_n} = 0,$$
we see how the form of Φ is recovered.

the expression $\omega_1 dx_1 + \ldots + \omega_m dx_m$ is not necessarily a perfect differential, for a special example to the contrary was shewn; but when the form $F = 0$ is made particular, not general, the condition may be satisfied, and an integral can then be deduced.

We proceed to consider the significance of the result as regards the astronomical problem.

283. Taking the case of the problem of three bodies, the equation $F = 0$ is

$$\prod_{a=1}^{8} (s - s_a) = 0,$$

where the eight roots s_a are the eight values of

$$\pm r_{12} \pm r_{23} \pm r_{31}.$$

There is thus the possibility of 28 equalities among the roots. Each of the three equations

$$r_{12} = 0, \quad r_{23} = 0, \quad r_{31} = 0,$$

provides four equalities; each of the three equations

$$r_{23} = r_{31}, \quad r_{31} = r_{12}, \quad r_{12} = r_{23},$$

provides two equalities; each of the three equations

$$r_{23} + r_{31} = 0, \quad r_{31} + r_{12} = 0, \quad r_{12} + r_{23} = 0,$$

provides two equalities; and each of the four equations

$$r_{12} \pm r_{23} \pm r_{31} = 0,$$

provides one equality: making up the total of 28. We consider the rationalised forms of these equations in turn.

As regards one of the first three, say $r_{12} = 0$, its rationalised form is

$$(x_1 - x_4)^2 + (x_2 - x_5)^2 + (x_3 - x_6)^2 = 0,$$

say

$$\xi_1^2 + \xi_2^2 + \xi_3^2 = 0,$$

in the notation of § 279. We have to give equal roots to t in

$$(\xi_1 + \eta_1 t)^2 + (\xi_2 + \eta_2 t)^2 + (\xi_3 + \eta_3 t)^2 = 0,$$

so that

$$(\xi_1^2 + \xi_2^2 + \xi_3^2)(\eta_1^2 + \eta_2^2 + \eta_3^2) = (\xi_1 \eta_1 + \xi_2 \eta_2 + \xi_3 \eta_3)^2.$$

Thus

$$\Phi = (\xi_1 \eta_2 - \xi_2 \eta_1)^2 + (\xi_2 \eta_3 - \xi_3 \eta_2)^2 + (\xi_3 \eta_1 - \xi_1 \eta_3)^2.$$

Now Φ is resoluble into two factors (§ 279), linear in η_1, η_2, η_3; but they are conjugate complex factors. As they are not real, they are to be rejected (§ 276) as a possible basis for an integral. Hence no function ψ_0, with the assigned properties, arises in connection with the equation $r_{12} = 0$.

The same result holds for each of the other two equations $r_{23} = 0$, $r_{31} = 0$, of the first set.

As regards one of the second three, say $r_{12} = r_{31}$, its rationalised form is

$$(x_1 - x_4)^2 + (x_2 - x_5)^2 + (x_3 - x_6)^2 = (x_1 - x_7)^2 + (x_2 - x_8)^2 + (x_3 - x_9)^2,$$

say

$$\xi_1^2 + \xi_2^2 + \xi_3^2 = \xi_4^2 + \xi_5^2 + \xi_6^2,$$

which also is the rational form of one of the third three, viz. $r_{12} + r_{31} = 0$. We are to give equal roots to t in

$$(\xi_1 + \eta_1 t)^2 + (\xi_2 + \eta_2 t)^2 + (\xi_3 + \eta_3 t)^2 = (\xi_4 + \eta_4 t)^2 + (\xi_5 + \eta_5 t)^2 + (\xi_6 + \eta_6 t)^2,$$

so that

$$\left\{ \sum_{r=1}^{r=3} \xi_r^2 - \sum_{q=4}^{q=6} \xi_q^2 \right\} \left\{ \sum_{r=1}^{r=3} \eta_r^2 - \sum_{q=4}^{q=6} \eta_q^2 \right\} = \left\{ \sum_{r=1}^{r=3} \xi_r \eta_r - \sum_{q=4}^{q=6} \xi_q \eta_q \right\}^2,$$

that is,

$$\Phi = \left\{ \sum_{r=1}^{r=3} \xi_r^2 - \sum_{q=4}^{q=6} \xi_q^2 \right\} \left\{ \sum_{r=1}^{r=3} \eta_r^2 - \sum_{q=4}^{q=6} \eta_q^2 \right\} - \left\{ \sum_{r=1}^{r=3} \xi_r \eta_r - \sum_{q=4}^{q=6} \xi_q \eta_q \right\}^2.$$

This expression Φ, of the second order in the quantities η, is not resoluble into linear factors[*]; hence no quantities ψ_0, with the assigned properties, arise through the equations $r_{12} \pm r_{31} = 0$.

Similarly no quantities ψ_0 arise through either of the other two equations of the second set, and none through either of the other two equations of the third set.

---

[*] This may be seen by taking $\Phi = 0$, regarding it as a quadratic in η_1, and solving the quadratic. If Φ were resoluble, being say

$$A\eta_1^2 + 2B\eta_1 + C = 0,$$

the radical $(B^2 - AC)^{\frac{1}{2}}$ would be a linear function of the other quantities η: that this may be the case, $B^2 - AC$, qua function of those quantities η, must be a perfect square. That this is not so, follows from considering the terms in η_2^2, η_3^2, $\eta_2 \eta_3$; they are

$$- \eta_2^2 (\xi_3^2 - r^2{}_{13})(r^2{}_{12} - r^2{}_{13}) - \eta_3^2 (\xi_2^2 - r^2{}_{13})(r^2{}_{12} - r^2{}_{13}) + 2\xi_2 \xi_3 \eta_2 \eta_3 (r^2{}_{12} - r^2{}_{13}),$$

or, dropping the factor $r^2{}_{13} - r^2{}_{12}$, they are

$$\eta_2^2 (\xi_3^2 - r^2{}_{13}) + \eta_3^2 (\xi_2^2 - r^2{}_{13}) - 2\xi_2 \xi_3 \eta_2 \eta_3,$$

which is not the perfect square of any linear function of η_2 and η_3.

As regards the last set of four equations, they have each the same rationalised form, viz.

$$(r^2_{12} - r^2_{13} + r^2_{23})^2 - 4r^2_{12} r^2_{23} = 0.$$

In the simplified form of this when $r_{23} = 0$, we fall back upon the last case: the deduced quantity Φ for that simpler form is not resoluble. *A fortiori*, when r_{23} is not zero, the deduced quantity Φ cannot be resoluble.

Hence it follows that, in the problem of three bodies, the critical equations $\Sigma = 0$ give rise to quantities Φ, which are either irresoluble or, if resoluble, give rise to complex factors. No function ψ_0, of the kind indicated at the beginning of § 279, exists in this case.

284. Similarly for the problem of n bodies. The only instances, in which the quantities Φ deduced from critical equations $\Sigma = 0$ prove resoluble, are those coming through equations

$$r^2_{\lambda\mu} = 0,$$

which correspond to the instances $r^2_{12} = 0$, $r^2_{23} = 0$, $r^2_{31} = 0$ in the problem of three bodies; the resolved factors are conjugate complex quantities, and the corresponding functions ψ_0 are to be rejected. All other instances give rise to quantities Φ which are irresoluble; and so there arise no functions ψ_0 for consideration.

SUMMARY OF RESULTS.

285. At this stage we shall now begin the detailed consideration of the equations in the astronomical problem. Except for the form of the equation connected with the irrationality s, the differential equations have been of a general type: and it remains to be seen how far the particular form of the equations may affect the results thus far attained. These results may be summarised as follows, enunciated in connection with the equations

$$\frac{dx_r}{dt} = y_r, \quad \frac{dy_r}{dt} = A_r(x_1, \dots, x_m), \qquad (r = 1, \dots, m),$$

(which are to be identified with the differential equations of the problem of n bodies), and with the equation that defines the irrational quantity in that problem.

Any algebraic integral of the differential equations, which does not involve t, can be derived by purely algebraical operations from integrals ϕ, characterised by the following properties :

(i)　　Each integral ϕ is a polynomial in y, the coefficients of the powers being rational functions of x and s; and no parametric constants occur in the expression for ϕ.

(ii)　　The integral ϕ remains unchanged, save as to a factor which is a power of κ, by the substitutions

$$x' = x\kappa^2, \quad s' = s\kappa^2, \quad y' = y\kappa^{1+2N} :$$

such integrals being called *homogeneous*.

(iii)　　When the aggregates of terms in ϕ, which are of the same dimensions in the variables y, are gathered together, ϕ can be expressed in the form

$$\phi_0 + \phi_2 + \phi_4 + \ldots,$$

where ϕ_2 is of dimensions less by two than ϕ_0, ϕ_4 of dimensions less by two than ϕ_2, and so on, in the variables y.

(iv)　　The aggregate ϕ_0 of terms in ϕ of the highest order in the variables y, is rational in the variables x and does not explicitly involve s; it is a polynomial in the quantities

$$x_2 y_1 - x_1 y_2, \quad x_3 y_1 - x_1 y_3, \quad \ldots, \quad x_m y_1 - x_1 y_m,$$

the coefficients in the polynomial being rational functions of y which, except for a power of y_1 as a possible denominator, are also integral functions of the variables y.

EQUATIONS OF THE ASTRONOMICAL PROBLEM.

286.　In connection with the differential equations as set out in § 264, it is convenient to use the symbols

$$\left. \begin{aligned} \frac{\partial}{\partial u} &= \sum_{a=1}^{n} \left(X_a \frac{\partial}{\partial x_a} + Y_a \frac{\partial}{\partial y_a} + Z_a \frac{\partial}{\partial z_a} \right) \\ \frac{\partial}{\partial v} &= \sum_{a=1}^{n} \left(A_a \frac{\partial}{\partial X_a} + B_a \frac{\partial}{\partial Y_a} + C_a \frac{\partial}{\partial Z_a} \right) \end{aligned} \right\},$$

so that the effect of $\frac{\partial}{\partial u}$ upon any quantity is to increase its order in the variables X, Y, Z by unity, and the effect of $\frac{\partial}{\partial v}$ is to

decrease that order by unity. We are concerned with homogeneous integrals ϕ, which are polynomials in X, Y, Z, and are
rational in the variables x and s; when ϕ is arranged in the form

$$\phi = \phi_0 + \phi_2 + \phi_4 + \dots,$$

where ϕ_0 is the aggregate of terms of the highest order in X, Y, Z,
then ϕ_0 is a polynomial in the variables x. It will be proved that
ϕ can be compounded from the classical integrals.

Since ϕ is an integral of the differential equations, we have

$$\frac{d\phi}{dt} = \frac{\partial\phi}{\partial u} + \frac{\partial\phi}{\partial v} = 0.$$

Arranging this equation, so that terms of the same dimensions in
the variables X, Y, Z are grouped together, we have

$$\frac{\partial\phi_0}{\partial u} + \left(\frac{\partial\phi_0}{\partial v} + \frac{\partial\phi_2}{\partial u}\right) + \left(\frac{\partial\phi_2}{\partial v} + \frac{\partial\phi_4}{\partial u}\right) + \dots = 0,$$

which must be identically satisfied. Hence

$$\frac{\partial\phi_0}{\partial u} = 0, \quad \frac{\partial\phi_0}{\partial v} + \frac{\partial\phi_2}{\partial u} = 0, \dots$$

Our immediate aim is the determination of the form of ϕ_0; it
will be found that the first two of these equations suffice for
the purpose.

The first equation is

$$\sum_{a=1}^{n} \left(X_a \frac{\partial\phi_0}{\partial x_a} + Y_a \frac{\partial\phi_0}{\partial y_a} + Z_a \frac{\partial\phi_0}{\partial z_a}\right) = 0,$$

a partial differential equation of the first order. The equations
subsidiary to the solution are

$$\frac{dx_1}{X_1} = \dots = \frac{dx_a}{X_a} = \frac{dy_a}{Y_a} = \frac{dz_a}{Z_a} = \dots = \frac{dz_n}{Z_n}$$

$$= \frac{dX_1}{0} = \dots = \frac{dX_a}{0} = \frac{dY_a}{0} = \frac{dZ_a}{0} = \dots = \frac{dZ_n}{0}.$$

The necessary $6n - 1$ integrals of this system are furnished by

$$X_1, Y_1, Z_1; \ X_2, Y_2, Z_2; \ \dots; \ X_n, Y_n, Z_n;$$

$$f_a = x_a X_1 - x_1 X_a, \quad \text{for } \alpha = 2, 3, \dots, n;$$

$$\left.\begin{array}{l} g_a = y_a X_1 - x_1 Y_a \\ h_a = z_a X_1 - x_1 Z_a \end{array}\right\}, \text{ for } \alpha = 1, 2, \dots, n.$$

From § 285, it follows that ϕ_0 is expressible as a polynomial in the variables f, g, h; and that, except possibly as to a power of X_1 for a denominator, the coefficients of the various powers of f, g, h in ϕ_0 are polynomials in X, Y, Z. When the form of ϕ_0 as a function of the variables x, y, z, X, Y, Z is known, this modified expression of ϕ_0 is obtained, by substituting for the variables x their values

$$x_r = \frac{f_r}{X_1} + x_1\frac{X_r}{X_1}, \quad y_r = \frac{g_r}{X_1} + x_1\frac{Y_r}{X_1}, \quad z_r = \frac{h_r}{X_1} + x_1\frac{Z_r}{X_1};$$

the variable x_1 then disappears from ϕ_0.

Let these same substitutions for the variables x be made in ϕ_2, which does not necessarily satisfy the equation $\frac{\partial\phi_2}{\partial u} = 0$. The variable x_1 does not then necessarily disappear from ϕ_2, as it does from ϕ_0; and we have

$$\frac{\partial\phi_2}{\partial u} = X_1\frac{\partial\phi_2}{\partial x_1}.$$

Thus

$$X_1\frac{\partial\phi_2}{\partial x_1} + \sum_{a=1}^{n}\left(A_a\frac{\partial\phi_0}{\partial X_a} + B_a\frac{\partial\phi_0}{\partial Y_a} + C_a\frac{\partial\phi_0}{\partial Z_a}\right) = 0,$$

where, in the terms to be summed, all the quantities are to be expressed in terms of the $3n$ variables $x_1, g_1, h_1; f_2, g_2, h_2; \dots;$ f_n, g_n, h_n; and the $3n$ variables X, Y, Z. Denoting the result by U, we have

$$X_1\frac{\partial\phi_2}{\partial x_1} + U = 0,$$

and therefore

$$\phi_2 = -\int\frac{U}{X_1}dx_1 + \phi_2',$$

where ϕ_2' is a solution of the equation

$$\frac{\partial\phi_2'}{\partial u} = 0.$$

Now ϕ_2 is a polynomial in X, Y, Z; and $\int\frac{U}{X_1}dx_1$ is certainly a polynomial in those variables; hence ϕ_2' must also be a polynomial in X, Y, Z, of the same order in them as ϕ_2, that is, of order less by two than ϕ_0. Moreover, ϕ_2' is a solution of the same equation as ϕ_0; so that we can regard ϕ_2' as an initial set of terms of highest order for a new function ϕ', that highest order being two less than in ϕ. Thus ϕ' is of the same character as ϕ; whatever is proved as to generic character of ϕ will hold also for ϕ'; and so there will be no loss of generality as regards ϕ if we neglect ϕ', in effect, if we take $\phi_2' = 0$. Then

$$\phi_2 = -\int \frac{U}{X_1}\, dx_1.$$

Now ϕ_2 is rational in the variables x, y, z, s; and therefore the subject of integration must be such as to render the integral a rational function of those variables.

287. Changing the variables in

$$U = \sum_{a=1}^{n}\left(A_a \frac{\partial \phi_0}{\partial X_a} + B_a \frac{\partial \phi_0}{\partial Y_a} + C_a \frac{\partial \phi_0}{\partial Z_a}\right)$$

from x_1, y_1, z_1; ...; x_n, y_n, z_n; to x_1, g_1, h_1; f_2, g_2, h_2; ...; f_n, g_n, h_n; we have a new expression for U in the form

$$U = \sum_{a=1}^{n}\left(A_a \frac{\partial \phi_0}{\partial X_a} + B_a \frac{\partial \phi_0}{\partial Y_a} + C_a \frac{\partial \phi_0}{\partial Z_a}\right)$$
$$+ \frac{\partial \phi_0}{\partial g_1}(y_1 A_1 - x_1 B_1) + \frac{\partial \phi_0}{\partial h_1}(z_1 A_1 - x_1 C_1)$$
$$+ \sum_{\lambda=2}^{n}\left\{\frac{\partial \phi_0}{\partial f_\lambda}(x_\lambda A_1 - x_1 A_\lambda) + \frac{\partial \phi_0}{\partial g_\lambda}(y_\lambda A_1 - x_1 B_\lambda) + \frac{\partial \phi_0}{\partial h_\lambda}(z_\lambda A_1 - x_1 C_\lambda)\right\},$$

where ϕ_0 is now supposed to be a function of the new variables f, g, h and of X, Y, Z.

When the values of A, B, C are substituted, the expression for U consists of a sum of terms, each of which is fractional; the denominator in each is of the form r^3_{kl}, the integers k and l changing from term to term. We select those which have $r^3_{1\rho}$ as their denominator; let their aggregate be denoted by $U_{1\rho} \div r^3_{1\rho}$. We also select those which have $r^3_{\sigma\tau}$ as their denominator; let their aggregate be denoted by $U_{\sigma\tau} \div r^3_{\sigma\tau}$. Clearly, we have

$$\rho = 2,\ 3,\ \dots,\ n; \quad \sigma,\ \tau = 2,\ 3,\ \dots,\ n.$$

It appears that, in the quantities $U_{1\rho}$ and $U_{\sigma\tau}$, certain combinations of terms occur that are symmetric in x, y, z; or in X, Y, Z; or in f, g, h; or in sets of these variables. To abbreviate the formulæ, a symbol S will be used, with the definition

$$SF(x_\lambda, f_\lambda, X_\lambda) = F(x_\lambda, f_\lambda, X_\lambda) + F(y_\lambda, g_\lambda, Y_\lambda) + F(z_\lambda, h_\lambda, Z_\lambda),$$

whatever be the function F; the symbol, in fact, expresses symmetrical summation for three variables, the sole exception being as regards g_1, h_1, because there is no variable f_1. With this understanding, we have

$$U_{1\rho} = m_\rho \left[\frac{\partial \phi_0}{\partial g_1} \{ y_1 (x_\rho - x_1) - x_1 (y_\rho - y_1) \} + \frac{\partial \phi_0}{\partial h_1} \{ z_1 (x_\rho - x_1) - x_1 (z_\rho - z_1) \} \right]$$

$$+ S \left\{ (x_\rho - x_1) \left(m_\rho \frac{\partial \phi}{\partial X_1} - m_1 \frac{\partial \phi}{\partial X_\rho} \right) \right\} + m_1 x_1 S (x_\rho - x_1) \frac{\partial \phi_0}{\partial f_\rho}$$

$$+ m_\rho (x_\rho - x_1) \sum_{\lambda=2}^{n} S x_\lambda \frac{\partial \phi_0}{\partial f_\lambda}$$

$$= \Theta_{1\rho} x_1^2 + \Phi_{1\rho} x_1 + \Psi_{1\rho},$$

on further substitution for the variables x, y, z in terms of f, g, h. The coefficient $\Theta_{1\rho}$ is given by

$$X_1^2 \Theta_{1\rho} = m_\rho \left\{ \frac{\partial \phi_0}{\partial g_1} (Y_1 X_\rho - X_1 Y_\rho) + \frac{\partial \phi_0}{\partial h_1} (Z_1 X_\rho - X_1 Z_\rho) \right\}$$

$$+ m_\rho (X_\rho - X_1) \sum_{\lambda=2}^{n} S X_\lambda \frac{\partial \phi_0}{\partial f_\lambda} + m_1 X_1 S (X_\rho - X_1) \frac{\partial \phi_0}{\partial f_\rho} ;$$

the explicit values of the coefficients $\Phi_{1\rho}$, $\Psi_{1\rho}$ are not required.

Similarly we have

$$U_{\sigma\tau} = S (x_\tau - x_\sigma) \left(m_\tau \frac{\partial \phi_0}{\partial X_\sigma} - m_\sigma \frac{\partial \phi_0}{\partial X_\tau} \right) - x_1 S (x_\tau - x_\sigma) \left(m_\tau \frac{\partial \phi_0}{\partial f_\sigma} - m_\sigma \frac{\partial \phi_0}{\partial f_\tau} \right)$$

$$= \Theta_{\sigma\tau} x_1^2 + \Phi_{\sigma\tau} x_1 + \Psi_{\sigma\tau},$$

where

$$X_1 \Theta_{\sigma\tau} = - S (X_\tau - X_\sigma) \left(m_\tau \frac{\partial \phi_0}{\partial f_\sigma} - m_\sigma \frac{\partial \phi_0}{\partial f_\tau} \right);$$

the explicit values of the coefficients $\Phi_{\sigma\tau}$, $\Psi_{\sigma\tau}$ are not required.

Now taking $f_{\alpha\beta}$ to denote $f_\alpha - f_\beta$ (where $f_{\alpha1} = f_\alpha$), $X_{\alpha\beta}$ to denote $X_\alpha - X_\beta$, and so for the other variables, we have

$$x_\rho - x_1 = \frac{1}{X_1} (f_{\rho 1} + x_1 X_{\rho 1}),$$

$$x_\sigma - x_\tau = \frac{1}{X_1} (f_{\sigma\tau} + x_1 X_{\sigma\tau});$$

so that, for our purpose,

$$\left. \begin{array}{l} r^2_{1\rho} = a_{1\rho} x_1^2 + 2 b_{1\rho} x_1 + c_{1\rho} \\ r^2_{\sigma\tau} = a_{\sigma\tau} x_1^2 + 2 b_{\sigma\tau} x_1 + c_{\sigma\tau} \end{array} \right\},$$

where the coefficients a, b, c do not involve x_1. Thus

$$U = \sum_{\rho=2}^{n} \frac{\Theta_{1\rho} x_1^2 + \Phi_{1\rho} x_1 + \Psi_{1\rho}}{(a_{1\rho} x_1^2 + 2b_{1\rho} x_1 + c_{1\rho})^{\frac{3}{2}}} + \sum_{\rho, \sigma = 2, 3, \ldots, n} \frac{\Theta_{\rho\sigma} x_1^2 + \Phi_{\rho\sigma} x_1 + \Psi_{\rho\sigma}}{(a_{\rho\sigma} x_1^2 + 2b_{\rho\sigma} x_1 + c_{\rho\sigma})^{\frac{3}{2}}}.$$

But we have

$$\phi_2 = -\int \frac{U}{X_1} dx_1,$$

all the quantities except x_1 being constant in the integration; so that the right-hand side is the integral of a number of terms of the form

$$\int \frac{\theta_0 x_1^2 + \theta_1 x_1 + \theta_2}{(ax_1^2 + 2bx_1 + c)^{\frac{3}{2}}} dx_1.$$

This integral contains a logarithmic term

$$\frac{\theta_0}{a^{\frac{3}{2}}} \log \left\{ x_1 + \frac{b}{a} + \left(x_1^2 + 2\frac{b}{a} x_1 + \frac{c}{a} \right)^{\frac{1}{2}} \right\};$$

and ϕ_2 is to be a rational function in the variables x and s, so that it cannot contain a logarithmic term. Hence each logarithmic term must disappear from the expression for ϕ_2: a condition that can be satisfied, only if

$$\theta_0 = 0$$

in every case. We therefore have

$$\Theta_{1\rho} = 0, \quad \Theta_{\sigma\tau} = 0,$$

for ρ, σ, $\tau = 2, 3, \ldots, n$.

Form of the Leading Aggregate of an Integral.

288. The conditions $\Theta_{1\rho} = 0$, $\Theta_{\sigma\tau} = 0$, impose limitations upon the form of ϕ_0: and manifestly they are a set of simultaneous partial differential equations, satisfied by ϕ_0 as the dependent variable. Since all the derivatives of ϕ_0, which occur in them, are taken with regard to f, g, h, and since the coefficients of all those derivatives involve only the variables X, Y, Z but not f, g, h, it follows that the Jacobian differential relations of co-existence

$$(\Theta_{1\rho}, \Theta_{\sigma\tau}) = 0$$

are satisfied identically. Hence the aggregate of the equations either constitutes a complete system or contains a complete system: to settle which of the alternatives is valid, it is sufficient to investigate how far the equations are independent of one another in linear algebraical combinations.

For this purpose, let

$$\frac{1}{m_\sigma} \frac{\partial \phi_0}{\partial f_\sigma} = F_\sigma, \text{ for } \sigma = 2, \ldots, n;$$

$$\frac{1}{m_\sigma} \frac{\partial \phi_0}{\partial g_\sigma} = G_\sigma, \quad \frac{1}{m_\sigma} \frac{\partial \phi_0}{\partial h_\sigma} = H_\sigma, \text{ for } \sigma = 1, 2, \ldots, n;$$

and let

$$F_\sigma - F_\tau = F_{\sigma\tau}, \quad X_\sigma - X_\tau = X_{\sigma\tau},$$

with corresponding symbols for G, H, Y, Z. Then the equations $\Theta_{\sigma\tau} = 0$ become

$$X_{\sigma\tau} F_{\sigma\tau} + Y_{\sigma\tau} G_{\sigma\tau} + Z_{\sigma\tau} H_{\sigma\tau} = 0,$$

holding for $\sigma, \tau = 2, 3, \ldots, n$.

A typical equation of the set $\Theta_{1\rho} = 0$ is

$$G_1 (Y_1 X_\rho - X_1 Y_\rho) + H_1 (Z_1 X_\rho - X_1 Z_\rho) + (X_\rho - X_1) \Upsilon$$
$$+ X_1 \{ (X_\rho - X_1) F_\rho + (Y_\rho - Y_1) G_\rho + (Z_\rho - Z_1) H_\rho \} = 0,$$

where

$$\Upsilon = \frac{1}{m_1} \sum_{\lambda=2}^{n} SX_\lambda \frac{\partial \phi_0}{\partial f_\lambda}$$
$$= \sum_{\lambda=2}^{n} \frac{m_\lambda}{m_1} (X_\lambda F_\lambda + Y_\lambda G_\lambda + Z_\lambda H_\lambda).$$

There is no quantity f_1 among the variables $f, g, h,$ and therefore there is no derivative $\frac{1}{m_0} \frac{\partial \phi_0}{\partial f_1}$; that is, the symbol F_1 does not occur among the set F, G, H. We introduce F_1 to denote a subsidiary quantity, defined by the equation

$$\Upsilon + X_1 F_1 + Y_1 G_1 + Z_1 H_1 = 0,$$

so that, in fact, it has the symmetric form

$$\sum_{\lambda=1}^{n} m_\lambda (X_\lambda F_\lambda + Y_\lambda G_\lambda + Z_\lambda H_\lambda) = 0,$$

on multiplication by m_1: and we use the symbol $F_{\rho 1}$ to denote $F_\rho - F_1$. Substituting $-(X_1 F_1 + Y_1 G_1 + Z_1 H_1)$ for Υ in the equation $\Theta_{1\rho} = 0$, we find

$$- X_1 \{ (X_\rho - X_1) F_1 + (Y_\rho - Y_1) G_1 + (Z_\rho - Z_1) H_1 \}$$
$$+ X_1 \{ (X_\rho - X_1) F_\rho + (Y_\rho - Y_1) G_\rho + (Z_\rho - Z_1) H_\rho \} = 0;$$

that is, on removing the factor X_1, the equation is

$$X_{\rho 1} F_{\rho 1} + Y_{\rho 1} G_{\rho 1} + Z_{\rho 1} H_{\rho 1} = 0,$$

for $\rho = 2, 3, \ldots, n$.

The whole system of equations satisfied by ϕ_0 thus is

$$X_{\lambda\mu}F_{\lambda\mu} + Y_{\lambda\mu}G_{\lambda\mu} + Z_{\lambda\mu}H_{\lambda\mu} = 0,$$

for $\lambda,\ \mu = 1,\ 2,\ \ldots,\ n$; we are required to find how many of them are linearly independent of one another, or, what is the same thing, we are required to select a set, in terms of which every other equation of the system is linearly expressible.

It is clear that the equations

$$S\,X_{21}F_{21} = 0, \quad S\,X_{31}F_{31} = 0,$$

are independent of one another. Since $X_{32} = X_{31} - X_{21},\ F_{32} = F_{31} - F_{21}$, we have

$$0 = S\,X_{32}F_{32}$$
$$= S\,X_{31}F_{31} + S\,X_{21}F_{21} - S\,X_{21}F_{31} - S\,X_{31}F_{21},$$

that is,

$$S\,X_{21}F_{31} + S\,X_{31}F_{21} = 0,$$

which is clearly independent of the first two. Treating the three equations for $\lambda = \lambda,\ \mu = 1,\ 2,\ 3$, in the same way as the equation $\lambda,\ \mu = 2,\ 3$ has been treated, we find

$$\left.\begin{array}{l} X_{\lambda 1}F_{\lambda 1} + Y_{\lambda 1}G_{\lambda 1} + Z_{\lambda 1}H_{\lambda 1} = 0 \\[4pt] X_{31}F_{\lambda 1} + Y_{31}G_{\lambda 1} + Z_{31}H_{\lambda 1} = -(X_{\lambda 1}F_{31} + Y_{\lambda 1}G_{31} + Z_{\lambda 1}H_{31}) \\[4pt] X_{21}F_{\lambda 1} + Y_{21}G_{\lambda 1} + Z_{21}H_{\lambda 1} = -(X_{\lambda 1}F_{21} + Y_{\lambda 1}G_{21} + Z_{\lambda 1}H_{21}) \end{array}\right\},$$

which potentially express $F_{\lambda 1},\ G_{\lambda 1},\ H_{\lambda 1}$ in terms of $F_{21},\ G_{21},\ H_{21},\ F_{31},\ G_{31},\ H_{31}$. The actual values of $F_{\lambda 1},\ G_{\lambda 1},\ H_{\lambda 1}$ are

$$\left.\begin{array}{rl} F_{\lambda 1} = & \xi Y_{\lambda 1} - \eta Z_{\lambda 1} \\[4pt] G_{\lambda 1} = -\xi X_{\lambda 1} & + \zeta Z_{\lambda 1} \\[4pt] H_{\lambda 1} = & \eta X_{\lambda 1} - \zeta Y_{\lambda 1} \end{array}\right\},$$

provided the quantities $\xi,\ \eta,\ \zeta$ satisfy the six equations

$$\left.\begin{array}{l} -Y_{31}\xi + Z_{31}\eta = -F_{31} \\[4pt] X_{31}\xi \qquad\ - Z_{31}\zeta = -G_{31} \\[4pt] -X_{31}\eta + Y_{31}\zeta = -H_{31} \end{array}\right\}, \qquad \left.\begin{array}{l} -Y_{21}\xi + Z_{21}\eta = -F_{21} \\[4pt] X_{21}\xi \qquad\ - Z_{21}\zeta = -G_{21} \\[4pt] -X_{21}\eta + Y_{21}\zeta = -H_{21} \end{array}\right\}.$$

These six equations are equivalent to three only, in virtue of

$$S\,X_{21}F_{21} = 0, \quad S\,X_{31}F_{31} = 0, \quad S\,X_{21}F_{31} + S\,X_{31}F_{21} = 0,$$

which are satisfied; hence $\xi,\ \eta,\ \zeta$ are uniquely determinate.

We now construct a set of equations by retaining those, which arise for

$$\lambda, \mu = 1, 2, 3, \text{ being three,}$$

$$\lambda = 4, 5, \ldots, n, \ \mu = 1, 2, 3, \text{ being } 3\,(n - 3).$$

This set contains $3n - 6$ equations: and no one of the set is a linear combination of any of the rest. Taking any other equation of the full original set, we have

$$\begin{aligned} S\,X_{\rho\sigma}F_{\rho\sigma} &= S\,(X_{\rho 1} - X_{\sigma 1})\,(F_{\rho 1} - F_{\sigma 1}) \\ &= S\,X_{\rho 1}F_{\rho 1} + S\,X_{\sigma 1}F_{\sigma 1} - S\,(X_{\sigma 1}F_{\rho 1} + X_{\rho 1}F_{\sigma 1}) \\ &= -\,S\,(X_{\sigma 1}F_{\rho 1} + X_{\rho 1}F_{\sigma 1}), \end{aligned}$$

in virtue of members of the retained set. Substituting the values of $F_{\rho 1}$, $G_{\rho 1}$, $H_{\rho 1}$, $F_{\sigma 1}$, $G_{\sigma 1}$, $H_{\sigma 1}$ as deduced from the retained set, we have the right-hand side equal to

$$-\,\xi\,(X_{\sigma 1}Y_{\rho 1} + X_{\rho 1}Y_{\sigma 1} - Y_{\sigma 1}X_{\rho 1} - Y_{\rho 1}X_{\sigma 1}) + \text{two similar terms}$$

$$= \xi\,.\,0 + \eta\,.\,0 + \zeta\,.\,0 = 0\,;$$

that is, the equation

$$S\,X_{\rho\sigma}F_{\rho\sigma} = 0$$

is satisfied, in virtue of the retained set. Hence the original system of equations is equivalent to a set of $3n - 6$ equations linearly independent of one another.

Taking now the set of equations thus retained, let F_1 be removed from them by the relation

$$\sum_{\lambda=1}^{n} m_\lambda\,(X_\lambda F_\lambda + Y_\lambda G_\lambda + Z_\lambda H_\lambda) = 0\,;$$

they become once more a set of partial differential equations determining ϕ_0. They are linearly independent of one another, and the Jacobian conditions of coexistence are satisfied; they therefore form a complete system, and the number of equations in the system is $3n - 6$. Now the total number of independent variables in that system of partial differential equations is

$$3n - 1, \text{ on account of } f,\ g,\ h,$$

$$+\,3n, \text{ on account of } X,\ Y,\ Z,$$

$= 6n - 1$ in all; consequently[*] the number of independent solutions is $6n - 1 - (3n - 6)$, that is, $3n + 5$. In terms of these $3n + 5$ solutions, any other can be expressed; that is, our quantity ϕ_0 is thus expressible, when the $3n + 5$ solutions are known.

* See Part I. of this work, § 38.

The retained equations involve derivatives with regard to f, g, h, but not with regard to X, Y, Z; it follows that $3n$ independent solutions are

$$X_1, Y_1, Z_1; \ldots; X_n, Y_n, Z_n.$$

Therefore five other solutions are required; they must involve the variables f, g, h. According to their form, will be the form of ϕ_0 when expressed in terms of them; as ϕ_0 is a polynomial in f, g, h, it is natural to seek for solutions which are linear in those variables, and therefore are linear in the variables x, y, z. But the classical integrals of our problem provide (§ 264) five integrals of the differential equations which have this character; and therefore the terms, of highest order in the variables X, Y, Z in these integrals, give five possible solutions of the equations for ϕ_0. As a matter of fact, the terms of highest order (it is unity in each case) constitute the whole of the integral; and thus five other solutions are provided by

$$A, B, C, A', B', C',$$

subject to the relation

$$L'A' + M'B' + N'C' = 0.$$

Thus ϕ_0 is expressible in terms of X, Y, Z, and of any five (or all six) of the quantities A, B, C, A', B', C'. We proceed to prove that

$$\phi_0 = G(A, B, C, A', B', C', X_1, Y_1, Z_1, \ldots, X_n, Y_n, Z_n),$$

where G is polynomial in its arguments.

289. In § 286 it was proved that ϕ_0 is a polynomial in the variables f, g, h, the coefficients in the polynomial being themselves polynomials in X, Y, Z, except for a possible power of X_1 as a denominator. We have

$$X_1 A = X_1 \sum_{a=1}^{n} m_a(y_a Z_a - z_a Y_a) = \sum_{a=1}^{n} m_a(g_a Z_a - h_a Y_a),$$

$$X_1 B = X_1 \sum_{a=1}^{n} m_a(z_a X_a - x_a Z_a) = \sum_{a=1}^{n} m_a(h_a X_a - f_a Z_a),$$

$$X_1 C = X_1 \sum_{a=1}^{n} m_a(x_a Y_a - y_a X_a) = \sum_{a=1}^{n} m_a(f_a Y_a - g_a X_a),$$

$$X_1 A' = N' \sum m_a g_a - M' \sum m_a h_a,$$

$$X_1 B' = L' \sum m_a h_a - N' \sum m_a f_a,$$

$$X_1 C' = M' \sum m_a f_a - L' \sum m_a g_a:$$

with the restriction that there is no variable f_1, which accordingly will be taken as zero in these expressions. If it is possible to express any five of the variables f, g, h (say g_1, h_1, f_2, g_2, h_2), in terms of five of the six quantities A, B, C, A', B', C' (say A, B, C, B', C'), and of the remaining variables f, g, h; and if the expressions are substituted in ϕ_0; then all the other variables f, g, h must disappear from the result, because ϕ_0 is expressible in terms of X, Y, Z, A, B, C, B', C' alone (§ 288).

All that is necessary to secure the possibility of such expression is, that the determinant of the coefficients of the five variables g_1, h_1, f_2, g_2, h_2 in the five equations shall not vanish. This determinant, as given by the foregoing equations for X_1A, X_1B, X_1C, X_1B', X_1C', is D_1, where

$$D_1 = \begin{vmatrix} m_1Z_1 & , & -m_1Y_1, & 0 & , & m_2Z_2 & , & -m_2Y_2 \\ 0 & , & m_1X_1 & , & -m_2Z_2, & 0 & , & m_2X_2 \\ -m_1X_1, & & 0 & , & m_2Y_2 & , & -m_2X_2, & 0 \\ 0 & , & m_1L' & , & -m_2N', & 0 & , & m_2L' \\ -m_1L', & & 0 & , & m_2M' & , & -m_2L', & 0 \end{vmatrix}$$

$$= m_1{}^2 m_2{}^3 L' (X_2 - X_1) \begin{vmatrix} L', & X_1, & X_2 \\ M', & Y_1, & Y_2 \\ N', & Z_1, & Z_2 \end{vmatrix},$$

which does not vanish when there are more than two quantities m_1, m_2. If however there be only two bodies, so that $L' = m_1 X_1 + m_2 X_2$, and similarly for M' and N', then D_1 would vanish: and the inference could not be made. We have already dealt with the case of two bodies (§ 264); and we may therefore assume that $n > 2$.

When the substitution for f_1, g_1, f_2, g_2, h_2 is effected upon ϕ_0, it takes the form

$$G_1 (A, B, C, B', C', X, Y, Z) D_1{}^{-q} X_1{}^{-\lambda},$$

where q and λ may be positive whole numbers, and G_1 is polynomial in its arguments. But instead of using the five equations to determine f_1, g_1, f_2, g_2, h_2, they could have been used to eliminate g_1, h_1, g_2, h_2, g_3; and then the form of ϕ_0 would have been

$$G_2 (A, B, C, B', C', X, Y, Z) D_2{}^{-q} X_1{}^{-\lambda},$$

where

$$D_2 = \begin{vmatrix} m_1 Z_1 , & -m_1 Y_1, & m_2 Z_2 , & -m_2 Y_2, & m_3 Z_3 \\ 0 , & m_1 X_1 , & 0 , & m_2 X_2 , & 0 \\ -m_1 X_1, & 0 , & -m_2 X_2, & 0 , & -m_3 X_3 \\ 0 , & m_1 L' , & 0 , & m_2 L' , & 0 \\ -m_1 L', & 0 , & -m_2 L', & 0 , & -m_3 L' \end{vmatrix}$$

$$= m_1{}^2 m_2{}^2 m_3 L'^2 (X_2 - X_1)$$
$$\times \{ Z_3 (X_2 - X_1) + Z_1 (X_3 - X_2) + Z_2 (X_1 - X_3) \}.$$

Similarly if the equations were used to eliminate f_1, g_1, f_3, g_3, h_3, the form of ϕ_0 would be

$$G_3 (A, B, C, B', C', X, Y, Z) D_3{}^{-q} X_1{}^{-\lambda},$$

where

$$D_3 = m_1{}^2 m_2{}^3 L' (X_3 - X_1) \begin{vmatrix} L', & X_1, & X_3 \\ M', & Y_1, & Y_3 \\ N', & Z_1, & Z_3 \end{vmatrix}.$$

The form of ϕ_0 resulting from the transformation is, of course, independent of the manner in which the equations are manipulated. The only factor common* to D_1, D_2, D_3 is L'; hence in the first form, G_1 is divisible by $(X_2 - X_1)(L', Y_1, Z_2)$: in the second form, G_2 is divisible by $(X_2 - X_1)(Z_3, X_2, 1)$: and in the third form, G_3 is divisible by $(X_3 - X_1)(L', Y_1, Z_3)$. Thus ϕ_0 is of the form

$$\bar{G}_1 (A, B, C, B', C', X, Y, Z) L'^{-q} X_1{}^{-\lambda},$$

where q and λ may be positive whole numbers.

If other five equations had been used, say those which involve A, B, C, A', B', then a form

$$\bar{G}_2 (A, B, C, A', B', X, Y, Z) M'^{-r} X_1{}^{-\mu}$$

would be obtained; and the other possible set of five would lead to a form

$$\bar{G}_3 (A, B, C, A', C', X, Y, Z) N'^{-s} X_1{}^{-\nu},$$

where r, μ, s, ν may be positive whole numbers.

290. First, it is clear that λ, μ, ν all are zero. For instead of proceeding from the form of ϕ_0 in § 286, which is a polynomial in f, g, h, and has a possible power of X_1 as a denominator, we can

---

* When there are only three bodies in the system, then D_1 and D_3 have a common factor (X_1, Y_2, Z_3), as well as L'; but that other common factor is not contained in D_2.

proceed from the earlier form of ϕ_0 in § 278, where it is a polynomial in x, y, z, and the coefficients in the polynomial are themselves polynomial in X, Y, Z. We should then eliminate five of the variables x, y, z, by using any five of the equations

$$A = \sum_{a=1}^{n} m_a (y_a Z_a - z_a Y_a),$$

$$B = \sum_{a=1}^{n} m_a (z_a X_a - x_a Z_a),$$

$$C = \sum_{a=1}^{n} m_a (x_a Y_a - y_a X_a),$$

$$A' = N' \sum_{a=1}^{n} m_a y_a - M' \sum_{a=1}^{n} m_a z_a,$$

$$B' = L' \sum_{a=1}^{n} m_a z_a - N' \sum_{a=1}^{n} m_a x_a,$$

$$C' = M' \sum_{a=1}^{n} m_a x_a - L' \sum_{a=1}^{n} m_a y_a;$$

and the remainder of the variables x, y, z would then disappear from ϕ_0. The determinants of the variables in these equations are the same as before, and the same argument applies; and now there is no question of a power of X_1 in the denominator. Hence $\lambda = 0$, $\mu = 0$, $\nu = 0$.

Secondly, it is to be expected that the three forms

$$\bar{G}_1 L'^{-q}, \quad \bar{G}_2 M'^{-r}, \quad \bar{G}_3 N'^{-s},$$

can be made one and the same, in virtue of the relation

$$L'A' + M'B' + N'C' = 0.$$

To effect this change, take the form

$$\bar{G}_1 (A, B, C, B', C', X, Y, Z) L'^{-q};$$

replace X_1, Y_1, Z_1, by their values, in terms of L', M', N' and the rest of the variables X, Y, Z; and denote the new expression by

$$H (A, B, C, B', C', L', M', N', X_2, ..., Z_n) L'^{-q}.$$

As H is a polynomial in its arguments, let it be expressed in ascending powers of L'; then the form of ϕ_0 is (say)

$$\phi_0 = \frac{H_0}{L'^m} + \frac{H_1}{L'^{m-1}} + \cdots,$$

where the quantities H_0, H_1, ... are polynomials that do not involve L'. Now ϕ_0 is not infinite when $L' = 0$; hence H_0 must vanish when $L' = 0$. But when $L' = 0$, then

$$B'M' + C'N' = 0;$$

hence H_0, which is polynomial in B', C', M', N', and vanishes with $B'M' + C'N'$, must contain that quantity as a factor, say

$$H_0 = \bar{H}_0 (B'M' + C'N') = - A'L'\bar{H}_0,$$

where $\bar{H}_0$ is a polynomial in the same arguments as H_0. Thus

$$\phi_0 = \frac{H_1 - A'\bar{H}_0}{L'^{m-1}} + \cdots$$

In the same way, we can shew that, if $m > 1$, then $H_1 - A'\bar{H}_0$ contains a factor $B'M' + C'N'$, and so is expressible in a form $- A'L'\bar{H}_1$, where $\bar{H}_1$ is a polynomial in the same quantities as H_0. Hence finally, we shall have no fractional terms; and the form of ϕ_0 clearly is

$$\bar{G}(A, B, C, A', B', C', L', M', N', X_2, ..., Z_n),$$

or say

$$G(A, B, C, A', B', C', X_1, ..., Z_n),$$

a function that is polynomial in each of its arguments.

Further Limitations on the Leading Aggregate of an Integral.

291. This limitation upon the form of ϕ_0 has arisen through the necessity of excluding logarithmic terms from

$$- \int \frac{U}{X_1} dx_1$$

in the expression for ϕ_2. We proceed now to consider the expression that actually can be obtained for ϕ_2; but instead of returning to the expression obtained in § 287, we substitute the form of ϕ_0, which has just been derived, in the equation

$$- \frac{\partial \phi_2}{\partial x_1} X_1 = \frac{\partial \phi_0}{\partial v}.$$

Now

$$\frac{\partial A}{\partial v} = \sum_{a=1}^{n} \left(A_a \frac{\partial A}{\partial X_a} + B_a \frac{\partial A}{\partial Y_a} + C_a \frac{\partial A}{\partial Z_a} \right) = 0,$$

and similarly for the quantities $\dfrac{\partial B}{\partial v}, \dfrac{\partial C}{\partial v}, \dfrac{\partial A'}{\partial v}, \dfrac{\partial B'}{\partial v}, \dfrac{\partial C'}{\partial v}$.

Hence, as

$$\phi_0 = G(A, B, C, A', B', C', X_1, \ldots, Z_n),$$

we have

$$\frac{\partial \phi_0}{\partial v} = \underset{\rho,\,\sigma=1,\,\ldots,\,n}{\Sigma\Sigma} \frac{\Phi_{\rho\sigma}}{r^3_{\rho\sigma}},$$

where

$$\Phi_{\rho\sigma} = S(x_\sigma - x_\rho)\left(m_\sigma \frac{\partial \phi_0}{\partial X_\rho} - m_\rho \frac{\partial \phi_0}{\partial X_\sigma}\right).$$

We introduce symbols $\xi_{\rho\sigma}, \eta_{\rho\sigma}, \zeta_{\rho\sigma}, x_{\rho\sigma}, y_{\rho\sigma}, z_{\rho\sigma}$, defined by the formulæ

$$x_\rho - x_\sigma = x_{\rho\sigma}, \quad m_\sigma \frac{\partial \phi_0}{\partial X_\rho} - m_\rho \frac{\partial \phi_0}{\partial X_\sigma} = \xi_{\rho\sigma},$$

and two similar pairs; and then we have

$$\frac{\partial \phi_2}{\partial x_1} X_1 = -\frac{\partial \phi_0}{\partial v}$$

$$= \underset{\rho,\,\sigma}{\Sigma\Sigma} \frac{x_{\rho\sigma}\xi_{\rho\sigma} + y_{\rho\sigma}\eta_{\rho\sigma} + z_{\rho\sigma}\zeta_{\rho\sigma}}{r^3_{\rho\sigma}},$$

where, to secure the significance of the left-hand side, the variables on the right must be made $x_1, f_1, g_1, f_2, \ldots, h_n$. To obtain ϕ_2, we must effect a quadrature with regard to x_1.

The quantities $\xi_{\rho\sigma}, \eta_{\rho\sigma}, \zeta_{\rho\sigma}$ do not involve x_1, when the new set of variables has been adopted; for x_1 does not then occur in ϕ_0. Hence, during the quadrature, $\xi_{\rho\sigma}, \eta_{\rho\sigma}, \zeta_{\rho\sigma}$ are effectively constant.

The expression in § 287 for $r_{\sigma\tau}$ is

$$r^2_{\sigma\tau} = a_{\sigma\tau}x_1^2 + 2b_{\sigma\tau}x_1 + c_{\sigma\tau},$$

where

$$X_1^2 a_{\sigma\tau} = SX^2_{\sigma\tau}, \quad X_1^2 b_{\sigma\tau} = SX_{\sigma\tau}f_{\sigma\tau}, \quad X_1^2 c_{\sigma\tau} = Sf^2_{\sigma\tau},$$

for $\sigma, \tau = 2, 3, \ldots, n$; and the form holds also for $r^2_{1\rho}$, if we take $f_{\rho 1} = f_\rho$, there being no quantity f_1. Thus

$$\phi_2 = -\int \frac{dx_1}{X_1} \frac{\partial \phi_0}{\partial v}$$

$$= \Sigma\Sigma\, S\, \frac{\xi_{\rho\sigma}}{X_1^2} \int \frac{f_{\rho\sigma} + x_1 X_{\rho\sigma}}{(a_{\rho\sigma}x_1^2 + 2b_{\rho\sigma}x_1 + c_{\rho\sigma})^{\frac{3}{2}}}\, dx_1.$$

Now

$$\frac{d}{dx}\left\{\frac{\alpha x + \beta}{(ax^2 + 2bx + c)^{\frac{1}{2}}}\right\} = \frac{x(b\alpha - a\beta) + c\alpha - b\beta}{(ax^2 + 2bx + c)^{\frac{3}{2}}};$$

hence taking

$$ab_{\rho\sigma} - \beta a_{\rho\sigma} = S\,\xi_{\rho\sigma}X_{\rho\sigma} \Big\}$$
$$\alpha c_{\rho\sigma} - \beta b_{\rho\sigma} = S\,\xi_{\rho\sigma}f_{\rho\sigma} \Big\}\,,$$

we have

$$\phi_2 = \Sigma\Sigma\,\frac{\alpha x_1 + \beta}{X_1^2 r_{\rho\sigma}}\,.$$

Having actually obtained a value for ϕ_2, we transform it back to the old variables x, y, z. We have

$$X_1^2 a_{\rho\sigma} = S X^2_{\rho\sigma},$$
$$X_1^2 b_{\rho\sigma} = X_1 S X_{\rho\sigma}x_{\rho\sigma} - x_1 S X^2_{\rho\sigma},$$
$$X_1^2 c_{\rho\sigma} = X_1^2 S x^2_{\rho\sigma} - 2x_1 X_1 S X_{\rho\sigma}x_{\rho\sigma} + x_1^2 S X^2_{\rho\sigma},$$

so that

$$X_1^2 (b^2_{\rho\sigma} - a_{\rho\sigma}c_{\rho\sigma}) = (S X_{\rho\sigma}x_{\rho\sigma})^2 - (S X^2_{\rho\sigma})(S x^2_{\rho\sigma}) = E_{\rho\sigma},$$

say. Also

$$(b^2_{\rho\sigma} - a_{\rho\sigma}c_{\rho\sigma})\,\alpha = b_{\rho\sigma}(S\xi_{\rho\sigma}X_{\rho\sigma}) - a_{\rho\sigma}(S\xi_{\rho\sigma}f_{\rho\sigma}) = P_{\rho\sigma} \Big\}$$
$$(b^2_{\rho\sigma} - a_{\rho\sigma}c_{\rho\sigma})\,\beta = c_{\rho\sigma}(S\xi_{\rho\sigma}X_{\rho\sigma}) - b_{\rho\sigma}(S\xi_{\rho\sigma}f_{\rho\sigma}) = Q_{\rho\sigma} \Big\}\,,$$

say; then

$$\phi_2 = \Sigma\Sigma\,\frac{\alpha x_1 + \beta}{X_1^2 r_{\rho\sigma}}$$
$$= \Sigma\Sigma\,\frac{x_1 P_{\rho\sigma} + Q_{\rho\sigma}}{E_{\rho\sigma}r_{\rho\sigma}}$$
$$= \Sigma\Sigma\,\frac{F_{\rho\sigma}}{E_{\rho\sigma}r_{\rho\sigma}}\,,$$

where

$$F_{\rho\sigma} = (x_1 b_{\rho\sigma} + c_{\rho\sigma})(S\xi_{\rho\sigma}X_{\rho\sigma}) - (x_1 a_{\rho\sigma} + b_{\rho\sigma})\,(S\xi_{\rho\sigma}f_{\rho\sigma})$$
$$= \left(S x^2_{\rho\sigma} - \frac{x_1}{X_1}S X_{\rho\sigma}x_{\rho\sigma}\right)(S\xi_{\rho\sigma}X_{\rho\sigma}) - \frac{1}{X_1}S X_{\rho\sigma}x_{\rho\sigma}$$
$$\times \{S\xi_{\rho\sigma}(X_1 x_{\rho\sigma} - x_1 X_{\rho\sigma})\}$$
$$= (S x^2_{\rho\sigma})(S\xi_{\rho\sigma}X_{\rho\sigma}) - (S X_{\rho\sigma}x_{\rho\sigma})(S\xi_{\rho\sigma}x_{\rho\sigma}).$$

The integral ϕ, which is under consideration, is a polynomial in the variables X, Y, Z. The quantity ϕ_0 is the aggregate of terms of highest order in those variables, and ϕ_2 is the aggregate of terms of the next highest order: that is, ϕ_2 is a polynomial in the variables X, Y, Z, so that

$$\Sigma\Sigma\,\frac{F_{\rho\sigma}}{E_{\rho\sigma}r_{\rho\sigma}}$$

must be a polynomial in X, Y, Z. The quantity $E_{\rho\sigma}$ is resoluble into a couple of factors, linear in $X_{\rho\sigma}$, $Y_{\rho\sigma}$, $Z_{\rho\sigma}$, with (conjugate) complex coefficients; and neither of the factors is a factor of any other quantity E. In order, therefore, to secure the polynomial character of ϕ_2, it is necessary that $F_{\rho\sigma}$ should be actually divisible by $E_{\rho\sigma}$, when the two quantities are regarded as functions of $X_{\rho\sigma}$, $Y_{\rho\sigma}$, $Z_{\rho\sigma}$. The conditions are easily obtained. Dropping the subscripts ρ and σ temporarily, we have

$$F = (\xi X + \eta Y + \zeta Z)(x^2 + y^2 + z^2) - (\xi x + \eta y + \zeta z)(Xx + Yy + Zz),$$

where ξ, η, ζ are (unknown) homogeneous functions of X, Y, Z; and F is to be a multiple of

$$E = (X^2 + Y^2 + Z^2)(x^2 + y^2 + z^2) - (Xx + Yy + Zz)^2$$

$$= \frac{1}{y^2 + z^2}[\{X(y^2 + z^2) - x(Yy + Zz) - ir(Yz - Zy)\}$$
$$\{X(y^2 + z^2) - x(Yy + Zz) + ir(Yz - Zy)\}],$$

where $r^2 = x^2 + y^2 + z^2$, the multiple being an integral function, so far as concerns the variables X, Y, Z. Hence when we make

$$X(y^2 + z^2) - x(Yy + Zz) - ir(Yz - Zy) = 0,$$

F must vanish; that is, substituting this value of X in $F = 0$, we have one relation linear in ξ, η, ζ. Similarly, by using the other factor of E in the same way, we should obtain another relation linear in ξ, η, ζ; and (owing to the difference of the coefficients in the factors of E) this relation will be different from the former. Consequently the two relations lead to

$$\frac{\xi}{\Xi} = \frac{\eta}{H} = \frac{\zeta}{Z},$$

as unique conditions, which secure that F is divisible by E. But the requirement as to divisibility is clearly satisfied by

$$\frac{\xi}{X} = \frac{\eta}{Y} = \frac{\zeta}{Z},$$

which accordingly are the conditions. Restoring the suffixes ρ and σ, we have

$$\frac{\xi_{\rho\sigma}}{X_{\rho\sigma}} = \frac{\eta_{\rho\sigma}}{Y_{\rho\sigma}} = \frac{\zeta_{\rho\sigma}}{Z_{\rho\sigma}};$$

and these equations must be satisfied for all the values

$$\rho, \sigma = 1, 2, \ldots, n.$$

They clearly are a set of partial differential equations of the first order, homogeneous and of the first degree in the derivatives, satisfied by

$$G(A, B, C, A', B', C', X_1, Y_1, Z_1, \ldots, Z_n),$$

which is the value of ϕ_0.

292. To construct the most general solution of the set, we proceed to obtain the complete system to which it belongs. For this purpose, let

$$\frac{\partial \phi_0}{\partial X_\tau} = p_\tau, \quad \frac{\partial \phi_0}{\partial Y_\tau} = q_\tau, \quad \frac{\partial \phi_0}{\partial Z_\tau} = r_\tau,$$

for $\tau = 1, 2, \ldots, n$; then two equations of the set are

$$\Theta = (m_\sigma p_\rho - m_\rho p_\sigma)(Y_\rho - Y_\sigma) - (m_\sigma q_\rho - m_\rho q_\sigma)(X_\rho - X_\sigma) = 0,$$

$$\Theta' = (m_\tau p_\rho - m_\rho p_\tau)(Y_\rho - Y_\tau) - (m_\tau q_\rho - m_\rho q_\tau)(X_\rho - X_\tau) = 0.$$

The Jacobian condition of coexistence of these two equations, being

$$\sum_{\lambda=1}^{n} \left\{ S \left(\frac{\partial \Theta}{\partial p_\lambda} \frac{\partial \Theta'}{\partial X_\lambda} - \frac{\partial \Theta}{\partial X_\lambda} \frac{\partial \Theta'}{\partial p_\lambda} \right) \right\} = 0$$

in general, is for the present instance

$$- m_\sigma (Y_\rho - Y_\sigma)(m_\tau q_\rho - m_\rho q_\tau) + (m_\sigma q_\rho - m_\rho q_\sigma) m_\tau (Y_\rho - Y_\tau)$$

$$- m_\sigma (X_\rho - X_\sigma)(m_\tau p_\rho - m_\rho p_\tau) + (m_\sigma p_\rho - m_\rho p_\sigma) m_\tau (X_\rho - X_\tau) = 0.$$

Using $\Theta = 0$, $\Theta' = 0$, to eliminate $m_\tau q_\rho - m_\rho q_\tau$, $m_\sigma q_\rho - m_\rho q_\sigma$, from this condition, and removing a factor

$$(X_\rho - X_\tau)(X_\rho - X_\sigma) + (Y_\rho - Y_\tau)(Y_\rho - Y_\sigma)$$

which does not vanish, we find

$$(m_\sigma p_\rho - m_\rho p_\sigma) m_\tau (X_\rho - X_\tau) - (m_\tau p_\rho - m_\rho p_\tau) m_\sigma (X_\rho - X_\sigma) = 0;$$

that is, in our former notation, we have

$$\frac{\xi_{\rho\sigma}}{m_\rho m_\sigma X_{\rho\sigma}} = \frac{\xi_{\rho\tau}}{m_\rho m_\tau X_{\rho\tau}}.$$

But the first fraction is equal to

$$\frac{\eta_{\rho\sigma}}{m_\rho m_\sigma Y_{\rho\sigma}}, \quad \frac{\zeta_{\rho\sigma}}{m_\rho m_\sigma Z_{\rho\sigma}};$$

and so for the second: hence all the $\frac{3}{2} n(n-1)$ quantities

$$\frac{\xi_{\rho\sigma}}{m_\rho m_\sigma X_{\rho\sigma}}, \quad \ldots$$

have a common value, say U'. We therefore have to consider the set of equations

$$\frac{1}{m_\rho m_\sigma} \xi_{\rho\sigma} = U' X_{\rho\sigma}, \quad \frac{1}{m_\rho m_\sigma} \eta_{\rho\sigma} = U' Y_{\rho\sigma}, \quad \frac{1}{m_\rho m_\sigma} \zeta_{\rho\sigma} = U' Z_{\rho\sigma},$$

for $\rho, \sigma = 1, 2, \ldots, n$: these certainly include the former set. From this new set, it is easy to construct a complete system, that is, a system, in which (i) the equations are linearly and algebraically independent of one another, and (ii) all the Jacobian conditions of coexistence of the members of the system are satisfied, either identically or in virtue of the system.

The equation

$$\frac{1}{m_\rho m_\sigma} \xi_{\rho\sigma} = U' X_{\rho\sigma}$$

can be written

$$\frac{p_\rho}{m_\rho} - \frac{p_\sigma}{m_\sigma} = U' (X_\rho - X_\sigma).$$

Obviously, the $\frac{1}{2} n (n-1)$ equations which occur are all satisfied in virtue of the $n-1$ equations, given by

$$\rho = 2, 3, \ldots, n-1, n; \ \sigma = 1.$$

Similarly the $\frac{1}{2} n (n-1)$ equations

$$\frac{1}{m_\rho m_\sigma} \eta_{\rho\sigma} = U' Y_{\rho\sigma}$$

are satisfied in virtue of the corresponding $n-1$ equations: and likewise for the $\frac{1}{2} n (n-1)$ equations

$$\frac{1}{m_\rho m_\sigma} \zeta_{\rho\sigma} = U' Z_{\rho\sigma}.$$

Hence the whole set of $\frac{3}{2} n (n-1)$ equations is satisfied in virtue of $3 (n-1)$ equations, which are linearly independent of one another. On removing the quantity U', so that the equations are

$$\frac{\dfrac{p_\rho}{m_\rho} - \dfrac{p_\sigma}{m_\sigma}}{X_\rho - X_\sigma} = \frac{\dfrac{q_\rho}{m_\rho} - \dfrac{q_\sigma}{m_\sigma}}{Y_\rho - Y_\sigma} = \frac{\dfrac{r_\rho}{m_\rho} - \dfrac{r_\sigma}{m_\sigma}}{Z_\rho - Z_\sigma}$$

for $\rho = 2, 3, \ldots, n-1$; $\sigma = 1$, we have a set of $3n - 4$ algebraically independent equations.

When the equations are taken in the form

$$(m_\sigma p_\rho - m_\rho p_\sigma) (Y_\rho - Y_\sigma) - (m_\sigma q_\rho - m_\rho q_\sigma)(X_\rho - X_\sigma) = 0,$$

that is,

$$\xi_{\rho\sigma} Y_{\rho\sigma} - \eta_{\rho\sigma} X_{\rho\epsilon} = 0,$$

all the Jacobian conditions of coexistence are satisfied. Hence the set of $3n - 4$ equations, thus constructed, is a complete system.

The equations involve partial derivatives of ϕ_0. The number of variables, independent for the present system, which enter into the expression for ϕ_0, is $3n + 5$, viz. A, B, C, A', B', C' (which are equivalent to five), $X_1, \ldots, Z_n$. As the complete system involves $3n - 4$ equations, it follows[*] that the number of independent solutions is 9, $= 3n + 5 - (3n - 4)$; and in terms of them, any other solution can be expressed.

In the complete system, the only derivatives of ϕ_0 that explicitly occur are those taken with regard to the variables X, Y, Z. Hence solutions as required are given by

$$A, B, C, A', B', C',$$

equivalent to five; and therefore four others are required, which must involve the variables X, Y, Z.

The classical integrals provide four others. Three of them are

$$L' = \overset{n}{\underset{a=1}{\Sigma}} m_a X_a, \quad M' = \overset{n}{\underset{a=1}{\Sigma}} m_a Y_a, \quad N' = \overset{n}{\underset{a=1}{\Sigma}} m_a Z_a,$$

because the quantity ϕ_0 in each case is the whole of the integral. The energy equation is

$$\tfrac{1}{2} \overset{n}{\underset{a=1}{\Sigma}} m_a (X_a^2 + Y_a^2 + Z_a^2) - \Sigma\Sigma \frac{m_\rho m_\sigma}{r_{\rho\sigma}} = \text{const.},$$

say

$$\phi = T - U;$$

then for this integral, $\phi_0 = T$, that is,

$$T = \tfrac{1}{2} \overset{n}{\underset{a=1}{\Sigma}} m_a (X_a^2 + Y_a^2 + Z_a^2),$$

is another integral of our system for ϕ_0. Thus four solutions are given by

$$L', M', N', T.$$

Bruns's Theorem.

293. The function ϕ_0 was proved (§ 291) to be polynomial in the quantities $A, B, C, A', B', C', X_1, Y_1, Z_1, \ldots, Z_n$; and we have just proved that it is expressible in terms of $A, B, C, A', B', C', L', M', N', T$. Hence when we proceed to eliminate some of the variables X, Y, Z, by means of the last set of quantities, the rest of those variables must also disappear. For this purpose, the first six quantities are ineffective: because each of them, and every combination of them, involve the variables x, y, z; and therefore

---

[*] Part I. of this work, § 38.

the requisite elimination can be carried out only by means of L', M', N', T. Evidently we shall be able to eliminate four of the variables X, Y, Z (say X_1, Y_1, Z_1, X_2) by means of them; and then Y_2, Z_2, X_3, ..., Z_n must disappear, owing to the later form of ϕ_0.

We have

$$m_1 Y_1 = M' - \eta, \quad m_1 Z_1 = N' - \zeta,$$

where η involves the rest of the variables Y, and ζ the rest of the variables Z. Also

$$m_1 X_1 + m_2 X_2 = L' - \xi,$$
$$m_1 X_1^2 + m_2 X_2^2 = 2T - \xi',$$

where ξ involves the rest of the variables X, and ξ' involves η, ζ and the rest of the variables X, Y, Z. The last two equations give irrational expressions for X_1 and X_2, the irrational quantity being

$$\{(m_1 + m_2)(2T - \xi') - (L' - \xi)^2\}^{\frac{1}{2}}.$$

Let the values thus obtained for X_1, Y_1, Z_1, X_2 be substituted in

$$G(A, B, C, A', B', C', X, Y, Z);$$

the quantities ξ, η, ζ, ξ' are to disappear. But ξ and ξ' can disappear, only if those combinations of the preceding irrational quantity, which actually occur, are rational. This being the case, it follows that such combinations are rational in L' and T. No one of the substitutions made in G admits the possibility of a fractional form; and therefore

$$\phi_0 = \mathfrak{G}(A, B, C, A', B', C', L, M, N, T),$$

where $\mathfrak{G}$ denotes a function that is polynomial in each of its arguments. When substitution takes place for the various arguments in terms of the variables x, y, z, X, Y, Z, the resulting expression for the function ϕ_0 is integral and homogeneous in the variables X, Y, Z; and it represents the aggregate of terms of the highest order in X, Y, Z, which are contained in some integral ϕ that is polynomial in those variables.

Now A, B, C, A', B', C', L, M, N, $T - U$ are the classical integrals of the differential equation; and any function of them is an integral, so that, if ϕ' denote

$$\mathfrak{G}(A, B, C, A', B', C', L, M, N, T - U),$$

then ϕ' is an integral, free from explicit occurrence of the time. Manifestly it is a polynomial in the variables X, Y, Z; and the aggregate of its terms of highest order in those variables is evidently ϕ_0. Hence

$$\phi - \phi'$$

is an integral of the differential equations, which is of the same character as ϕ. Also, since

$$\phi = \phi_0 + \phi_2 + \phi_4 + \dots,$$

$$\phi' = \phi_0 + U\frac{\partial \phi_0}{\partial T} + \tfrac{1}{2}U^2\frac{\partial^2 \phi_0}{\partial T^2} + \dots,$$

it follows that the order of the aggregate of terms of the highest order in X, Y, Z, which occur in $\phi - \phi'$, is two less than the order of ϕ_0; and it may be more than two less, if $\phi_2 = U\frac{\partial \phi_0}{\partial T}$. In other words, $\phi - \phi'$, while it is of the same polynomial character as ϕ in X, Y, Z, is of lower order than ϕ. Denoting it by $\bar{\phi}$, we have

$$\phi = \phi' + \bar{\phi};$$

so that ϕ can be composed algebraically from the classical integrals, if $\bar{\phi}$, of order at least two lower than ϕ in X, Y, Z, can be so composed.

Moreover, $\bar{\phi}_0$ is of the same character as ϕ_0. For we have

$$\bar{\phi}_0 = \phi_2 - U\frac{\partial \phi_0}{\partial T},$$

or, if we reintroduce the part ϕ_2' omitted in § 286, we have

$$\bar{\phi}_0 = \phi_2' + \phi_2 - U\frac{\partial \phi_0}{\partial T},$$

where ϕ_2' is of the same character as ϕ_0. Now, by §§ 291, 292, we have

$$\phi_2 = \Sigma\Sigma \frac{1}{r_{\rho\sigma}} m_\rho m_\sigma U'$$

$$= UU'.$$

Also

$$U' = \frac{1}{m_\rho m_\sigma X_{\rho\sigma}}\left(m_\sigma \frac{\partial \phi_0}{\partial X_\rho} - m_\rho \frac{\partial \phi_0}{\partial X_\sigma}\right)$$

$$= \frac{1}{m_\rho m_\sigma X_{\rho\sigma}}(m_\sigma m_\rho X_\rho - m_\rho m_\sigma X_\sigma)\frac{\partial \phi_0}{\partial T}$$

$$= \frac{\partial \phi_0}{\partial T};$$

and therefore
$$\bar{\phi}_0 = \phi_2',$$
shewing that $\bar{\phi}_0$ is of the same character as ϕ_0. Hence the preceding argument applies to $\bar{\phi}$.

Repeated a sufficient number of times to each such function of decreased order as it arises, it shews that ϕ can be composed algebraically from the classical integrals if this property holds for an integral, that is of the first order in X, Y, Z (which is the case when ϕ_0 is of odd order), or does not involve X, Y, Z (which is the case when ϕ_0 is of even order).

As regards an integral of the first order in X, Y, Z, the aggregate of terms of the first order (being the highest order) is, by our preceding investigation, a polynomial in A, B, C, A', B', C', L', M', N', T. But the polynomial cannot involve T, for then it would be of the second order in X, Y, Z; that is, ϕ_0 is a linear combination of A, B, C, A', B', C', L', M', N', with constant coefficients. This linear combination is the whole of the integral; and it is composed of the classical integrals.

As regards an integral which does not involve X, Y, Z, we have proved that such integrals do not exist (§ 273): they are merely constants. The reduction of ϕ to expression in terms of the classical integrals would be effected at the preceding stage.

The proviso is satisfied in each case: and therefore we have the theorem:

Every algebraical integral of the differential equations of the problem of three bodies, or of the problem of more than three bodies, can be compounded by purely algebraical processes from the classical integrals.

One remark may be added. In the course of § 289, it was pointed out that the argument did not apply in case there were only two bodies: the problem is therefore omitted from the enunciation. It is known (§ 264) that the theorem does not apply.

General Remarks in Conclusion.

294. The preceding method, which has proved effective for the differential equations of celestial dynamics, does not seem directly applicable to the investigation of the algebraical integrals,

and of the conditions of their existence, for a general equation $F(w'', w', w, z) = 0$ of the second order or for one of higher order.

The briefest review of the discussion of the corresponding questions even for a general equation of the first order, as merely begun in §§ 141—144, is sufficient to shew that the analogous method at once lands the enquiry among considerations, which are connected with rational transformation of surfaces in ordinary space and of configurations in hyperspace. Such considerations, important in themselves, lead rather to descriptive than to functional properties of the differential equations ; and interesting as are the results obtained* by Picard, Painlevé, and other writers, they belong to a region of ideas, which lie outside the scope and the method of treatment of the portion of the subject discussed in this volume. Accordingly these results, and kindred matter, will not be developed here : the discussion of ordinary differential equations that are not linear is concluded at this stage.

It must not however be supposed that all the important questions, arising in the line of main development of this subject, have been dealt with. In addition to the omission of the theory of algebraic integrals, already indicated, other omissions that may be mentioned are the theory of reducibility, Poincaré's theory of periodic solutions of various kinds possessed by simultaneous equations of the first order, Picard's asymptotic solutions, the whole range of Lie's group-theory as applied to ordinary equations, and the geometrical properties of curves implied by one equation and by systems of equations. In order to prevent inordinate expansion, it was found convenient to impose initial limitations upon the class of questions that would be investigated : my purpose has been to secure an ample discussion of those branches of the subject which lie within the range selected.

---

* For references, see p. 276, note, of this volume.

INDEX TO PART II.

(Part II occupies volumes II and III of the present work. The figures refer to the pages in these volumes. The Table of Contents at the beginning of volume II and of volume III may also be consulted.)

For EU product safety concerns, contact us at Calle de José Abascal, 56–1°,
28003 Madrid, Spain or eugpsr@cambridge.org.

www.ingramcontent.com/pod-product-compliance
Ingram Content Group UK Ltd.
Pitfield, Milton Keynes, MK11 3LW, UK
UKHW010852090126
466816UK00011B/179